Solid-State Fermentation Bioreactors

David A. Mitchell · Nadia Krieger
Marin Berovič (Eds.)

Solid-State Fermentation Bioreactors

Fundamentals of Design
and Operation

With 194 Figures and 32 Tables

 Springer

Dr. David A. Mitchell
Federal University of Paraná
Department of Biochemistry
and Molecular Biology
P.O. Box 19046
81531-990 Curitiba-PR, Brazil
davidmitchell@ufpr.br

Dr. Nadia Krieger
Federal University of Paraná
Department of Chemistry
P.O. Box 19081
81531-990 Curitiba-PR, Brazil
nkrieger@ufpr.br

Dr. Marin Berovič
University of Ljubljana
Department of Chemical, Biochemical
and Environmental Engineering
Hajdrihova 19
1001 Ljubljana, Slovenia
marin.berovic@Uni-Lj.si

Library of Congress Control Number: 2006922414

ISBN-10 3-540-31285-4 Springer Berlin Heidelberg New York
ISBN-13 978-3-540-31285-7 Springer Berlin Heidelberg New York
e-ISBN 3-540-31286-2
DOI 10.1007/3540312854

Springer is a part of Springer Science+Business Media
springer.com

© Springer-Verlag Berlin Heidelberg 2006
Printed in Germany

Typesetting and Production: LE-TEX, Jelonek, Schmidt & Vöckler GbR, Leipzig, Germany
Coverdesign: design&production, Heidelberg, Germany

Printed on acid-free paper   31/3100/YL – 5 4 3 2 1 0

Preface

Although solid-state fermentation (SSF) has been practiced for many centuries in the preparation of traditional fermented foods, its application to newer products within the framework of modern biotechnology is relatively restricted. It was considered for the production of enzymes in the early 1900s and for the production of penicillin in the 1940s, but interest in SSF waned with the advances in submerged liquid fermentation (SLF) technology. The current dominance of SLF is not surprising: For the majority of fermentation products, it gives better yields and is easier to apply. It is notoriously difficult to control the fermentation conditions in SSF; these difficulties are already apparent at small scale in the laboratory and are exacerbated with increase in scale. However, there are particular circumstances and products for which SSF technology is appropriate. For example, a desire to reuse solid organic wastes from agriculture and food processing rather than simply discarding them leads naturally to the use of SSF. Further, some microbial products, such as fungal enzymes and spores, amongst others, are produced in higher yields or with better properties in the environment provided by SSF systems.

With recognition of this potential of SSF, a revival of interest began in the mid-1970s. However, the theoretical base for SSF bioreactor technology only began to be established around 1990. Before this, there were many examples of SSF bioreactors, especially those used in the *koji* industry, but there was little or no information about the efficiency of heat and mass transfer processes within them. The work that has been carried out over the last 15 years is sufficient to establish a general basis of engineering principles of SSF bioreactors. This book brings together this work in order to provide this basis. It makes the key point that, given the complexity of SSF systems, efficient performance of SSF bioreactors will only be achieved through: (1) the use of mathematical models in making design and operating decisions for bioreactors and (2) The application of control theory.

Before proceeding, we must point out that we are quite aware of the potential problems that might be used by our use of the word "fermentation". In this book we use it not in its metabolic sense but rather in its more general sense of "controlled cultivation of microorganisms". Although several terms are used to denote this fermentation technique, the most common by far is "solid-state fermentation".

This book focuses on SSF bioreactors. It does not aim to introduce SSF itself. We assume that readers interested in learning about SSF bioreactors are familiar with SSF processes themselves. Even if not, a reader who understands the basic principles of SLF processes and SLF bioreactor design will be able to understand this book. In any case, readers requiring a general background regarding SSF can consult books or review articles (e.g., see the Further Reading section of Chap. 1).

Even with this focus on SSF bioreactors, the book deliberately addresses general issues and concepts. Specific examples are given to illustrate concepts, but the book neither considers all types of bioreactors that have been used nor presents all mathematical models that have been developed. We do not attempt to present all the engineering know-how so far generated for SSF bioreactors. Rather, we aim to introduce the fundamental concepts and ideas.

The main audience intended for this book is the researcher/worker in SSF who is currently developing an SSF process with the intention of eventually commercializing it. Our aim is to give this reader a broad overview of what is involved in designing a bioreactor and optimizing its performance.

We recognize that many readers may not have the necessary background to set up and solve mathematical models of bioreactor performance. This book does not attempt to teach the necessary modeling skills. Such a task would require a lengthy treatise on various mathematical and engineering fundamentals. A basic understanding of differential and integral calculus will help readers to understand various of the chapters, although it is by no means necessary to be an expert.

After reading this book, the "non-engineering reader" should:

- understand qualitatively the importance of the various mass transfer, heat transfer and biological phenomena that are important in SSF systems, and the interactions amongst these various phenomena;
- understand what mathematical models of bioreactors can do. If you understand what models can and cannot do, then even if you do not have the skills to develop a model yourself, you will know when it is appropriate to seek the help of someone with such modeling skills (a "modeler");
- be able to "talk the same language" as the "modeler". In other words, you should be able to define clearly for the modeler what you wish to do, and you should be able to understand the questions that the modeler poses. In this way you can interact with modelers, even if they have no experience with SSF.

This book should also be useful for readers with modeling skills but who are working in SSF for the first time. In a succinct way, it outlines the important phenomena and the basic principles of SSF bioreactor design and operation.

We welcome comments, suggestions and criticisms about this book. Our aim is to help you to understand SSF bioreactors better. We would appreciate knowing just how well we have achieved this aim. The addresses of the editors and authors are given after the Table of Contents.

November 2005

David Mitchell   Nadia Krieger   Marin Berovič

Acknowledgements

As leader of the editorial team and the main contributing author of the book, I would first like to thank my PhD supervisors, Paul Greenfield and Horst Doelle. You set me on a path that I have found both challenging and interesting over the last twenty years or so. In fact, I still remember the moment, in mid-1984, when I decided to do my PhD in the area of solid-state fermentation. Paul Greenfield said to me "I have heard of this area called solid-state fermentation, I think that you can make some contributions in the area". Well Paul, you were right, I have managed something. Thanks! Of course, there is still much to do.

I must also thank my co-editors and co-authors. This book would never have been written without your input. From each of you I have learnt something about solid-state fermentation. Further, I recognize that I am indebted to many colleagues who, while not being co-authors, have helped me to understand solid-state fermentation better. I will not cite names because the list is enormous. It includes not only colleagues with whom I have interacted personally, but also colleagues who have published papers in the area of solid-state fermentation that have helped me to develop my understanding of this area.

This book, in part, represents a synthesis of work undertaken by my research group and supported by various funding agencies. I am indebted to these agencies for funding my research over the last 15 years or so. I received two grants to work on solid-state fermentation bioreactors from the "Australian Research Council Small Grants Scheme". Since my move to Brazil, I have received funding from several state and federal granting bodies. These include (1) the "Araucaria Foundation" (Fundação Araucária), a research agency of the state of Paraná; (2) the Brazilian National Council for Scientific and Technological Development (CNPq, Conselho Nacional de Desenvolvimento Científico e Tecnológico) and (3) the Brazilian-Argentinean Biotechnology Committee (CBAB, Comitê Brasileiro-Argentino de Biotechnologia), for which the funds originated from the Brazilian Ministry of Science and Technology (MCT, Ministério de Ciência e Tecnologia) and were administered by CNPq. CNPq has also been kind enough to award me a research scholarship.

Finally, thanks are due to the Springer staff, especially Marion Hertel, Beate Siek, and Joern Mohr, for their patience and guidance.

David Mitchell

Contributing Authors

Prof. Eduardo Agosin
Department of Chemical and Bioprocess Engineering
Pontificia Universidad Católica de Chile
P.O. Box 306, Santiago 22, Post Code 6904411, Chile
E-mail: agosin@ing.puc.cl

Prof. Marin Berovič
Department of Chemical, Biochemical and Environmental Engineering
Faculty of Chemistry and Chemical Technology, University of Ljubljana
Askerceva 9, Ljubljana 1000, Slovenia
E-mail: marin.berovic@Uni-Lj.si

Dr. Mario Fernández
Deparment of Sciences and Engineering, Universidad de Talca
Camino Los Niches km 1, Curicó, Chile
E-mail: mafernandez@utalca.cl

Dr. Matthew T. Hardin
Division of Chemical Engineering, University of Queensland
St Lucia 4072, Australia
E-mail: matth@cheque.uq.edu.au

Dr. Lilik Ikasari
Food Division, Quest International Indonesia
Jl. Raya Jakarta-Bogor Km. 35, Cimanggis, Kab. Bogor, West Java, Indonesia
E-mail: lilik.ikasari@questintl.com

Dr. Morteza Khanahmadi
Agricultural Engineering Research Section
Isfahan Center for Research of Agricultural Science & Natural Resources
Amirieh, Agriculture Blvd, Isfahan, 81785-199, Iran
E-mail: khanahmadi@yahoo.com

Dr. Nadia Krieger
Department of Chemistry, Universidade Federal do Paraná
Cx. P. 19081 Centro Politécnico, Curitiba 81531-990, Paraná, Brazil
E-mail: nkrieger@ufpr.br

Dr. Luiz Fernando L. Luz Junior
Department of Chemical Engineering, Universidade Federal do Paraná
Cx. P. 19011 Centro Politécnico, Curitiba 81531-990, Paraná, Brazil
E-mail: luzjr@ufpr.br

Dr. David A. Mitchell
Department of Biochemistry and Molecular Biology
Universidade Federal do Paraná
Cx. P. 19046 Centro Politécnico, Curitiba 81531-990, Paraná, Brazil
E-mail: davidmitchell@ufpr.br

Dr. Montira Nopharatana
Department of Food Engineering
King Mongkut's University of Technology Thonburi,
91 Prachauthit Rd., Tungkru, Bangkok 10140, Thailand
E-mail: montira.nop@kmutt.ac.th

Dr. J. Ricardo Pérez-Correa
Department of Chemical and Bioprocess Engineering
Pontificia Universidad Católica de Chile
P.O. Box 306, Santiago 22, Post Code 6904411, Chile
E-mail: perez@ing.puc.cl

Dr. Luis B. Ramos Sánchez
Department of Chemical Engineering, University of Camagüey.
Circunvalación Norte, km 5 1/2, s/n, Camagüey. CP 74650. Camagüey, Cuba
E-mail: lramos@qui.reduc.edu.cu

Dr. Penjit Srinophakun
Department of Chemical Engineering, Kasetsart University
P.O.Box 1032, Kasetsart Post Office, Bangkok 10903, Thailand
E-mail: fengpjs@ku.ac.th

Dr. Deidre M. Stuart
School of Environmental Sciences and Natural Resource Management
University of New England, Armidale, NSW 2350, Australia
E-mail: deidre.stuart@pobox.une.edu.au

Ms. Graciele Viccini MSc
Department of Biochemistry and Molecular Biology
Universidade Federal do Paraná
Cx. P. 19046 Centro Politécnico, Curitiba 81531-990, Paraná, Brazil
E-mail: gvic@pop.com.br

Dr. Oscar F. von Meien
UN-RIO/ST/EISA – PETROBRÁS, Av. Gen. Canabarro 500 - 5° ad.,
Maracanã, Rio de Janeiro, RJ 20271-900, Brasil
E-mail: meienov@petrobras.com.br

Contents

13 Appropriate Levels of Complexity for Modeling SSF Bioreactors 179
David A. Mitchell, Luiz F.L. Luz Jr, Marin Berovič, and Nadia Krieger

14 The Kinetic Sub-model of SSF Bioreactor Models: General Considerations...191
David A. Mitchell and Nadia Krieger

20.4 Solids-to-Air Heat and Mass Transfer Coefficients Within Beds..........283
20.5 Bed-to-Headspace Transfer Coefficients..284
20.6 Conclusions 289
Further Reading...289

21 Bioreactor Modeling Case Studies: Overview 291
David A. Mitchell
21.1 What Can the Models Be Used to Do?..291
21.2 Limitations of the Models...292
21.3 The Amount of Detail Provided about Model Development...............293
21.4 The Order of the Case Studies 294

22 A Model of a Well-mixed SSF Bioreactor 295
David A Mitchell and Nadia Krieger
22.1 Introduction 295
22.2 Synopsis of the Model 295
    22.2.1 The System, Equations, and Assumptions 295
    22.2.2 Values of Parameters and Variables...301
22.3 Insights the Model Gives into the Operation of Well-Mixed
    Bioreactors 303
    22.3.1 Insights into Operation at Laboratory Scale 303
    22.3.2 Insights into Operation at Large Scale 307
    22.3.3 Effect of Scale and Operation on Contributions to Cooling of
        the Solids 310
22.4 Conclusions on the Operation of Well-Mixed Bioreactors...................312
Further Reading...314

23 A Model of a Rotating-Drum Bioreactor 315
David A. Mitchell, Deidre M. Stuart, and Nadia Krieger
23.1 Introduction 315
23.2 A Model of a Well-Mixed Rotating-Drum Bioreactor 315
    23.2.1 Synopsis of the Mathematical Model and its Solution 315
    23.2.2 Predictions about Operation at Laboratory Scale 320
    23.2.3 Scale-up of Well-Mixed Rotating-Drum Bioreactors...................325
23.3 What Modeling Work Says about Rotating-Drum Bioreactors
    Without Axial Mixing..328
23.4 Conclusions on the Design and Operation of Rotating-Drum
    Bioreactors 329
Further reading 330

24 Models of Packed-Bed Bioreactors 331
*David A. Mitchell, Penjit Srinophakun, Oscar F. von Meien,
Luiz F.L. Luz Jr, and Nadia Krieger*
24.1 Introduction 331
24.2 A Model of a Traditional Packed-Bed Bioreactor 331
    24.2.1 Synopsis of the Mathematical Model and its Solution 333

Abbreviations

A/D	analog-digital
ASFB	air-solid fluidized bed
a_w	water activity
CER	CO_2 evolution rate
COU	cumulative O_2 uptake
CRDB	continuous rotating drum bioreactor
CSSFB	continuous-flow solid-state fermentation bioreactor
CSTB	continuous stirred-tank bioreactor
CTFB	continuous tubular flow bioreactor
DM	dry matter
D/A	digital-analog
DMC	dynamic matrix control
FCV	flow control valve
GA_3	gibberellic acid
GC	gas chromatography
GC/MS	gas chromatography coupled with mass spectrometry
GPM	gallons per minute
HEPA	high efficiency particulate air
HPLC	high-performance liquid chromatography
IBM	International Business Machines
IDS	initial dry solids
IDM	initial dry matter
INRA	*Institut National de la Recherche Agronomique* (National Agronomic Research Institute)
ISFET	ion sensitive field effect transistor
IR	infrared
IWC	initial water content
k_Fa	biofilm conductance (used to characterize the efficiency of O_2 transfer between the gas and biofilm phases in SSF)
k_La	overall mass transfer coefficient (used to characterize the efficiency of O_2 transfer between the gas and liquid phases in SLF)
L/D	length to diameter ratio

MPC	model predictive control
NLMPC	nonlinear model predictive control
ODE	ordinary differential equation
OUR	O_2 uptake rate
PDE	partial differential equation
PI	proportional/integral
PID	proportional/integral/derivative (as defined in Chap. 27.2.2)
PLC	Programmable Logic Controller
PUC	Pontificia Universidad Católica
RTD	resistance temperature detector
RQ	respiratory quotient
SI	*Système International* (international metric system)
SLF	submerged liquid fermentation
SSF	solid state fermentation (as defined in Chap. 1)
t_{90}	time for the biomass to reach 90% of its maximum value
TC	thermocouple
TDR	time domain reflectometry
T_{opt}	optimum temperature
$T_{subscript}$	temperature, with the meaning as indicated by the subscript
UV	ultraviolet
vvm	volume per volume per minute
WC	water content
wt	weight
X	dry biomass
Z-N	Ziegler and Nichols (in relation to controller tuning rules)

Notation

Please note that, due to the fact that different models use different nomenclature and units, the nomenclature is covered chapter-by-chapter. In most cases the symbols are also explained where they first appear in the text.

Chapter 3

Pr	productivity (kg h^{-1} m^{-3})
$t_{process}$	time between successive harvests (h)
$V_{bioreactor}$	bioreactor volume (m^3)
$X_{harvest}$	amount of biomass (or product) at the time of harvesting (kg)
$X_{initial}$	amount of biomass (or product) at zero time (kg)

Chapter 4

$T_{subscript}$	temperature of phase or subsystem indicated by subscript (°C)

Chapter 5

H	superficial velocity of the air (m s^{-1})
$T_{subscript}$	temperature of phase or subsystem indicated by subscript (°C)
V_Z	bed height (m)

Chapter 6

C	O$_2$ concentration in the surrounding atmosphere (g cm^{-3})
D	effective diffusivity of O$_2$ in the bed (cm^2 h^{-1})
D_c	critical tray depth (cm)
k	thermal conductivity of the bed (W m^{-1} °C^{-1})
N_{Bi}	Biot number
R_Q	volumetric heat production rate (W m^{-3})
R_X	overall growth rate (kg-dry-biomass m^{-3} h^{-1})
R_{XM}	maximum growth rate (g-dry-biomass cm^{-3}-bed h^{-1})
T_a	surrounding air temperature (°C)
T_s	bed surface temperature (°C)
X	biomass density (kg-dry-biomass m^{-3})
X_{max}	maximum possible biomass density (kg-dry-biomass m^{-3})
Y_{XO}	yield coefficient of biomass from O$_2$ (g-dry-biomass g-O$_2$$^{-1}$)
z	spatial coordinate as a dimensionless fraction of the total bed height
Z	total bed height (m)
α	coefficient for bed-to-air heat transfer at the bed top (W m^{-2} °C^{-1})
α_b	coefficient for bed-to-air heat transfer at the bed bottom (W m^{-2} °C^{-1})

μ	specific growth rate parameter (h^{-1})
Θ	temperature difference between the bottom of bed and the tray surface when no heat transfer through the bottom of the tray (°C)
μ_{FO}	fractional specific growth rate based on O_2 (dimensionless)
μ_{FT}	fractional specific growth rate based on temperature (dimensionless)
μ_{max}	maximum value that the specific growth rate parameter can have (h^{-1})

Chapter 8

$A_{subscript}$	area, with meaning indicated by subscript (m^2)
D	drum diameter (m)
F_{mix}	volumetric exchange rate between the dead and plug-flow regions relative to the drum volume and mean residence time (dimensionless)
h	coefficient for bed-to-headspace heat transfer $(W\ m^{-2}\ °C^{-1})$
N_C	critical rotational speed (rpm)
R_B	ratio of exposed surface area of the bed to the bed volume (m^{-1})
R_{conv}	rate of convective heat removal to the headspace gases (W)
$T_{subscript}$	temperature of phase or subsystem indicated by subscript (°C)
$V_{subscript}$	volume, with meaning indicated by subscript (m^3)
θ_ω	angle subtended at the center of the drum by the bed surface for fractional filling ω (radians)
ω	fractional filling of the drum $(m^3$-bed m^{-3}-total-bioreactor-volume)

Chapter 11

F	mass flow of fresh solids $(kg\ h^{-1})$
f_m	solids flow through well-mixed region $(kg\ h^{-1})$
f_p	solids flow through plug-flow region $(kg\ h^{-1})$
f_R	recycled solid-flow $(kg\ h^{-1})$
M	overall mass of solids in the bioreactor (kg)
M_m	mass of solids in the well-mixed region (kg)
M_p	mass of solids in the plug-flow region (kg)
X	biomass content in product and recycle streams $(g\ kg\text{-dry-matter}^{-1})$
X_{max}	maximum possible biomass content $(g\ kg\text{-dry-matter}^{-1})$
X_o	initial biomass content in the fresh feed stream $(g\ kg\text{-dry-matter}^{-1})$
$X_o{'}$	biomass content after mixing the fresh feed and recycle streams $(g\ kg\text{-dry-matter}^{-1})$
α	fraction of the flow that passes through the plug-flow region (f_p/f_m)
β	fraction of the "in-bioreactor" mass in the plug-flow region (M_p/M_m)
γ	recycle ratio (dimensionless)
μ	specific growth rate parameter (h^{-1})

Chapter 12

a	constant in the double-Arrhenius equation (h^{-1})
A	area across which heat transfer takes place (m^2)
b	constant in the double-Arrhenius equation (dimensionless)
C_{Pair}	heat capacity of dry air $(J\ kg\text{-dry-air}^{-1}\ °C^{-1})$

C_{PB}	overall bed heat capacity (J kg^{-1} °C^{-1})
C_{Pvapor}	heat capacity of water vapor (J kg-vapor^{-1} °C^{-1})
$Ea1, Ea2$	constants in the double-Arrhenius equation (J mol^{-1})
F	air flow rate on a dry basis (kg-dry-air h^{-1})
h	heat transfer coefficient (J m^{-2} s^{-1} °C^{-1})
H	outlet air humidity (kg-vapor kg-dry-air^{-1})
H_{in}	inlet air humidity (kg-vapor kg-dry-air^{-1})
M	total bed mass (kg)
R	universal gas constant (J mol^{-1} K^{-1})
S	dry substrate concentration (kg-dry-substrate m^{-3}).
t	time (h)
T	temperature of the substrate bed and the outlet air (°C)
T_{in}	inlet air temperature (°C)
T_{surr}	temperature of the surroundings (°C)
W	water content on a dry basis (kg-H$_2$O kg-dry-substrate^{-1})
X	amount of biomass in the bioreactor (kg)
X_{max}	maximum possible amount of biomass in the bioreactor (kg)
Y_Q	yield of metabolic heat from growth (J kg-biomass^{-1})
μ	specific growth rate parameter (h^{-1})
ΔH_{vap}	enthalpy of vaporization of water (J kg-H$_2$O^{-1})

Chapter 13
A	area for heat transfer as indicated by subscript (m^2)
a_w	water activity as indicated by subscript
D_S	diffusivity of the substrate (m^2 h^{-1})
h	heat transfer coefficient as indicated by subscript (J h^{-1} m^{-2} °C^{-1})
H	humidity of phase indicated by subscript (kg-vapor kg-dry-air^{-1})
k	mass transfer coefficient (kg m^{-2} h^{-1})
K_S	saturation constant of the Monod equation (g L^{-1})
r	radial position (m)
S	substrate concentration (g L^{-1})
$S|_r$	substrate concentration at radial position r (g L^{-1})
t	time (h)
T	temperature as indicated by subscript (°C)
X	biomass concentration (g L^{-1})
$X|_r$	biomass concentration at radial position r (g L^{-1})
Y_{XS}	yield of biomass from substrate (g-dry-biomass g-substrate^{-1})
μ_{max}	specific growth rate parameter (h^{-1})

Chapter 14
A	constant of the deceleration growth equation (dimensionless)
C	biomass content (basis not specified)
C_m	maximum biomass content (basis not specified)
C_o	initial biomass content (basis not specified)
C_{XA}	biomass content related to dry matter at zero time (g-dry-biomass g-initial-dry-solids^{-1})

C_{XM}	biomass content related to fresh matter at time of sampling (g-dry-biomass g-moist-solids^{-1})
C_{XR}	biomass content related to dry matter at time of sampling (g-dry-biomass g-dry-solids^{-1})
C_{XW}	biomass content related to fresh matter at zero time (g-dry-biomass g-initial-moist-solids^{-1})
D	mass of dry solids (g)
D_o	initial mass of dry solids (g)
k	growth equation parameter (g-dry-biomass h^{-1} for the linear equation, h^{-1} for the deceleration equation)
M	mass of moist solids (g)
M_o	initial mass of moist solids (g)
S	mass of dry residual substrate (g)
S_o	initial mass of dry residual substrate (g)
t	time
W	mass of water (g)
W_o	initial mass of water (g)
X	mass of dry biomass (g)
μ	specific growth rate parameter (h^{-1})

Chapter 15

C_{XR}	biomass content, relative basis (g-dry-biomass g-dry-solids^{-1})
C_{XA}	biomass content, absolute basis (g-dry-biomass g-initial-dry-solids^{-1})
C_{CA}	concentration of a biomass component (mg-component g-initial-dry-solids^{1})
C_F	content of biomass component within the biomass (mg-component g-biomass^{-1})
d_i	dry mass of sample removed from the "ith" flask for determination of the moisture content (g)
D_i	dry mass of solids in the sample removed for biomass determination (g-dry-solids)
d_o	dry mass of sample removed at zero time (g)
G_x	biomass glucosamine content (mg-glucosamine mg-dry-biomass^{-1})
IDS_i	dry substrate initially added to the "ith" flask (g)
IWC	initial water content, wet basis (% by mass)
m_i	wet mass of sample removed from the "ith" flash for determination of the moisture content (g)
m_o	wet mass of sample removed at zero time (g)
M_i	fresh mass of the sample removed for biomass determination (g-moist-solids)
M_{oi}	mass of substrate initially added to the "ith" flask (g)
WC_i	moisture content of the "ith" flask at the time of sampling, wet basis (% by mass)
X	mass of biomass or a component (g)
λ	lag time (h)

Chapter 16

a_o to a_4 fitting parameters for Eq. (16.14)

a_{ws} water activity of the solid substrate phase

A parameter of the deceleration equation (dimensionless)

A fitting parameter of the double-Arrhenius equation (h^{-1})

A_d frequency factor for the death reaction (h^{-1})

A_g frequency factor for the growth reaction (h^{-1})

A_D frequency factor for denaturation reaction (dimensionless)

A_S frequency factor for synthesis reaction (dimensionless)

b fitting parameter of Eq. (16.16)

B fitting parameter of the double-Arrhenius equation (dimensionless)

C_m maximum biomass content (g-biomass 100-g-dry-matter^{-1})

C_{XA} biomass content, absolute basis (g-dry-biomass g-initial-dry solids^{-1})

C_{XAM} maximum biomass content, absolute basis (g-dry-biomass g-initial-dry-solids^{-1})

C_{XAO} initial biomass content, absolute basis (g-dry-biomass g-initial-dry-solids^{-1})

C_{XAD} absolute concentration of dead biomass (g-dry-biomass g-initial-dry solids^{-1})

C_{XAT} absolute concentration of total biomass, i.e., both viable and dead (g-dry-biomass g-initial-dry solids^{-1})

C_{XAV} absolute concentration of viable biomass (g-dry-biomass g-initial-dry solids^{-1})

C_{XR} biomass content, relative basis (g-dry-biomass g-dry-solids^{-1})

D_1 to D_4 fitting parameters of Eq. (16.17)

D total dry mass of solids in the bioreactor (kg)

D_o initial total dry mass of solids in the bioreactor (kg)

E_{a1}, E_{a2} fitting parameters of the double-Arrhenius equation (J mol^{-1})

E_{ad} activation energy for the death reaction (J mol^{-1})

E_{ag} activation energy for the growth reaction (J mol^{-1})

E_{aS} activation energy for synthesis reaction (J mol^{-1})

E_{aD} activation energy for denaturation reaction (J mol^{-1})

f specific growth rate as a fraction of the specific growth rate under optimum conditions (dimensionless)

f_T specific growth rate as a fraction of the specific growth rate under optimum conditions, based on temperature (dimensionless)

f_T specific growth rate as a fraction of the specific growth rate under optimum conditions based on water activity (dimensionless)

F state of the intracellular "essential enzyme pool" (dimensionless)

F_1 to F_3 fitting constants of Eq. (16.15)

k growth equation parameter (g-dry-biomass g-initial-dry solids^{-1} h^{-1} in the linear equation, h^{-1} in the deceleration equation)

k_d specific death rate coefficient (h^{-1})

k_D coefficient of the denaturation reaction (h^{-1})

k_S coefficient of the autocatalytic synthesis reaction (h^{-1})

m_S maintenance coefficient (kg-dry-substrate kg-dry-biomass $^{-1}$ h^{-1})

r_d	death rate (g-dry-biomass g-initial-dry solids^{-1} h^{-1})
R	universal gas constant (J mol^{-1} °C^{-1})
S	total dry mass of residual substrate (kg)
t	time (h)
T	temperature (°C).
T_{max}	maximum temperature for growth (°C)
T_{min}	minimum temperature for growth (°C)
T_{opt}	optimum temperature for growth (°C)
X	total dry mass of biomass (kg)
X_{max}	maximum total dry mass of biomass (kg)
Y_{XS}	true growth yield (kg-dry-biomass kg-dry-substrate^{-1})
μ	specific growth rate parameter (h^{-1})
$\mu_{measured}$	measured value of the specific growth rate (h^{-1})
μ_{opt}	specific growth rate parameter under optimal growth conditions (h^{-1})
μ_T	specific growth rate parameter as a function of temperature (h^{-1})
μ_W	specific growth rate parameter as a function of water activity (h^{-1})

Chapter 17

$a_{w(subscript)}$	water activity of phase or subsystem indicated by subscript
CER	carbon dioxide evolution rate (mol-CO_2 h^{-1})
COU	cumulative O_2 uptake (mol-O_2)
C_{in}	inlet O_2 concentration (typically %volume)
C_{out}	outlet O_2 concentration (typically %volume)
C_{XA}	absolute biomass concentration (kg-dry-biomass kg-dry-solids^{-1})
D	inactivation parameter used in Eqs. (17.8) and (17.9)
D_o	initial mass of dry solids within the bioreactor (kg)
F	dry air flow rate (L h^{-1})
L	initial particle length (m)
l_c	residual particle length at time t (m)
m_A	maintenance coefficient for production or consumption of the species indicated by subscript (kg-A kg-dry-biomass^{-1} h^{-1})
m_c	maintenance coefficient for CO_2 (mol-CO_2 kg-dry-biomass^{-1} h^{-1})
m_d	fitting constant in Eq. (17.9)
m_o	maintenance coefficient for O_2 (mol-O_2 kg-dry-biomass^{-1} h^{-1})
m_N	maintenance coefficient for nutrient (kg-nutrient kg-dry-biomass^{-1} h^{-1})
m_P	coefficient for product formation related to maintenance metabolism (kg-product kg-dry-biomass^{-1} h^{-1})
m_Q	maintenance coefficient for heat production (J kg-dry-biomass^{-1} h^{-1})
m_S	maintenance coefficient for residual dry substrate (kg-dry-substrate kg-dry-biomass^{-1} h^{-1})
m_W	maintenance coefficient for water production (kg-H_2O kg-dry-biomass^{-1} h^{-1}).
OUR	oxygen uptake rate (mol-O_2 h^{-1})
P	mass of product (kg)
P_o	mass of product present at time zero (kg)

$r_{subscript}$	overall rate of change in a species, with the particular species being indicated by the subscript (kg h^{-1} or mol h^{-1})
r_C	overall rate of CO_2 production (mol-CO_2 h^{-1})
r_N	overall rate of nutrient consumption (kg-nutrient h^{-1})
r_P	overall rate of product formation (kg h^{-1})
r_O	overall rate of O_2 consumption (mol-O_2 h^{-1})
r_Q	overall rate of metabolic waste heat production (J h^{-1})
r_W	overall rate of metabolic water production (kg-H_2O h^{-1}
t	time (h)
t_d	time at which inactivation kinetics appear (h)
t_r	time at which inactivation kinetics disappear (h)
T	time for complete particle degradation in Eq. (17.10)
X	total mass of biomass within the bioreactor (kg-dry-biomass)
X_m	maximum possible biomass, logistic equation (kg-dry-biomass)
X_o	initial biomass (kg-dry-biomass)
Y_{AB}	stoichiometric relationship between two species as indicated by subscripts A and B (kg-A kg-B^{-1})
Y_{CX}	yield of CO_2 from biomass (mol-CO_2 kg-dry-biomass^{-1})
Y_{PX}	yield of product from growth (kg-product kg-dry-biomass^{-1})
Y_{QC}	yield of heat from CO_2 production (J kg-$CO_2$$^{-1}$)
Y_{QX}	yield of heat from the growth reaction (J kg-dry-biomass^{-1})
Y_{WX}	yield of water from the growth reaction (kg-H_2O kg-dry-biomass^{-1})
Y_{XN}	yield of biomass from a nutrient (kg-dry-biomass kg-nutrient^{-1})
Y_{XO}	yield of biomass from O_2 (kg-dry-biomass mol-$O_2$$^{-1}$)
Y_{XS}	yield of biomass based on overall residual dry substrate (kg-dry-biomass kg-dry-substrate^{-1})
λ	fractional particle length (i.e., l_c/L in Eq. (17.10)) (dimensionless)
λ	time at which product begins to be produced (h), used in Eq. (17.14)
μ	specific growth rate parameter (h^{-1})

Chapter 18

$A_{subscript}$	area of the transfer surface indicated by subscript (m^2)
$a_{wsolid}*$	water activity for solids to be in equilibrium with the gas phase
$a_{w(subscript)}$	water activity of phase or subsystem indicated by subscript
C_{Pbed}	overall heat capacity of the bed (J kg^{-1} °C^{-1})
$C_{Psubscript}$	heat capacity of phase or subsystem indicated by subscript (J kg^{-1} °C^{-1})
D	total mass of dry solids in the bed (kg-dry-solids)
G	mass flux of dry air (kg-dry-air m^{-2} h^{-1})
$h_{subscript}$	heat transfer coefficient as indicated by subscript (J h^{-1} m^{-2} °C^{-1})
hA	global heat transfer coefficient (J h^{-1} °C^{-1})
H_{sat}	saturation humidity (kg-vapor kg-dry-air^{-1})
$H_{subscript}$	humidity of a subsystem or phase as indicated by subscript (kg-vapor kg-dry-air^{-1})
k	thermal conductivity (J m^{-1} h^{-1} °C^{-1})
k_w	mass transfer coefficient for water (kg-H_2O m^{-2} h^{-1})
m_{bed}	mass of the bed (kg)

M_{water}	overall mass of water in the bed (kg)
r_Q	rate of metabolic heat production (J h^{-1})
r_W	rate of metabolic water production (kg h^{-1})
$R_{subscript}$	rate of a mass transfer phenomenon that involves water (kg-H$_2$O h^{-1})
$Q_{subscript}$	rate of a heat transport phenomenon as indicated by subscript (J h^{-1})
t	time (h)
$T_{subscript}$	temperature of phase or subsystem indicated by subscript (°C)
V_Z	air superficial velocity (m h^{-1})
W	water content of the bed (kg-H$_2$O kg-dry-solids^{-1})
W_{sat}	water content that the solids would have if they were in equilibrium with the gas phase, dry basis (kg-H$_2$O kg-dry solid^{-1})
x	distance, usually horizontal distance within the phase (m)
z	distance, usually vertical (axial) distance within the phase (m)
λ	enthalpy of vaporization of water (J kg-H$_2$O^{-1})
$\rho_{subscript}$	density of phase or subsystem indicated by subscript (kg m^{-3})

Chapter 19

a	fitting parameter of the Antoine equation
a_{wg}	gas phase water activity
a_{ws}	water activity of the solids
b	fitting parameter of the Antoine equation
c	fitting parameter of the Antoine equation
$C_{Psubscript}$	heat capacity of phase or subsystem indicated by subscript (J kg^{-1} °C^{-1})
d	fitting parameter of the Antoine equation
H	humidity (kg-vapor kg-dry-air^{-1})
H	saturation humidity (kg-vapor kg-dry-air^{-1})
M_g	gas molecular weight (kg mol^{-1})
$m_{subscript}$	mass of the item indicated by subscript (g or kg)
n	number of moles (mol)
P	pressure (Pa)
P_w	vapor pressure of water (Pa)
P_{sat}	saturation vapor pressure of water (Pa)
R	universal gas constant (J mol^{-1} K^{-1})
S	shrinkage factor (m^3-dry-bed m^{-3}-moist-bed)
T	temperature (°C)
T_K	temperature (K)
T_s	solids temperature (°C)
V_P	specific packed volume on a dry basis (m^3 kg-dry-matter^{-1})
$V_{subscript}$	volume of phase or subsystem indicated by subscript (L or m^3)
w_i	mass fraction contributed by component "i"
W	solids water content, dry basis (kg- H$_2$O kg-dry-solids^{-1})
ε	bed porosity (dimensionless)
λ	enthalpy of vaporization of water (J kg-H$_2$O^{-1})
ρ_b	bed packing density (g L^{-1} or kg m^{-3})
$\rho_{subscript}$	density of phase or subsystem indicated by subscript (g L^{-1} or kg m^{-3})

Chapter 20

A	area for transfer (m^2)
A_g	cross-sectional area of headspace normal to gas flow (m^2)
c_{air}	dimensionless air humidity (as defined by Eq. (20.11))
c_{bed}	dimensionless saturation water vapor concentration
$C_{Psubscript}$	heat capacity of phase or subsystem indicated by subscript (J kg^{-1} °C^{-1})
C_V	dimensionless constant associated with the bed viscosity
d	particle diameter (m)
D	bioreactor diameter (m)
f	porosity factor (dimensionless)
F	inlet air flow rate (kg-dry-air s^{-1})
g	gravitational acceleration (m s^{-2})
G	air flux through the bed (kg-air m^{-2} s^{-1})
h	maximum height of the bed (m)
ha	"volumetric" overall heat transfer coefficient (J s^{-1} m^{-3} °C^{-1})
$h_{subscript}$	heat transfer coefficient, as indicated by subscript (J s^{-1} m^{-2} °C^{-1})
H_{sat}	saturation humidity (kg-vapor kg-dry-air^{-1})
$H_{subscript}$	humidity of phase indicated by subscript (kg-vapor kg-dry-air^{-1})
ka	scaled water mass transfer coefficient (s^{-1})
k_b	thermal conductivity of the bed (J h^{-1} m^{-1} °C^{-1} or J s^{-1} m^{-1} °C^{-1})
k_{wall}	thermal conductivity of the wall (J s^{-1} m^{-1} °C^{-1})
K	secondary variable calculated by Eq. (20.14)
Ka	"volumetric" overall mass transfer coefficient (kg-dry-solids s^{-1} m^{-3})
L	bioreactor length (m)
L_{wall}	wall thickness (m)
M	percentage moisture content, wet basis (% by mass)
N	rotational speed (revolutions per second)
P	pressure (Pa)
Pe_{eff}	effective Peclet number
R_w	scaled overall water transfer rate (s^{-1})
s	mobile layer thickness (m)
S	fraction of the critical speed (dimensionless)
t_c	time of contact between the solid particles and the bioreactor wall (s)
$T_{subscript}$	temperature of phase or subsystem indicated by subscript (°C)
u_P	average particle velocity (m s^{-1})
W	solids water content (kg-H$_2$O kg-dry-solids^{-1})
α_b	thermal diffusivity of the bed (m^2 h^{-1})
γ	dynamic angle of repose of the solids (degrees)
δ	diffusivity of water vapor in air (m^2 s^{-1})
ρ_b	bed density (kg m^{-3}-bed)

Chapter 22

Also see Tables 22.1 and 22.2.
The model converts all parameters and variables to a consistent set of units.

A	area for heat transfer across bioreactor side wall (m^2)
A_1 to A_4	fitting parameters of the double-Arrhenius equation (Eq. (22.1))

a_{wg}	gas phase water activity
a_{wgin}	inlet air water activity
a_{wgo}	initial gas phase water activity
$a_{wg}{}^*$	outlet gas water activity set point for triggering water addition
a_{ws}	water activity of the solids
a_{wso}	initial water activity of the solids phase
b_o	initial biomass content (kg-biomass kg-initial-dry-solids^{-1})
b_m	maximum biomass content (kg-biomass kg-initial-dry-solids^{-1})
B	mass of bioreactor wall (kg)
C_{Pb}	heat capacity of bioreactor body (J kg^{-1} °C^{-1})
C_{Pg}	heat capacity of dry gas (J kg^{-1} °C^{-1})
C_{Pm}	heat capacity of dry matter (J kg^{-1} °C^{-1})
$C_{Pv,}$	heat capacity of water vapor (J kg^{-1} °C^{-1})
C_{Pw}	heat capacity of liquid water (J kg^{-1} °C^{-1})
D	bioreactor diameter (m)
D_1 to D_4	fitting parameters of Eq. (22.2)
F_{in}	flow rate of dry air at the air inlet (kg-dry-air s^{-1})
$fold$	fold increase in the solids-to-gas heat and mass transfer coefficients
G	mass of dry air held in the inter-particle spaces (kg)
ha	"volumetric" overall heat transfer coefficient (J s^{-1} m^{-3} °C^{-1})
h_{bw}	bioreactor-to-cooling-water heat transfer coefficient (J s^{-1} m^{-2} °C^{-1})
h_{gb}	gas-to-bioreactor heat transfer coefficient (J s^{-1} m^{-2} °C^{-1})
h_{sb}	solids-to-bioreactor heat transfer coefficient (J s^{-1} m^{-2} °C^{-1})
H	gas phase humidity (kg-vapor kg-dry-air^{-1})
H_B	bioreactor height (m)
H_{in}	inlet air humidity (kg-vapor kg-dry-air^{-1})
J	proportional gain (dimensionless)
Ka	"volumetric" overall mass transfer coefficient (kg-dry-solids s^{-1} m^{-3})
L	thickness of the bioreactor wall (mm)
M	total mass of dry solids in the bioreactor (kg)
M_o	initial mass of dry solids in the bioreactor (kg)
P	overall pressure in the bioreactor (mm Hg)
R	universal gas constant (J mol^{-1} °C^{-1})
S_o	initial mass of dry substrate in the bed (kg)
t	time (h)
$Type$	type of relation of growth with solids water activity
T_b	bioreactor body temperature (°C)
T_g	gas phase temperature (°C)
T_{in}	inlet air temperature (°C)
T_{opt}	optimum temperature for growth (°C)
T_s	solids temperature (°C)
$T_{setpoint}$	set point temperature for the cooling water control scheme (°C)
T_{sys}	initial temperature of the system (°C)
T_w	cooling water temperature (°C)
V_{bed}	volume of the bed within the bioreactor (m^3)
vvm	volumes of air per bed volume per minute (m^3-air (m^3-bed)$^{-1}$ min^{-1})

W	water content on a dry basis (kg-H_2O kg-dry-solids^{-1})
W_o	initial water content on a dry basis (kg-H_2O kg-dry-solids^{-1})
W_{sat}	water content (dry basis) that the solids would have if they were in equilibrium with the gas phase (kg-H_2O kg-dry-solids^{-1})
X	total amount of biomass in the bioreactor (kg)
X_m	maximum amount of biomass in the bioreactor (kg)
X_o	initial amount of biomass in the bioreactor (kg)
Y_{QX}	yield of heat from growth (J kg-biomass^{-1})
Y_{XS}	yield of biomass from dry substrate (kg-biomass kg-dry-substrate^{-1})
Y_{WX}	yield of water from growth (kg-H_2O kg-biomass^{-1})
ε	effective porosity (void fraction) of the substrate bed (dimensionless)
λ	enthalpy of vaporization of water (J kg-H_2O^{-1})
μ	specific growth rate parameter (h^{-1})
μ_{FT}	fractional growth rate based on temperature (dimensionless)
μ_{FW}	fractional growth rate based on water (dimensionless)
μ_{opt}	specific growth rate parameter under optimal growth conditions (h^{-1})
μ_T	specific growth rate parameter as a function of temperature (h^{-1})
μ_W	specific growth rate parameter as a function of water activity (h^{-1})
ρ_a	density of the air phase (kg-dry-air m^{-3})
ρ_b	density of the bioreactor wall (kg m^{-3})
ρ_S	density of dry solid particles (kg m^{-3})

Chapter 23
Also see Tables 23.1 and 23.2.
The model converts all parameters and variables to a consistent set of units.

a_{wb}	water activity of the bed
a_{win1}	water activity of the inlet air when $T \le T_{opt}$
a_{win2}	water activity of the inlet air when $T > T_{opt}$
a_{wSP}	bed water activity set point
A_1 to A_4	fitting parameters of the double-Arrhenius equation (Eq. (22.1))
A_{bh}	area of contact between the bed and the headspace (m^2)
A_{bw}	area of contact between the bed and the bioreactor wall (m^2)
A_g	cross sectional area of the headspace normal to the gas flow (m^2)
A_{hw}	area of contact between the headspace and the bioreactor wall (m^2)
A_{ws}	area of contact between the bioreactor wall and the surroundings (m^2)
b_o	initial biomass concentration (kg-biomass kg-initial-dry-substrate^{-1})
b_m	maximum possible biomass concentration (kg-biomass kg-initial-dry-substrate^{-1})
B	total mass of bioreactor wall (kg)
C_{pb}	heat capacity of the bioreactor wall (J kg^{-1} °C^{-1})
C_{pg}	heat capacity of dry air (J kg^{-1} °C^{-1})
C_{pm}	heat capacity of dry matter (J kg^{-1} °C^{-1})
C_{pv}	heat capacity of water vapor (J kg^{-1} °C^{-1})
C_{pw}	heat capacity of liquid water (J kg^{-1} °C^{-1})
D	bioreactor diameter (m)

F	dry air flow through the headspace (kg-dry-air min^{-1})
G	mass of dry gas in the headspace (kg)
h_{bh}	bed-to-headspace heat transfer coefficient (J s^{-1} m^{-2} °C^{-1})
h_{bw}	bed-to-wall heat transfer coefficient (J s^{-1} m^{-2} °C^{-1})
h_{hw}	headspace-to-bed heat transfer coefficient (J s^{-1} m^{-2} °C^{-1})
h_{ws}	wall-to-surroundings heat transfer coefficient (J s^{-1} m^{-2} °C^{-1})
H	headspace humidity (kg-vapor kg-dry-air^{-1})
H_{in}	inlet gas humidity (kg-vapor kg-dry-air^{-1})
k_w	bed to headspace water mass transfer coefficient (kg-dry-solids s^{-1} m^{-2})
L	bioreactor length (m)
m_Q	maintenance coefficient for heat production (J kg-dry-biomass s^{-1})
m_S	maintenance coefficient for substrate (kg-dry-substrate s^{-1} kg-biomass^{-1})
m_W	maintenance coefficient for water production (kg-H$_2$O kg-dry-biomass s^{-1})
M	total mass of dry solids in the bed (kg)
n	fold-increase in transfer rates due to mixing
P	overall pressure in the bioreactor (mm Hg)
R	universal gas constant (J mol^{-1} °C^{-1})
t	time (h)
T_b	bed temperature (°C)
T_h	headspace temperature (°C)
T_{in}	inlet air temperature (°C)
T_{opt}	optimum temperature for growth (°C)
T_s	temperature of the surroundings (°C)
T_w	bioreactor wall temperature (°C)
$Type$	type of relation of growth with solids water activity
vvm	rate of dry air flow (m^3-air (m^3-bioreactor)$^{-1}$ min^{-1})
W	water content of the bed, dry basis (kg-H$_2$O kg-dry-solids^{-1})
W_o	initial water content of the bed, dry basis (kg-H$_2$O kg-dry-solids^{-1})
W_{sat}	water content (dry basis) that the solids would have if they were in equilibrium with the headspace gases (kg-H$_2$O kg-dry-solids^{-1})
X	total mass of dry biomass in the bed (kg-dry-biomass)
X_m	maximum possible mass of dry biomass in the bed (kg-dry-biomass)
X_o	initial mass of dry biomass in the bed (kg-dry-biomass)
Y_Q	metabolic heat yield coefficient (J kg-biomass^{-1})
Y_W	metabolic water yield coefficient (kg-H$_2$O kg-biomass^{-1})
Y_{XS}	biomass yield from dry substrate (kg-biomass kg-dry-substrate^{-1})
λ	enthalpy of vaporization of water (J kg-H$_2$O^{-1})
ρ_a	density of the air phase (kg-dry-air m^{-3})
ρ_b	overall density of the solid bed, wet basis (kg-wet-solids m^{-3})
μ	specific growth rate parameter (h^{-1})
μ_{opt}	specific growth rate parameter under optimum conditions (h^{-1})
$\%fill$	percentage of the drum volume occupied by the solid bed

Chapter 24

Also see Table 24.1.

The model converts all parameters and variables to a consistent set of units.

A_1 to A_4	fitting parameters of the double-Arrhenius equation (Eq. (22.1))
b	biomass concentration (kg-biomass kg-dry-substrate^{-1})
b_m	maximum biomass concentration (kg-biomass kg-dry-substrate^{-1})
C_{pa}	air heat capacity (J kg^{-1} °C^{-1})
C_{pb}	bed heat capacity (J kg^{-1} °C^{-1})
C_{ps}	substrate heat capacity (J kg^{-1} °C^{-1})
f	estimate of dH_{sat}/dT (kg-vapor kg-dry-air^{-1} °C^{-1})
h	overall bed-to-wall heat transfer coefficient in the Zymotis bioreactor (W m^{-2} °C^{-1})
H	bioreactor height (m)
H_B	height of the bed in the Zymotis bioreactor (m)
H_{sat}	saturation air humidity (kg-H$_2$O kg-dry-air °C^{-1})
k_a	air thermal conductivity (W m^{-1} °C^{-1})
k_b	bed thermal conductivity (W m^{-1} °C^{-1})
k_s	substrate thermal conductivity (W m^{-1} °C^{-1})
K	proportional gain factor (dimensionless)
L	spacing between plates in the Zymotis bioreactor (cm)
R	universal gas constant (J mol^{-1} °C^{-1})
S	dry substrate concentration (kg-dry-substrate m^{-3}-bed)
t	time (h)
t_{90}	time for the average biomass concentration to reach 90% of the maximum biomass concentration (h)
T	bed temperature (°C)
T_a	inlet air temperature for the Zymotis bioreactor (°C)
T_{in}	inlet air temperature for the packed bed (°C)
T_o	initial bed temperature (°C)
T_{opt}	optimum temperature for growth (°C)
T_{out}	outlet air temperature (°C)
T_w	cooling water temperature (°C)
T^*	temperature at the top of the bed, halfway between the heat transfer plates in the Zymotis bioreactor (°C)
V_Z	air superficial velocity (cm s^{-1})
x	horizontal distance within the repeating unit of the substrate bed within the Zymotis bioreactor (m)
X	biomass concentration (kg-biomass kg-substrate^{-1})
X_m	maximum possible biomass concentration (kg-biomass kg-substrate^{-1})
X_o	initial biomass concentration (kg-biomass kg-substrate^{-1})
Y_{bs}	yield of biomass from substrate (kg-dry-biomass kg-dry-substrate^{-1})
Y_Q	yield of metabolic heat from growth (J kg-dry-biomass^{-1})
Y_{Wb}	yield of water from biomass (kg-H$_2$O kg-dry-biomass^{-1})
z	vertical position (m)
ε	porosity (void fraction) of the bed (dimensionless)
λ	enthalpy of vaporization of water (J kg-H$_2$O^{-1})

μ	specific growth rate parameter (h^{-1})
μ_{opt}	value of the specific growth rate parameter at the optimum temperature for growth (h^{-1})
ρ_a	air phase density (kg-dry-air m^{-3})
ρ_b	bed density (kg m^{-3})
ρ_s	solid particle density (kg m^{-3})

Chapter 25
Also see Tables 25.1 and 25.2.
The model converts all parameters and variables to a consistent set of units.

a_{wg}	outlet gas water activity
a_{wgin}	inlet air water activity
$a_{wg}*$	outlet gas water activity set point
a_{ws}	water activity of the solids
a_{wso}	initial water activity of the solids
A_1 to A_4	fitting parameters of the double-Arrhenius equation (Eq. (22.1))
b	biomass concentration (kg-biomass kg-dry-solids^{-1})
b_o	initial biomass concentration (kg-biomass kg-dry-solids^{-1})
b_m	maximum biomass concentration (kg-biomass kg-dry-solids^{-1})
C_{pg}	heat capacity of dry air (J kg^{-1} °C^{-1})
C_{pv}	heat capacity of water vapor (J kg^{-1} °C^{-1})
C_{ps}	heat capacity of the dry solids (J kg^{-1} °C^{-1})
C_{pw}	heat capacity of liquid water (J kg^{-1} °C^{-1})
G	inlet air flux (kg-dry-air s^{-1} m^{-2})
H	gas phase humidity (kg-vapor kg-dry-air^{-1})
ha	"volumetric" overall heat transfer coefficient (J s^{-1} m^{-3} °C^{-1})
H_{in}	inlet air humidity (kg-vapor kg-dry-air^{-1})
Ka	"volumetric" overall mass transfer coefficient (kg-dry-solids s^{-1} m^{-3})
P	air pressure within the bioreactor (mm Hg)
R	universal gas constant (J mol^{-1} °C^{-1})
S	total dry solids per cubic meter (kg-dry-solids m^{-3})
S_o	initial dry substrate per cubic meter (kg-dry-solids m^{-3})
t	time (h)
t_{mix}	time taken by the mixing event (h)
T_g	gas phase temperature (°C)
T_{in}	inlet air temperature (°C)
T_s	solid phase temperature (°C)
T_{so}	initial temperature of the solid phase (°C)
T_{opt}	optimum temperature for growth (°C)
$Type$	type of relation of growth with solids water activity
W	water content of the bed, dry basis (kg-H_2O kg-dry-solids^{-1})
W_o	initial water content of the bed, dry basis (kg-H_2O kg-dry-solids^{-1})
W_{sat}	water content (dry basis) that the solids would have if they were in equilibrium with the gas in the void spaces (kg-H_2O kg dry solids^{-1})
Y_{BS}	yield of biomass from substrate (kg-biomass kg-dry-substrate^{-1})
Y_Q	yield of metabolic heat from growth (J kg-biomass^{-1})

Y_{WB}	yield of metabolic water from growth (kg-H_2O kg-biomass^{-1})
z	axial position (m)
Z	height of the bioreactor (m)
ε	porosity (void fraction) of the bed (dimensionless)
λ	enthalpy of vaporization of water (J kg-H_2O $^{-1}$)
μ	specific growth rate parameter (h^{-1})
μ_{FT}	fractional growth rate based on temperature (dimensionless)
μ_{FW}	fractional growth rate based on water (dimensionless)
μ_{mix}	fractional value of specific growth rate parameter during mixing (dimensionless)
μ_{opt}	specific growth rate parameter under optimal conditions (h^{-1})
ρ_s	density of the dry substrate particles (kg-dry-substrate m^{-3}-substrate)
ρ_g	density of the gas phase in the bed (kg-dry-air m^{-3})

Chapter 26

A, B, C	matrices
f	frequency (Hz)
N	number of measurements between two time instants
Q	known covariance of the process noise vector
R	known covariance of the measurement noise vector
u_k	vector of manipulated variables
v_k	pure random vector representing the measurement noise
w_k	pure random vector representing the process noise
y	a variable
y_i	value of the variable measured at time i
y_k	vector of measured variables
$\tilde{y}_N$	average value of the N measurements of a variable
α	gain inverse

Chapter 27

A	amplitude of the process output oscillations
C	control horizon
$e(t)$	error computed at time t
$F1$	cold fluid inlet flow rate
K_c	proportional gain
K_{cu}	ultimate gain
P	oscillation period
P	prediction horizon
Pu	ultimate period
r_k	vector of respective set points,
t	time
$T1$	cold fluid outlet temperature
u	the controller output
u_1, u_2	the two possible process input values
$u(t)$	controller output

u_L	lower bound on input
u_U	upper bound on input
W_D^y	diagonal matrix with weights that penalize the output deviations from the set points
$W_D^{\Delta u}$	diagonal matrix with weights that penalize the output deviations from the control movements
$\hat{y}_k$	vector of predicted plant outputs at time interval k
y_L	lower bound on output
y_U	upper bound on output
τ_d	derivative time
τ_i	integral time
Δu_k	vector of control moves

Chapter 28

$H_{subscript}$	humidity of phase indicated by subscript (kg-vapor kg-dry-air^{-1})
$T_{subscript}$	temperature of phase or subsystem indicated by subscript (°C)

Chapter 29

dH_{sat}/dT	change in the water-carrying capacity of air with a change in temperature (kg-vapor kg-dry-air^{-1} °C^{-1})
F	air flow-rate (m^3 h^{-1})
p	fan operating pressure (cm-H$_2$O)
P	fan power consumption (kW)
R_Q	maximum heat generation rate (kJ h^{-1})
Q_{rem}	capacity of the air to remove heat from the bed (kJ kg-air^{-1} °C^{-1})
V_b	bed volume (m^3)
X_{max}	maximum biomass content (kg-dry-biomass kg-dry-substrate^{-1})
W_{air}	required mass flow rate of air (kg h^{-1})
Y_q	metabolic heat yield coefficient (J kg-dry-biomass^{-1})
λ	enthalpy of evaporation of water (kJ kg-H$_2$O^{-1})
ρ_b	substrate packing density (kg-dry-substrate m^{-3})
μ_{max}	maximum value of the specific growth rate parameter (h^{-1})
ΔT	maximum allowable rise in air temperature (°C)

1 Solid-State Fermentation Bioreactor Fundamentals: Introduction and Overview

David A. Mitchell, Marin Berovič, and Nadia Krieger

1.1 What Is "Solid-State Fermentation"?

Solid-state fermentation (SSF) involves the growth of microorganisms on moist solid particles, in situations in which the spaces between the particles contain a continuous gas phase and a minimum of visible water. Although droplets of water may be present between the particles, and there may be thin films of water at the particle surface, the inter-particle water phase is discontinuous and most of the inter-particle space is filled by the gas phase. The majority of the water in the system is absorbed within the moist solid-particles (Fig. 1.1(a)). More detail about the spatial arrangement of the system components is given in Chap. 2.

In fact, here we follow the nomenclature proposed by Moo-Young et al. (1983) where the more general term "solid-substrate fermentation" is used to denote any type of fermentation process that involves solids, including suspensions of solid particles in a continuous liquid phase and even trickling filters (Fig. 1.1(b)). Therefore solid-state fermentation is classified as one type of solid-substrate fermentation. In this book we concentrate specifically on solid-state fermentation systems, in the manner that we defined them in the first paragraph.

The aim of the present section is not to give an in-depth explanation of all the characteristics of SSF systems, nor to compare SSF with submerged liquid fermentation (SLF). The further reading section at the end of this chapter gives some sources of general background information for readers who do not have much familiarity with SSF systems. Here we will give only a very broad summary of some of the main points:

- The majority of SSF processes involve filamentous fungi, although some involve bacteria and some involve yeasts.
- SSF processes may involve the pure culture of organisms, or the culture of several pure strains inoculated simultaneously or sequentially, while in some processes a "self-selected" microflora arises from the original microflora (e.g., in composting) or from a specially prepared traditional inoculum.
- The majority of SSF processes involve aerobic organisms. Note that we use the word "fermentation" in this book in the sense of its more general meaning, that

is, "the controlled cultivation of organisms" (the SSF literature uses the word fermentation in this sense).

- The substrates used in SSF processes are often products or byproducts of agriculture, forestry or food processing. Typically the source of nutrients comes from within the particle, although there are some cases in which nutrients are supplied from an external source. Usually a polymer gives the solid structure to the particle and this polymer may or may not be degraded by the microorganism during the fermentation. There are also some cases in which artificial or inert supports are used, with a nutrient solution absorbed within the matrix.

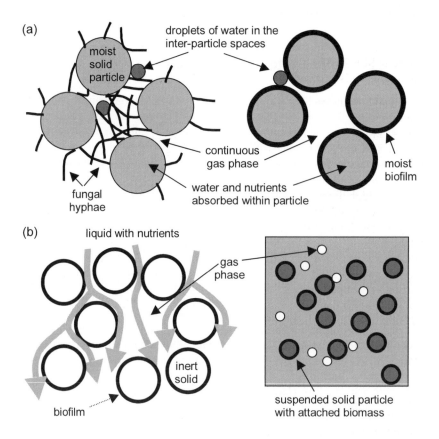

Fig. 1.1. The defining features of solid-state fermentation (SSF) systems (following the terminology of Moo-Young et al. 1983). **(a)** The arrangement of moist solid particles and the continuous gas phase in SSF systems involving a filamentous fungus (left-hand side) and a unicellular organism (right-hand side). **(b)** Other systems that involve growth on solids, but which are not defined as SSF due to the large amount of water in the inter-particle spaces. The left-hand diagram represents a trickling-filter type system while the right-hand diagram represents a suspension or slurry system

Much of this book will assume that we are working with pure cultures of aerobic filamentous fungi, to produce a specific product. In this case, there is a definite set of optimum conditions for growth of the process organism and product formation by it. Therefore this book does not consider composting, which is a specific application of SSF in which it is desirable for the temperature to vary during the process. Of course, with this and other important differences, such as the use of undefined mixed cultures, composting has its own literature, which is not directly relevant to the type of SSF process in which we are interested.

1.2 Why Should We Be Interested in SSF?

The environment that the organism experiences in SSF is different from that experienced in SLF. In SLF it is relatively easy to control the conditions to which the process organism is exposed:

- the fungal hyphae are bathed in a liquid medium and do not run the risk of desiccation;
- temperature control is typically not overly difficult, such that the organism is exposed to a constant temperature throughout its growth cycle;
- the availability of O_2 to the biomass can be controlled reasonably well at a particular level of saturation of the medium (although this can become very challenging in high density cultures);
- the availability of the nutrients to the organism can be controlled within relatively narrow limits if desired, through the feeding of nutrient solutions (at least in those processes in which soluble carbon and energy sources are provided);
- although shear forces do occur within mechanically stirred bioreactors, the nature and magnitude of these forces are well understood and it is possible to use bioreactors that provide a low-shear environment, if the organism is highly susceptible to shear damage, such as bubble columns or air lift bioreactors;
- pH control is relatively easy to provide.

In contrast, the environment in SSF can be quite stressful to the organism. For example:

- fungal hyphae are exposed to an air phase that can desiccate them;
- temperatures can rise to values that are well above the optimum for growth due to the inadequate removal of waste metabolic heat. In other words, the temperature to which the organism is exposed can vary during the growth cycle;
- O_2 is typically freely available at the surface of the particle, however, there may be severe restrictions in the supply of O_2 to a significant proportion of the biomass that is within a biofilm at the surface or penetrating into the particle;
- the availability of nutrients to the organism may be poor, even when the average nutrient concentration within the substrate particle, determined after homogenizing a sample of fermenting solid particles, is high. In other words, there tend to be large concentration gradients of nutrients within the particles;

- movement of the particles of the solid substrate can cause impact and shear damage. In the case of fungal processes the hyphae can suffer severe damage;
- it may be difficult to provide pH control.

Also, due to the different physical natures of the two systems, namely the presence of solid-air interfaces in SSF, growth morphologies of mycelial organisms, in terms of hyphal extension and branching patterns, may be quite different between SSF and SLF. This can be linked to different patterns of expression of genes, including those for several potential biotechnological products (Ishida et al. 2000).

These, and other differences, mean that SLF is an "easier" system with which to work. The ease of using SLF is greater still when substrate handling is considered. For example, it is much simpler and cheaper to pump liquids from one place to another than to move solids and it is easier to sterilize a large volume of liquid than a large volume of solids (in either batch or continuous sterilization mode). Given all these potential difficulties, for both the operator and the microorganism, it would appear that SLF should be the fermentation method of choice. In fact, in the majority of cases it is! However, there are certain instances in which, despite being more problematic, SSF may be appropriate:

- when the product needs to be in a solid form (e.g., fermented foods);
- when a particular product is only produced under the conditions of SSF or, if produced in both SLF and SSF, is produced in much higher levels in SSF. For example, certain enzymes are only induced in SSF and some fungi only sporulate when grown in SSF, in which the hyphae are exposed directly to an air phase. If it is desired to use genetically unmodified organisms in a process for the production of such a product, then SSF may be the only option;
- when the product is produced in both SLF and SSF, but the yield is much higher in SSF. For example, *Monascus* pigment and many fungal spores are produced in much higher yields in SSF;
- when socio-economic conditions mean that the fermentation process must be carried out by relatively unskilled workers. Some SSF processes can be relatively resistant to being overtaken by contaminants;
- when the product is produced in both SSF and SLF, but the product produced in SSF has desirable properties which the product produced in SLF lacks. For example, spore-based fungal biopesticides produced in SSF processes are usually more resistant to adverse conditions than those produced in SLF, and are therefore more effective when spread in the field;
- when it is imperative to use a solid waste in order to avoid the environmental impacts that would be caused by its direct disposal. This is likely to become an increasingly important consideration as the ever-increasing population puts an increasing strain on the environment.

1.3 What Are the Current and Potential Applications of SSF?

The considerations raised in the previous section have meant that SSF technology has been used for many centuries. Some examples of traditional SSF processes are:

- tempe, which involves the cultivation of the fungus *Rhizopus oligosporus* on cooked soybeans. The fungal mycelium binds the soybeans into a compact cake, which is then fried and eaten as a meat substitute. This fermented food is quite popular in Indonesia;
- the *koji* step of soy sauce manufacture, which involves the cultivation of the fungus *Aspergillus oryzae* on cooked soybeans. During the initial SSF process of 2 to 3 days, the fungal mycelium not only covers the beans but also secretes a mixture of enzymes into them. The fermented beans are then transferred into brine, in which, over a period of several months, the enzymes slowly degrade the soybeans, leaving a dark brown sauce.
- ang-kak, or "red rice", which involves the cultivation of the fungus *Monascus purpureus* on cooked rice. The fungus produces a dark red pigment. At the end of the fermentation the red fermented rice is dried and ground, with the powder being used as a coloring agent in cooking.

Beyond this, over the last three decades, there has been an upsurge in interest in SSF technology, with research being undertaken into the production of a myriad of different products, including:

- enzymes such as amylases, proteases, lipases, pectinases, tannases, cellulases, and rennet;
- pigments;
- aromas and flavor compounds;
- "small organics" such as ethanol, oxalic acid, citric acid, and lactic acid;
- gibberellic acid (a plant growth hormone);
- protein-enriched agricultural residues for use as animal feeds;
- animal feeds with reduced levels of toxins or with improved digestibility;
- antibiotics, such as penicillin and oxytetracycline;
- biological control agents, including bioinsecticides and bioherbicides;
- spore inocula (such as spore inoculum of *Penicillium roqueforti* for blue cheese production).

There is also research into the use of microorganisms growing in SSF conditions to mediate processes such as:

- decolorization of dyes;
- biobleaching;
- biopulping;
- bioremediation.

These processes commonly use waste products or byproducts of agriculture and food processing, selected as appropriate to favor growth of the producing organism and formation of the desired product. Such wastes and byproducts include wheat bran, rice bran, oil-press cakes, apple pomace, grape pomace, banana peels, citrus peels, wheat straw, rice straw, coffee pulp, citrus pulp, sugar beet pulp, coffee husk, and sugar beet molasses. Sometimes higher-value agricultural and food materials are used, such as granular milk curds, fodder beets, rice, and cassava meal. Recently there has also been some interest in the use of inert supports impregnated with nutrient solutions; at times natural inert supports such as sugar cane bagasse have been used, at other times artificial supports have been used, such as polyurethane foam cubes.

Note that the list presented above highlights only a small proportion of the overall activity in the development of SSF processes. Various reviews have been published on the applications of SSF, including details of the organisms and substrates used and the current chapter does not intend to repeat the information presented in these reviews. Readers with further interest should consult the reference section at the end of the chapter.

1.4 Why Do We Need a Book on the Fundamentals of SSF Bioreactors?

So if solid-state fermentation has such potential, why is it not a more widely used technology? Why are there relatively few "large-scale success stories" such as exemplified by the *koji* step of soy sauce production? Of course, part of the problem has already been touched upon in Sect. 1.2: Our inability to control conditions may well put a stress on the organism that causes it to produce a useful product in large quantities; however, too much stress may reduce yields and even kill the organism.

For SSF to be a more widespread technology, we need to know how to apply it, when appropriate, at both small scale (in "domestic" industries) and large scale (that is, involving large quantities in bioreactors). There is a lot of know-how related to the production of traditional fermented foods that involve SSF, which allows us to operate small-scale processes well. However, with the exception of certain success stories, SSF has not found widespread application at large scale. Why?

One of the problems is that we do not have the knowledge to translate success of one large-scale process (e.g., soy sauce *koji*) into the success of other large-scale processes. Here we are specifically talking about the question of "How do we design and operate large-scale bioreactors in such a manner as to have a profitable process". Our success with large-scale soy sauce *koji* does not necessarily translate into success with products that have lower profit margins.

Unlike SLF, for SSF we do not have a broad general theory or tools for designing and optimizing the operation of large-scale bioreactors. Of course, in both SLF and SSF each particular process can have its peculiarities, so a general theory does not mean that technology can be directly transferred from one process to another, but such a general theory does help by allowing one to focus on those peculiarities. The "theoretical foundations of SLF technology" (by which we really mean "the application of quantitative or engineering principles") began to be established in the late 1940s, and have been continuously extended and refined since then. In comparison, the engineering principles of SSF bioreactors only began to be developed around the late 1980s. Before then, it seems that the large-scale *koji* processes must have been developed over time through trial-and-error and experience, although it is also possible that soy sauce companies do have a good fundamental engineering know-how, but do not publish it in the general literature.

The consequence of the lack of these "theoretical foundations of SSF technology" is that, despite a very large upsurge of interest since the late 1970s (as judged by the increase in the number of publications on the topic of SSF in the scientific literature), there have been relatively few process that, having shown promise in the laboratory, have managed to leave the laboratory and be established as large-scale commercial processes. These processes perform well in the laboratory, where it is a trivial problem to provide O_2 and remove heat from the bed, but when attempts are made to establish large-scale processes, it is found to be impossible to control important process parameters, such as the temperature, within acceptable limits.

However, we have now reached a stage where our understanding is sufficient for it to be appropriate to bring together the theoretical foundations of SSF technology. This is what we aim to do in this book. However, our intention is not to be comprehensive in the sense of presenting all the engineering know-how so far generated for SSF bioreactors. Rather, we aim to introduce the fundamental concepts and ideas. An understanding of these fundamentals will provide the basis for readers to progress to the more advanced principles that are currently being established and published in the literature.

Beyond bringing the fundamental principles together, the book aims to provide a guide, based on current knowledge, about how best to design and operate the various different types of SSF bioreactor. Hopefully, it will stimulate further research into the area of SSF bioreactor performance.

The main argument of this book is that we need to apply a "biochemical engineering approach" to the problem of designing and optimizing the operation of SSF bioreactors. By a "biochemical engineering approach", we mean:

- the quantitative characterization of the key phenomena responsible for controlling bioreactor performance;
- the mathematical description of these phenomena within models intended to guide bioreactor design and operation;
- undertaking this characterization and description at an appropriate level of complexity, with the appropriate level depending on the balance between the usefulness of the mathematical tools in improving process performance and the mathematical and experimental difficulty in obtaining the functioning model.

Essentially, we are saying that it is necessary to develop mathematical models of the important phenomena, and use them as tools within experimental programs for bioreactor development.

If we do in fact achieve our aim of stimulating the development of SSF bioreactor technology, we can foresee a future time in which, in the development of any particular microbial fermentation product, both SSF and SLF will be considered, and the most promising of the two will be selected. SSF will not simply be ignored due to the lack of "know-how", as is currently the case in many parts of the world, especially those that do not have traditional fermented foods that are produced using SSF.

1.5 How Is this Book Organized?

As shown by Fig. 1.2, this book can be seen as consisting of five different parts. The subsections that follow give an overview of the argument that is developed within each of these parts.

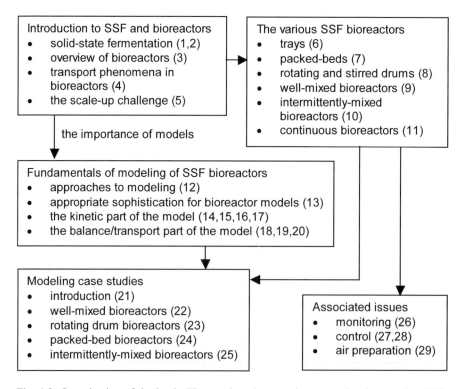

Fig. 1.2. Organization of the book. The numbers in parentheses are the chapters in which the various topics are covered

1.5.1 Introduction to Solid-State Fermentation and Bioreactors

Chapters 2 to 5 show the complexity of the task of designing efficient large-scale SSF bioreactors. Much of this complexity derives from the fact that the performance of an SSF bioreactor is the result of a complex interaction between biological and transport processes. Further, not only do the rates of these processes vary over time, but also processes such as heat and mass transfer involve several different phases within the bioreactor. Chapter 2 gives an overview of the phases present in an SSF bioreactor and the interaction between the biological and transport phenomena during the fermentation.

Chapter 3 then gives an overview of the various bioreactor types that have been used in SSF. The aim is not to give an exhaustive description of all design variations, but rather to recognize that it is useful to classify the many bioreactors into four groups, based on the manner in which they are aerated and agitated.

An understanding of the transport phenomena that occur in SSF bioreactors is essential in order to appreciate the difficulty of designing and operating efficient large-scale SSF bioreactors. Chapter 4 therefore introduces the important transport phenomena. This is done in a qualitative manner, with the quantitative aspects being covered later in the book. With this basis, Chap. 5 then explains the scale-up problem, or in other words, how limitations in mass and heat transfer mean that it is not appropriate to take a successfully operating laboratory-scale bioreactor and then simply design a geometrically identical larger version. This will not lead to a successfully operating large-scale bioreactor. Within this chapter, it becomes clear that mathematical models that combine the various biological and physical phenomena are essential as tools to guide bioreactor design and the optimization of bioreactor operation. The basic principles of these models and their application to particular bioreactors take up much of the latter part of the book.

1.5.2 Introduction to the Various Classes of SSF Bioreactors

Chapters 6 to 10 describe the various types of SSF bioreactors that have been used in batch-mode. For this purpose, they are divided into groups based on the aeration and agitation strategies. Classical bioreactors with these groups include tray bioreactors, packed-bed bioreactors, rotating drum bioreactors, and well-mixed or intermittently-mixed bioreactors with forced aeration. For each class of bioreactors the basic design and operating features are described, as well as several of the possible variations in these features. These chapters also relate information from the SSF literature about how these various bioreactors perform, highlighting the relative ease or difficulty of controlling conditions within the bioreactor and thereby of obtaining high productivity or not.

Continuous operation of SSF bioreactors is a subject that has received relatively little attention in the SSF literature. Chapter 11 describes the various ways in which SSF bioreactors can be operated in continuous mode, and also undertakes a preliminary analysis of bioreactor performance in this mode of operation. However, it will be clear that this is an area that needs much more attention.

1.5.3 Fundamentals of Modeling of SSF Bioreactors

Chapters 12 to 20 cover various aspects that are fundamental to an understanding of how to model SSF bioreactors. Chapter 12 starts with an overview of how modeling is undertaken, outlining a series of steps. The first of these steps involves making a decision about what degree of complexity is desired in the model, with more complex models potentially being more useful tools than simpler models, but also requiring much greater effort and sophistication, not only in the formulation and solution of the model equations, but also in the experimental work necessary to determine the various parameters that appear in the model. Chapter 13 then applies this question to mathematical models of SSF bioreactors, arguing that currently the best strategy is to develop and use so-called "fast-solving" models.

Chapter 12 also makes it clear that any model of an SSF bioreactor can be thought of as being comprised of two sub-models, a kinetic sub-model that describes the growth of the microorganism and a balance/transport sub-model that describes the various physical phenomena within the bioreactor. Chapter 13 argues that if a fast-solving model is desired, the kinetic sub-model should be quite simple and should not attempt to describe the dependence of the growth rate on nutrient concentrations, in order to avoid the necessity of describing simultaneous diffusion and reaction phenomena within the substrate particle.

The various steps in establishing appropriate equations for the kinetic sub-model are presented in Chaps. 14 to 17. Chapter 14 presents some basic considerations, highlighting one of the intrinsic difficulties faced in SSF systems, namely the difficulty in determining the amount of biomass in the system, which is especially problematic when the process organism is a filamentous fungus. It also presents the equations that are typically used to describe growth profiles in SSF and how the parameters of these equations can be determined by regression. Chapter 15 describes experimental systems and approaches that you can use to establish the growth profile for your own SSF system. Chapter 16 then shows how the equations should be written within the bioreactor model: Whereas the regression analysis of the growth profile is undertaken with the integral form of an equation, the equation must appear in a differential form within the kinetic sub-model. Chapter 16 also shows how the effect of the loss of dry matter from the system in the form of CO_2 can be taken into account in the kinetic sub-model. The equations developed in Chap. 16 involve various growth parameters that are in fact functions of the local conditions experienced by the microorganism, such as temperature and water activity. Chapter 17 shows how experiments can be undertaken and analyzed in order to establish appropriate correlations that give the value of the growth parameters for any given combination of local conditions. However, it will be obvious in this chapter that this is an area that needs further development.

The balance/transport sub-model is addressed in Chaps. 18 to 20. Chapter 18 introduces the concept of balance equations, showing how they include terms to describe the various transport phenomena that occur within and between subsystems within the bioreactor. The basic mathematical expressions used in these terms are presented. These expressions contain various parameters and physical constants, the values of which must be known in order to solve the bioreactor model. Chapter 19 describes these parameters and indicates how they might be determined. The balance/transport model also contains various heat and mass transfer coefficients. Chapter 20 describes various correlations that have been used and also lists some typical values that have been reported in the SSF literature.

Note that these chapters are written at a level intended for non-engineers. This section will not teach non-engineers all the skills that are needed for writing and solving models of SSF bioreactors. However, if you are not an engineer, these chapters will help you to understand the issues involved and this will greatly enrich your interaction with engineers during the bioreactor design process.

1.5.4 Modeling Case Studies of SSF Bioreactors

After a brief introduction in Chap. 21, Chaps. 22 to 25 present case studies in which fast-solving models are used to explore the design and operation of various SSF bioreactors. These include well-mixed bioreactors with forced aeration (Chap. 22), rotating drum bioreactors (Chap. 23), packed-bed bioreactors (Chap. 24), and intermittently-mixed forcefully-aerated bioreactors (Chap. 25). The case studies ask and answer questions such as "What aeration rate will be needed in order to control the bed temperature adequately in a large-scale bioreactor?".

These models, although still needing various improvements, can already be used as useful tools in the process of designing SSF bioreactors. The programs that are used in these chapters are available to readers from a web site. Details of this site and of the use of these programs are given in the Appendix.

1.5.5 Key Issues Associated with SSF Bioreactors

The last section of the book addresses several key issues in the operation of SSF bioreactors. Chapter 26 describes various process variables that we might like to monitor during the fermentation and gives suggestions for equipment that might be used to do this. It also addresses the question of data filtering, which is essential in order to eliminate random noise from the measured data.

Of course, one of the reasons that we might like to monitor the fermentation is to be able to undertake control actions in order to maintain the conditions in the bioreactor as near as possible to the optimum conditions for growth and product formation. Process control is a complex science. Chapter 27 introduces the basic principles of process control, at a level aimed for the non-engineer, although it is impossible to do this without presenting at least a few complicated mathematical equations! Chapter 28 then describes how control schemes can be applied to SSF bioreactors. It will become clear that this is an area that is still quite rudimentary and needs much more development.

Finally, a key step in the operation of an SSF bioreactor is the supply of air at an appropriate flow rate, temperature, and humidity. Chapter 29 describes how the air preparation system can be designed to do this and various related issues such as the selection of the air blower and the need for filtration. It will become clear that it is not an easy task to adjust the flow rate, temperature, and humidity of the air, independently, without building highly sophisticated systems. It presents a case study of the development of an air preparation system for a pilot-scale SSF bioreactor.

1.5.6 A Final Word

Solid-state fermentation bioreactor technology is still developing. We hope that this book stimulates you either to apply the principles presented to the design of a bioreactor for your own SSF process or even to contribute to development of the technology itself!

Further Reading

General features and applications of SSF
Doelle HW, Mitchell DA, Rolz CE (eds) (1992) Solid substrate cultivation. Elsevier Applied Science, London

A broad overview of solid-state fermentation
Mitchell DA, Berovic M, Krieger N (2002) Overview of solid state bioprocessing. Biotechnol Ann Rev 8:183–225

Physiological advantages that make SSF interesting for the production of certain products
Holker U, Hofer M, Lenz J (2004) Biotechnological advantages of laboratory-scale solid-state fermentation with fungi. Applied Microbiology and Biotechnology 64:175–186

Applications of SSF
Pandey A, Soccoll CR, Mitchell D (2000) New developments in solid-state fermentation: I – Bioprocesses and products. Process Biochemistry 35:1153–1169

2 The Bioreactor Step of SSF: A Complex Interaction of Phenomena

David A. Mitchell, Marin Berovič, Montira Nopharatana, and Nadia Krieger

2.1 The Need for a Qualitative Understanding of SSF

As argued in Chap. 1, mathematical models of bioreactor operation will be important tools in the development of bioreactors for solid-state fermentation (SSF) processes. These mathematical models must describe quantitatively the various phenomena within the SSF process that can potentially limit the performance of the bioreactor. One of the key early steps in modeling is to identify what these phenomena are, and to unite them in a qualitative description of the system, at an appropriate level of detail (an idea that will be developed further in Chap. 13). The current chapter provides a basis for this by describing SSF processes qualitatively, from several different perspectives. The current chapter presents:

- An overview of SSF processes.
- The physical structure of SSF systems.
- The phenomena occurring in SSF processes, including phenomena occurring at the microscale (the scale of the individual particle) and phenomena occurring at the macroscale (the scale of the bioreactor, looking at the substrate bed as a whole).

This chapter will make it very clear that the system is highly complex, and it will therefore be obvious that it is only with mathematical models that we can manage to understand the complex system behavior that stems from the combination of microscale and macroscale phenomena. Also, it will be clear that different phenomena will limit the performance of the process at different times during the fermentation, and that the relative importance of the various phenomena will depend on characteristics of the particular organism, substrate, and bioreactor that are used in a particular process. This understanding can lay the foundation for improvements in process performance.

2.2 The General Steps of an SSF Process

At the most general level, the major processing steps of an SSF process are no different from those of a submerged liquid fermentation (SLF) process, with which we assume that the reader has a general familiarity. These processing steps include (Fig. 2.1):

- Inoculum preparation
- Substrate preparation
- Bioreactor preparation
- Inoculation and loading
- Bioreactor operation
- Unloading
- Downstream processing
- Waste disposal

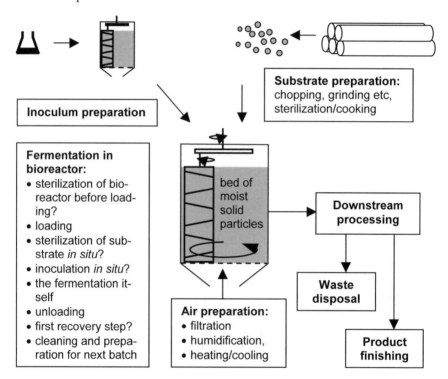

Fig. 2.1. An overview of an SSF process operated in batch mode. Note that the details can vary from process to process. For example, the substrate might either be sterilized within the bioreactor or sterilized before being added. At this level of detail, a solid-state fermentation process is no different from an SLF process. However, the details of the bioreactor and how it is designed and operated vary significantly between SSF and SLF

In the development of a process, attention must be given to all these steps. Some of the issues that need to be addressed in the various process steps are briefly mentioned below:

Substrate preparation. The substrate may need to be cut, milled, cracked, or granulated in order to obtain particles of an appropriate size. It may be necessary to add water and nutritional supplements or to cook or pre-treat the substrate to increase the availability of nutrients. The substrate might be sterilized, or at least pasteurized, outside the bioreactor. Alternatively, it may be possible and preferable to do this step with the substrate inside the bioreactor.

Inoculum preparation. The type and method of inoculum preparation depends on the microorganism involved. Many SSF processes involve filamentous fungi and therefore spore-based inocula may be used. The aim of this step is to develop an inoculum of sufficient size and high viability. The inoculum can often be prepared in one of various forms. For a fungal fermentation it may be possible to produce a suspended mycelial inoculum by SLF, or to undertake a solid-state fermentation followed either by suspension of spores in a liquid or by drying and grinding of the solid to produce a powder than can be used as the inoculum.

Bioreactor preparation. The bioreactor must be cleaned after the previous fermentation, and may need to be sterilized before addition of the substrate, although, as noted above, in some cases it might be appropriate to sterilize the substrate inside the bioreactor.

Inoculation and loading. The inoculation step may occur either prior to loading or after loading. If the substrate bed cannot be mixed within the bioreactor, inoculation must be done outside the bioreactor. If the bed can be mixed, then the best method of inoculation might be to spray the inoculum as a mist over the bed as it is being mixed. If the substrate is pasteurized or sterilized and inoculated outside the bioreactor, it may be necessary to undertake the loading step quite carefully in order to prevent or at least minimize the entry of contaminants. At large scale, loading will need to be mechanically assisted.

Bioreactor operation. Much attention will be paid to this step later in the book. The details will depend on the specific bioreactor design, however, the general task is to manipulate various operating variables, such as the flow rate and temperature of the inlet air, the bed mixing speed, and the cooling water temperature, in order to control key fermentation parameters, such as bed temperature and water activity, at the optimum values for growth and product formation.

Unloading. In some cases a leaching or drying step is undertaken within the bioreactor, in other cases the product recovery steps are undertaken outside of the bioreactor. In any case, solids must eventually be removed from the bioreactor. At large scale, unloading will need to be mechanically assisted.

Downstream processing. Depending on the process, either the whole of the fermented solids represents the product or a specific product is recovered from the solids and then purified. In the latter case, the extraction of the product from the solids represents a step in SSF processes that is not necessary in SLF processes. However, after extraction, the general principles of downstream processing are similar for both SSF and SLF.

Waste disposal. SSF is often suggested as a means of minimizing the impact of waste solid organic materials by preventing their being dumped in the environment. In some cases the whole solid is used as the product, for example, as a food or animal feed, but in others there will be a solid residue that must be disposed of adequately.

2.3 The Bioreactor Step of an SSF Process

The bioreactor step is a key step in an SSF process. It is in this step that the bioconversion takes place. More details about bioreactor operation will be given later. At this point it is only necessary to understand the general features of a typical SSF bioreactor and how it might be operated (Fig. 2.2). The bioreactor has two important functions:

- to hold the substrate bed and provide a barrier against both the release of the organism to the surroundings and the contamination of the substrate bed by organisms in the surroundings.
- to control, to the degree that is possible, the key environmental conditions, such as the bed temperature and water activity, at values which are optimal for growth and product formation by the microorganism.

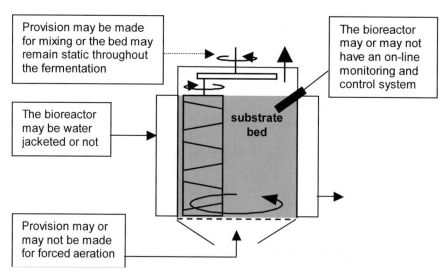

Fig. 2.2. A generalized diagram of an SSF bioreactor. Not all SSF bioreactors have all the features shown here. Details of the design of the various bioreactor types used in SSF are given in Chaps. 6 to 11

It is not possible simply to set the environmental conditions within the substrate bed at the desired value. The growth of the organism will tend to change the environmental conditions away from the optimal values and we must then intervene in order to try to bring them back to the optimum (Fig. 2.3). However, we can only manipulate a limited number of variables that are external to the bioreactor, called "operating variables". For example, we can change the agitation regime (if the bioreactor is agitated), the temperature, flow rate, and humidity of the inlet air (if the bioreactor is aerated), the addition of solutions or substances or the temperature and flow rate of the cooling water (if the bioreactor has cooled heat-transfer surfaces). The success of these external interventions in bringing the bed conditions back to the optimum values depends on the efficiency of the heat and mass transport processes within the substrate bed.

2.4 The Physical Structure of SSF Bioreactor Systems

In order to understand the phenomena occurring within an SSF bioreactor, it is necessary to understand the physical arrangement of the various phases within the system, since the various phenomena occur within and between these phases. We can choose two different levels of detail to examine the physical structure of the system, the macroscale and the microscale, as shown in the following subsections.

2.4.1 A Macroscale View of the Phases in an SSF Bioreactor

From a macroscale perspective, the bioreactor contains three phases (Fig. 2.4(a)):

- the bioreactor wall;
- a headspace full of gas, the extent of which depends on the bioreactor type;
- a substrate bed, composed of particles and air within the inter-particle spaces.

The bioreactor wall is important as a barrier. It should be a complete barrier to mass transfer. Matter can only enter the bioreactor through holes in this wall (addition ports, sampling ports, loading/unloading ports). It is a partial barrier to energy transfer. Energy can cross this boundary by conduction, at a rate that depends on the thermal properties of the material from which the bioreactor was constructed. Note that energy can also be stored in this wall, this storage being manifested as an increase in the temperature of the wall.

The headspace has functions such as:

- allowing the air that leaves the bed to reach the air outlet of the bioreactor. Foaming will typically not be a problem in SSF bioreactors, so the headspace does not play the role in foam control that it does in SLF bioreactors. In some SSF bioreactors, in which the solid particles within the bed are suspended in an air stream (gas-solid fluidized-beds), the headspace allows room for bed expansion and for disengagement of particles and air;

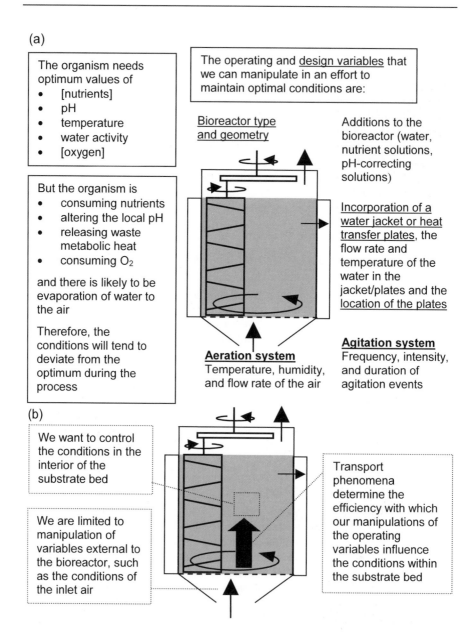

Fig. 2.3. The difficulty of controlling the conditions within an SSF bioreactor. **(a)** The organism changes the values away from the optimum values for growth and product formation, and we have only a limited number of operating variables, involving manipulations external to the substrate bed, with which we can attempt to bring the conditions back to the optimal values. **(b)** Transport phenomena within the bed determine the effectiveness with which any manipulation of the operating variables can control the conditions in the interior of the bed

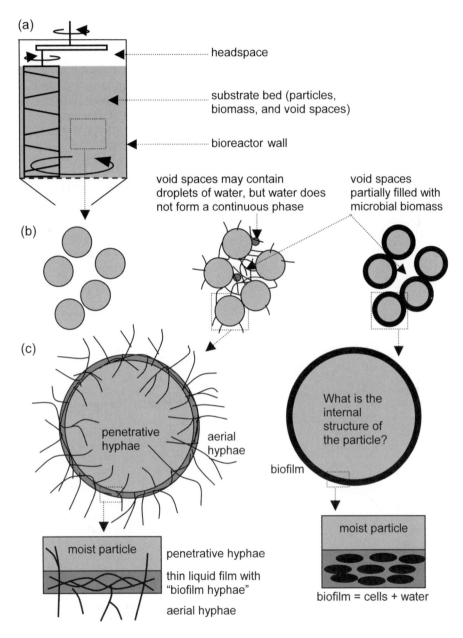

Fig. 2.4. The phases present within an SSF bioreactor. **(a)** Macroscale view; **(b)** Microscale view. From left to right the diagrams represent uninoculated substrate, the growth of a filamentous fungus and of a unicellular organism, such as a yeast or bacterium; **(c)** Greater detail of the microscale, showing a transverse section through the particles

- allowing space for particle movement in mixed bioreactors;
- bringing air to the bed surface in cases where the air is circulated through the headspace rather than being blown forcefully through the bed (such as the rotating and stirred drum bioreactors described in Chap. 8).

The bed is the site of the bio-reaction, that is, the site where the microorganism grows. It will be seen in more detail in the microscale view in the next section. As Fig. 2.5 shows, it is quite possible to have macroscale temperature, moisture, and gas concentration gradients across the bed, these resulting from mass and heat transfer processes that will be discussed in Chap. 4.

2.4.2 A Microscale Snapshot of the Substrate Bed

The physical appearance of the substrate bed changes during the process as the microorganism grows. The processes responsible for this change are described in Sect. 2.5. This section gives a "snapshot" view of the fermentation in the middle of the growth process. Of course, the substrate bed is quite complex in structure. As shown in Fig. 2.4(b), it contains a three-dimensional arrangement of substrate particles, inter-particle spaces, and microbial biomass.

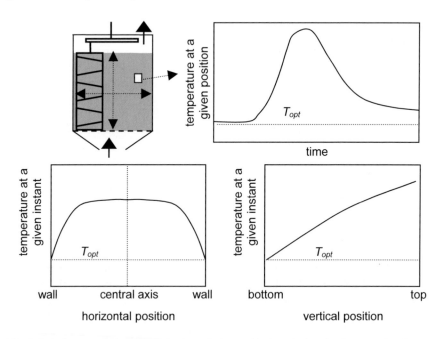

Fig. 2.5. At large scale in SSF there are variations with time during the fermentation for any given position and variations with position across the bioreactor at any given time. Temperature gradients are shown here as an example. Therefore the aim is to minimize temporal and spatial deviations from the optimal conditions (in this case, deviations from the optimum temperature for growth, T_{opt})

The size and shape of the substrate particles, along with the manner in which the bed is packed, will determine the sizes of the inter-particle spaces and the degree of continuity between them. The substrate particles are moist and will have a thin liquid film at their surface.

The microbial biomass is distributed as biofilms on the surfaces of the substrate particles, in the case that the process organism is unicellular, or as a network of hyphae, in the case that the process organism is a fungus (Fig. 2.4(b)).

The inter-particle spaces are filled with gas. For growth of a unicellular organism, the inter-particle spaces are well defined, and the biofilm is treated as part of the particle. For processes involving fungi, a network of aerial hyphae grows into the inter-particle spaces and the boundary between the inter-particle space and the particle is located at the surface of the thin liquid film that surrounds the particle. Note that even when the inter-particle spaces appear to be completely filled with hyphae, there is still a gas phase within the network, since the mycelial structure prevents the hyphae from occupying more than 34% of the available volume (Auria et al. 1995). In true SSF systems there will be no or very little liquid water within the inter-particle spaces, although small droplets may be held within the network of aerial hyphae of fungi.

Examining the particle with an even greater degree of detail, such as a cross-section through a substrate particle, we would typically see:

- that the substrate particle contains one or more types of macromolecule that confer the "solid" structure. The polymer or polymers that confer this structure may or may not be degraded by the microorganism during the process. If they are degraded, then the structure and properties of the particle at the end of the fermentation will be different from those of the original substrate particle. Substrate properties depend on the source of the substrate particle and how it was prepared, and this has consequences for hyphal penetration and accessibility of the nutrients, factors that can affect process performance. As a simple example, grains or pieces of stems have a cellular structure at the microscopic level, meaning that cell walls are present. On the other hand, particles made by granulating flours or meals have a more amorphous structure.
- that there is a spatial distribution of biomass (Fig. 2.4(c)). In the case of a unicellular microorganism, the biofilm is restricted to the exposed surfaces of the particle. The intercellular spaces are filled with water, giving the biofilm the consistency of a thick paste. In the case of a filamentous fungus, it is possible to distinguish aerial, "biofilm", and penetrative hyphae. Penetrative hyphae are those that have penetrated into the moist solid matrix. Aerial hyphae are those that are in direct contact with the air in the inter-particle spaces. Biofilm hyphae are those that are above the solid surface but are submerged in the liquid film at the particle surface. Depending on the extent of this liquid film, which might be stabilized by the presence of the hyphal network, the biofilm hyphae may represent a significant proportion of the overall biomass.
- that the particle surface, the location of which is used to define biofilm and penetrative hyphae, may be indistinct, especially if the organism attacks the polymer that gives structure to the substrate particle. For fungal growth, the

highest biomass concentration would typically be just above and just below the particle surface, where both nutrients from the substrate and O_2 from the gas phase are most readily available simultaneously.

- that, if we could visualize specific chemical components, we would see gradients in protons (i.e., in the pH), enzymes, polymers, hydrolysis products, other nutrients and gases within the substrate particle. During the rapid growth period the O_2 gradient is quite steep, with the O_2 concentration falling from a high value at the outer surface of the biofilm to essentially zero at 100 μm under the surface of the biofilm (Oostra et al. 2001). The substrate concentration gradient is typically in the other direction, such that the concentrations of soluble nutrients near the surface of the biofilm are quite low.

- that the substrate particle is moist. Water within the particle might be free water or involved in capillary sorption or hydration of macromolecules. There is a water film at the particle surface, the thickness of which will depend on the biomass properties and the water content of the particles. Note that the continuity of the surface water film with the liquid phase within the substrate particle and the continuity of the water phase within the particle itself depend on the substructure of the particle, given that intact cell walls disrupt continuity. This has consequences for molecular diffusion within the particle.

2.5 A Dynamic View of the Processes Occurring

In order to describe SSF systems as dynamic systems, it is useful to use two different time scales. A time scale of seconds to minutes is useful for describing the dynamics of the various biological and transport processes. On the other hand, a time scale of hours to days is useful for describing the gross changes in the system throughout the whole fermentation.

2.5.1 A Dynamic View with a Time Scale of Seconds to Minutes

In this view we unfreeze the snapshot of the fermentation described in Sect. 2.4.2 and concern ourselves with phenomena that occur on the timescale of seconds to minutes. It is again convenient to consider the macroscale and the microscale separately.

The dominant transport processes at the macroscale will depend on the bioreactor and the way it is operated. At this scale, the bed will typically be treated as a single pseudo-homogeneous phase, that is, as a single phase that has the average properties of the solid and air phases that comprise it. Figure 2.6(a) shows the various heat and mass transfer phenomena that occur within and between the various phases that were identified in Fig. 2.4(a), for a well-mixed, forcefully-aerated bioreactor. It is typically important to describe these transport phenomena mathematically in mass and energy balance equations; this topic receives detailed attention in Chap. 18 and is not discussed in detail here.

Figure 2.6(b) shows the microscale transport processes that occur in a typical SSF process in which a fungus grows aerobically using a polymer as its main carbon and energy source. At this scale, the particle and the inter-particle air are treated as different subsystems. Many of the transport processes shown are largely unaffected by the bioreactor and the way it is operated, that is, they are intrinsic to SSF systems due to the presence of the solid phase. At the substrate preparation stage, it might be possible to improve the efficiency of the inter-particle processes. For example, cooking may weaken or disrupt cell walls, reducing the barrier to penetration and diffusion, and may also hydrate polymers, making them more accessible to enzymes. In addition, the use of small particle sizes will reduce the distance over which diffusion must occur. However, these manipulations cannot entirely eliminate the importance of the intra-particle diffusion in SSF processes.

These processes include mass transfer processes such as:

- the diffusion of O_2, CO_2, and water vapor within static regions of the gas phase and their convective movement in regions of air flow, with the extent of static and flowing regions depending on whether the bed is forcefully aerated or not. Note that, even if air is blown forcefully through the bed, static layers of air are formed around any solid surfaces such as particle surfaces or hyphae;
- the diffusion of O_2, CO_2, water, nutrients, protons, products, and enzymes within the biofilm phase and the substrate particle;
- exchanges of O_2, CO_2, and water vapor between the various phases. Note that evaporation is typically treated as a phase change within the bed at the macroscale, whereas with a microscale view it is treated as a transfer between subsystems.
- Also, within the particle there will be the reaction of enzymes with their substrates. This is especially important in the context of SSF where the major carbon and energy source is quite often a macromolecule.

There will be various biological phenomena:

- translocation of nutrients within hyphae;
- growth, including processes such as the extension of hyphae or the expansion of a biofilm. In either case the biomass occupies volume that was previously occupied by either gas or substrate;
- physiological responses to the environment. Stress responses may be especially important in SSF, due to the combination of low water, low O_2 inside the particle, and high bed temperatures;
- genetic response mechanisms, such as induction and repression;
- cell death.

Compared to our understanding of these processes in SLF processes, relatively little is known about SSF. This is due to the fact that cell physiology is more difficult to study in SSF than in SLF. In particular, the well-mixed continuous-culture technique, which is a powerful tool in the study of microbial physiology in SLF, cannot be applied to SSF.

2.5.2 A Dynamic View with a Time Scale of Hours to Days

The preceding section gave little idea of the significant changes that occur over the whole fermentation. This section highlights the changes that occur on the time-scale of hours or days as the fermentation proceeds. The detailed description that follows, which is illustrated in Figs. 2.7 and 2.8, is for a process involving the aerobic growth of a fast-growing fungus on a polymeric carbon and energy source, starting with a spore-based inoculum. In this example, the substrate bed remains static during the whole process. Figure 2.9 shows the situation with the growth of a biofilm of a unicellular microorganism.

The early stages of the process. A fungal process typically begins with the mixing of a spore inoculum with cooked substrate particles. Each particle initially has a number of spores attached to it, with the number and uniformity of distribution of spores on each particle and amongst different particles depending on how the inoculation was done. The spores must germinate, which may take as long as 10 hours. During this time, the substrate bed might need to be warmed to ensure that the temperature is optimal for germination, although the necessity for this depends on how close the ambient temperature is to the required temperature. The various spores germinate at different times. Once each spore germinates, a germ tube extends away from the spore and branches to give daughter hyphae, which extend and then branch again, to give an expanding microcolony.

Fig. 2.6. (facing page) Various phenomena that occur within an SSF bioreactor during rapid growth of a fungus on a polymeric carbon and energy source. The example is for a well-mixed forcefully aerated bioreactor. **(a)** Macroscale phenomena. Energy is also stored within subsystems, which is evident as an increase in temperature. This is undesirable since it represents a deviation from the optimum temperature from growth or product formation. A major challenge in bioreactor design and operation is to minimize such temperature deviations. **(b)** Microscale phenomena. The example is for the growth of a fungus on the surface of a particle containing a polymeric carbon and energy source. Key [1] Diffusion of O_2 and CO_2 in regions of static gas layers; [2] Consumption of O_2 by aerial hyphae with release of CO_2; [3] Transfer of O_2 and CO_2 across the liquid layer at the particle surface; [4] Diffusion of O_2 and CO_2 within the particle; [5] Uptake of O_2 and release of CO_2 by hyphae submerged within an aqueous environment; [6] Release of hydrolytic enzymes by the biomass; [7] Diffusion of enzymes; [8] Reaction of enzymes with polymers to release soluble hydrolysis products; [9] Diffusion of soluble hydrolysis products within the substrate particle; [10] Uptake of soluble hydrolysis products by the biomass; [11] Translocation within the aerial hyphae; [12] Release of metabolic water from respiration; [13] Uptake of water for new biomass; [14] Diffusion of water within the substrate particle; [15] Evaporation of water from the liquid film at the particle surface; [16] Diffusion of water vapor in static gas layers; [17] Diffusion of soluble nutrients and their uptake by the biomass; [18] Release and diffusion of metabolic products; [19] Growth by extension and branching of hyphae

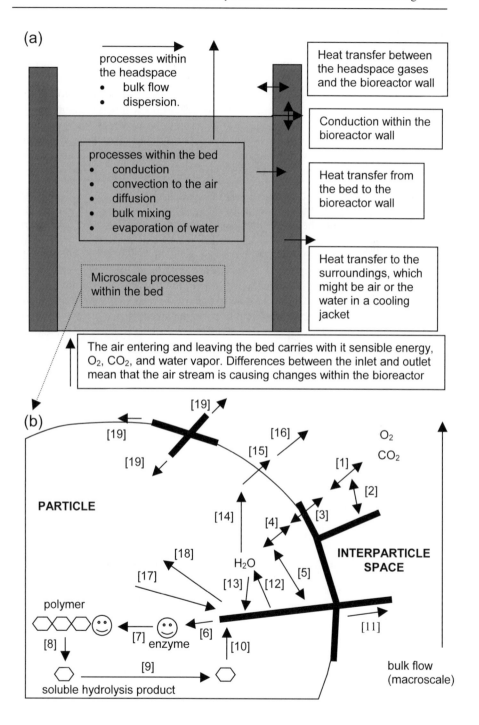

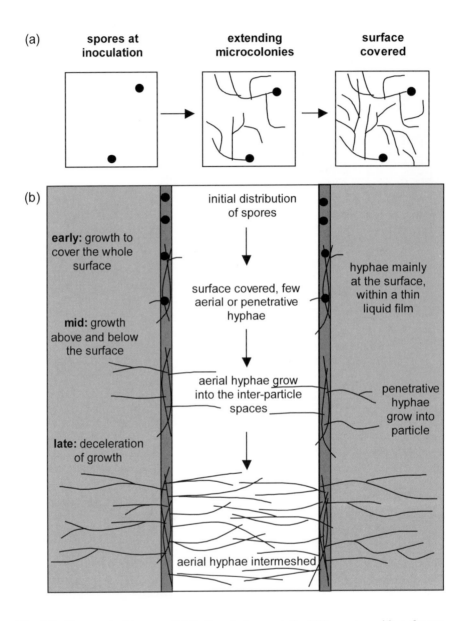

Fig. 2.7. Changes in biomass distribution during a static SSF process with a fungus. **(a)** Growth to cover the particle surface during the early phases of the fermentation, shown with an overhead view of the particle surface. **(b)** Development of aerial and penetrative hyphae during the fermentation, shown with a side view of a cut through two particles with an air space between them

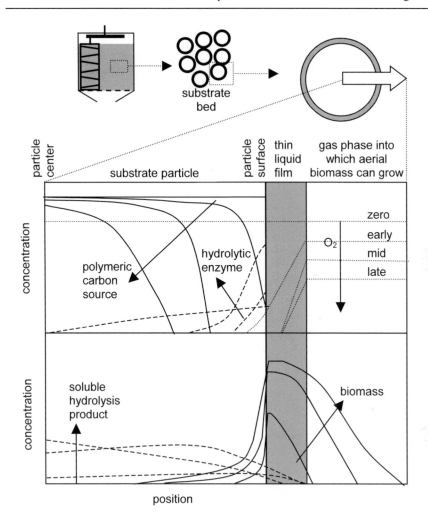

Fig. 2.8. Changes that take place during the fermentation with respect to concentration profiles along a radius that extends through the particle into the inter-particle gas phase. The example is given for the growth of a filamentous fungus on a polymeric carbon source, in a situation where the physical structure of the particle is derived from a second, inert polymer, such that the position of the surface does not change. The arrows show the direction of change during the fermentation. The initial concentrations of enzyme and soluble hydrolysis products in the substrate are zero and the initial biomass concentration is typically so small as to be negligible when spore inocula are used. The relative size of the thin liquid film at the particle surface is exaggerated in this diagram. Also, as a simplification, enzyme is assumed to be secreted only at the particle surface. It is assumed that O_2 concentrations fall in the inter-particle spaces, although the extent to which this is true will depend on where in the bioreactor this analysis is done. Key: Upper diagram of concentration versus position (———) polymeric carbon source, (- - -) hydrolytic enzyme, (·······) O_2. Lower diagram of concentration versus position (———) biomass, (- - -) soluble hydrolysis product

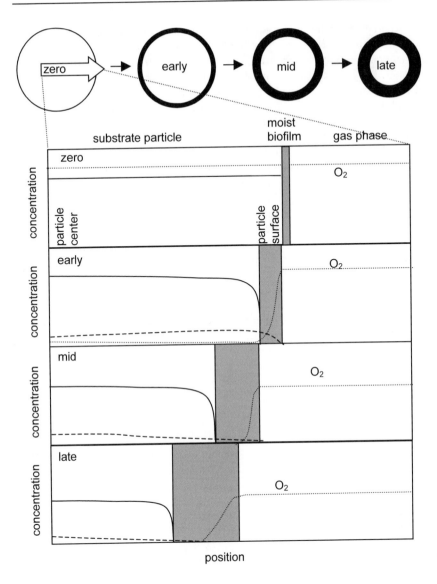

Fig. 2.9. Changes that take place during the fermentation with respect to concentration profiles along a radius that extends through the particle into the inter-particle gas phase. The example is given for the growth of a biofilm of a unicellular microorganism on a polymeric carbon source, in a situation where this polymer also provides the physical structure of the particle, such that the particle shrinks during the fermentation. For clarity, enzyme concentrations are not shown. In reality, the interface between the biomass and the substrate particle may be less distinct than is indicated here. This figure is based on modeling studies undertaken by Rajagopalan and Modak (1995) and Rajagopalan et al. (1997). Key: (——) Polymer concentration; (- - -) soluble hydrolysis product; (·······) O_2. For clarity, enzyme profiles are not shown, but are similar to those shown in Fig. 2.8

The original extension of the germ tube is fueled by reserves in the spores, but the continued growth depends on nutrients from the substrate. In the case that the carbon source is a polymer, this requires the secretion of the appropriate enzyme or enzymes. Enzymes diffuse away from the site of secretion into the particle. The speed of diffusion depends on the size of the enzyme and on the internal structure of the substrate particle. The enzymes begin hydrolyzing the polymer, and the soluble hydrolysis products then diffuse through the substrate. Oxygen consumption causes diffusion of O_2 through the static gas layer to the biomass and any initial O_2 within the substrate also diffuses to the biomass.

Soon hyphae from neighboring microcolonies meet one another, which causes negative interactions between the extending hyphae at the tips. For example, hyphae may change their growth direction or even cease to extend. During this time some hyphae will also have extended above the surface of the liquid film and others will have penetrated into the substrate (Fig. 2.7). During these very early stages, there is a sufficiently high O_2 concentration within the substrate to support this penetration: given the low biomass, the rate of O_2 uptake is low and diffusion can replenish O_2 reasonably effectively. Also due to the low biomass, the overall rate of heat production is very low, so it may still be necessary to warm the bed to provide the optimum temperature for growth.

So, early in an SSF process, growth is essentially biologically limited. Growth occurs at the maximum specific growth rate at which the organism is capable of growing on a solid surface at the prevailing temperature, pH, and water activity, although the extent to which this is true depends on how quickly enzymes are produced to liberate hydrolysis products from polymers. This period of biologically-limited growth can potentially be quite short once active growth has begun, possibly of the order of 2 to 10 hours, depending on the process.

The mid-stages of the process. The situation quickly changes as the biomass density increases, since the overall growth rate increases as the biomass increases, causing increases in the rates of growth-associated activities, such as nutrient and O_2 consumption and heat production. The consumption of O_2 and nutrients by the fungus decreases their concentrations in the immediate environment of the biomass (Fig. 2.8), and these changes typically occur more rapidly than O_2 and nutrients diffuse towards the biomass. The nutrient and O_2 concentrations within the biofilm continue to fall until they reach concentrations that are sufficiently low to decrease the growth rate. In this case, the process is limited by mass transfer.

During this phase the biomass density per unit surface area of substrate increases. The biomass may continue to penetrate into the substrate, although this might be relatively slow due to O_2 limitations. The production of aerial hyphae may contribute significantly to the overall increase in biomass density, but the density of the biomass in the biofilm may also increase. In an unmixed bed, the fungal hyphae form a network within the inter-particle spaces. Depending on the strength and density of this network, the substrate bed may be bound into a compact "cake". In a bed that is agitated, even if only intermittently, these aerial hyphae may be squashed onto the surface of the particle by the mixing, and may be damaged sufficiently to reduce growth. Mixing may also prevent sporulation, by damaging the developing aerial conidiophores before sporulation begins. Typically the mycelium squashed onto the particle surface is surrounded by a liquid film and is therefore considered as biofilm biomass. In this case the situation is closer to that presented in Fig. 2.9.

During this phase the rate of heat production soon exceeds the rate at which heat can be removed, such that the temperature of the substrate bed rises (Fig. 2.10). It continues to rise as long as the overall rate of heat production is greater than the overall rate of heat removal. Under operating conditions that have typically been used in SSF bioreactors, even if the inlet air and water jacket temperatures are maintained at the optimum temperature for growth, the temperature may increase over a period of several hours to values 10 to 20°C above the optimum, at least in some regions of the bioreactor. This is typically sufficient to affect growth deleteriously.

Therefore, during the mid-phases of an SSF process, growth can be limited by unfavorably high temperatures, or low concentrations of nutrients, soluble hydrolysis products or O_2. The major limiting factor depends on the growth rate of the microorganism, the properties of the substrate and the type of bioreactor used and how it is operated. Even for a single organism-substrate-bioreactor system, it is possible for different factors to be limiting at different times or even at the same time but at different locations within the bed.

If the polymer being hydrolyzed by the fungus is a structural polymer of the particle, then the particle properties will change. If other structural polymers are present that are not attacked, then the particle may simply lose strength. In many cases the size of the particle diminishes as the polymer is hydrolyzed. This might be accompanied by shrinkage of the bed, that is, a decrease in either or both of the bed height and width. With a fungal fermentation the "cake" may pull away from the walls. Note that these changes can affect the bulk scale transport processes. For example, the filling of the inter-particle spaces with biomass can cause increased pressure drops in a static bed with forced aeration (Fig. 2.10).

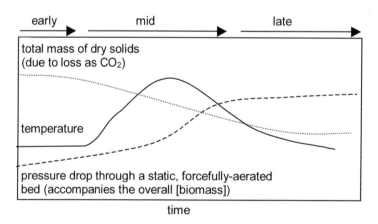

Fig. 2.10. Important variations throughout the whole fermentation at the macroscale

The latter stages of the process. Continued stress on the fungus due to high temperatures, low O_2 or lack of nutrients may trigger processes such as sporulation, termination of cell growth or death. As a result, the growth decelerates, and the rate of heat production falls. As the heat production rate falls, the temperature of the substrate bed falls.

This period may be quite important if the desired product is spores or a secondary metabolite, but for other products the process might typically be harvested at the onset of this phase. Depending on the process, this phase can be relatively short, consisting of a few hours, or quite long, consisting of days to weeks, such as can occur with the production of secondary metabolites or spores.

2.6 Where Has this Description Led Us?

This qualitative analysis of the physical nature of SSF bioreactor systems and the phenomena occurring within them has led us to the situation where we can say:

- There is a very wide array of phenomena occurring, and many of these can potentially limit growth. The limiting factor will depend on the microbe-substrate system, the bioreactor used and how it is operated, and the stage of the fermentation;
- The system is highly complex. It is so complex that this chapter, although it describes most of the important processes, finds it difficult to give a clear picture of the interactions between all the phenomena that are occurring simultaneously. The best approach to understanding the complexity of the interactions is to combine the various phenomena within a mathematical model;
- The inter-particle diffusion processes can be considered intrinsic to SSF, and there is little that can be done to affect them in the way that the bioreactor is operated. For example, little can be done to prevent O_2 from being exhausted within a small depth below the surface of substrate particles, and therefore O_2 limitation for at least some of the biomass can be considered to be an intrinsic characteristic of SSF systems;
- In bioreactor operation, the best that we can do to minimize overheating and O_2 limitations is to provide a gas environment close to the substrate particle that will lead to high rates of heat removal and high rates of oxygen supply. The operating variables available to try to achieve this are the rate at which air is supplied, the state (temperature, degree of saturation etc.) in which the air is supplied, and the frequency with which the bed is agitated.

Further Reading

Substrate preparation for SSF
Mitchell DA, Targonski Z, Rogalski J, Leonowicz A (1992) Substrates for processes. In: Doelle HW, Mitchell DA, Rolz CE (eds) Solid Substrate Cultivation. Elsevier Applied Science, London, pp 29–52

Downstream processing in SSF processes
Lonsane BK, Kriahnaiah MM (1992) Product leaching and downstream processing. In: Doelle HW, Mitchell DA, Rolz CE (eds) Solid Substrate Cultivation. Elsevier Applied Science, London, pp 147–171

Qualitative observations of the development of the microorganism during SSF
Mitchell DA, Greenfield PF, Doelle HW (1990) Mode of growth of *Rhizopus oligosporus* on a model substrate in solid-state fermentation. World J Microbiol Biotechnol 6:201–208

3 Introduction to Solid-State Fermentation Bioreactors

David A. Mitchell, Marin Berovič, and Nadia Krieger

3.1 Introduction

This book is about the design and operation of bioreactors for SSF. The current chapter briefly introduces the various bioreactor types, which will be described in more detail in Chaps. 6 to 11. Chapter 4 considers the heat and mass transfer phenomena that occur within bioreactors in a qualitative manner, while Chap. 5 shows how these heat and mass transfer phenomena are intimately linked to the question of how to design a large-scale SSF bioreactor; this discussion will highlight the need for mathematical models as tools in the scale-up process (Fig. 3.1).

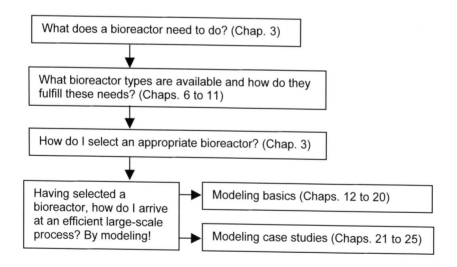

Fig. 3.1. The manner in which this book addresses the bioreactor design task

This book is written assuming that you are doing laboratory studies on a particular application of SSF, and the results give you confidence that you can develop a commercial process, for which you require a large-scale bioreactor. After undertaking the necessary kinetic characterization studies outlined in Chaps. 14 to 17, your task is to choose an appropriate bioreactor type, to design and build it, and then to operate it optimally. Clearly, the principles developed in this book are also appropriate for the situation in which you already have a process operating in a large-scale bioreactor, but in which the bioreactor was not designed based on engineering principles, and therefore you need to optimize its operation.

Note that in this book we are concentrating on the bioreactor itself. In all bioreactors there are various auxiliary operations and equipment. We do not cover bioreactor loading, unloading, and sterilization, but given the importance of the aeration system, Chap. 29 considers how it is possible to provide air at a desired flow rate, temperature, and humidity, and to change these conditions during the process. Chapters 26 and 28 consider the monitoring of the bioreactor and implementation of control schemes.

3.2 Bioreactor Selection and Design: General Questions

In taking a process that operates well in the laboratory and establishing a commercial process, the first step will be to identify what type of bioreactor will be suitable at large scale. Many of the laboratory experiments may have been undertaken in erlenmeyer flasks or thin packed-beds, but larger versions of these will probably not be appropriate for large-scale production.

Many considerations must be kept in mind when selecting a bioreactor, but a key question is: "What criterion do I use to compare different bioreactors and different operating conditions in order to be able to end up with the best system possible for my particular process?" Obviously the best criterion to use is the economic performance of the process. However, SSF processes have not been analyzed sufficiently to enable accurate estimates of capital and operating costs for new processes. In fact, at present the only way to compare the economic performance of bioreactors would be to build and operate a full-scale version of each bioreactor and record their capital and operating costs. It is likely to be some time before it will be possible to use economic performance as a criterion to guide bioreactor selection.

In the absence of sufficient information about the economics of SSF processes, the aim should then be to maximize the productivity of the bioreactor, in terms of product formation, which might be biomass or a metabolite. In other words, the criterion is the rate of production in kg of product per m^3 of bioreactor volume. Of course, if the substrate bed is not homogeneous, it will be necessary to calculate the productivity based on the evolution of the "volume-weighted biomass (or product)" curve. What is being sought is the combination of bioreactor operating strategy and harvesting time that will give the greatest value for:

$$Pr = \frac{X_{harvest} - X_{initial}}{t_{process} \cdot V_{bioreactor}},$$

(3.1)

where Pr is the productivity (kg h^{-1} m^{-3}), $X_{harvest}$ is the amount of biomass (or product) at the time of harvesting (kg), $X_{initial}$ is the amount of biomass (or product) at zero time (kg), $t_{process}$ is the overall process time (the time between successive harvests, h), and $V_{bioreactor}$ is the bioreactor volume (m^3).

Therefore bioreactor selection will be guided by the answers to several key questions about factors that affect the productivity. These are discussed in the following subsections.

3.2.1 The Crucial Questions

Possibly the three most important initial questions are:

- To what degree is the microorganism, or the desired form of the final product, affected deleteriously by agitation?
- How fast does the organism grow and how sensitive is it, and product formation by it, to increases in temperature?
- What are the aeration requirements of the system?

The answers to these questions will influence decisions about the type of aeration, mixing, and heat removal mechanisms that the large-scale bioreactor must have. Of course, these considerations are interconnected and affect the ability to control the macroscale variables of the process.

To what degree is the microorganism, or the desired form of the final product, affected deleteriously by agitation? Bioreactors can either be completely static, intermittently agitated, or continuously agitated. Frequent or continuous agitation would be desirable if it were tolerated, because it aids bulk transport of heat and O$_2$, improving the ability to control the conditions within the bed. Further, evaporative cooling of the bed can dry it out to water activities that restrict growth, meaning that it is often desirable to add water during the fermentation. It is only feasible to add water while the bed is being mixed. However, agitation can also affect the process deleteriously. It may damage hyphae in fungal-based processes, which might adversely affect growth and product formation. Conversely, it may be desired that the final product be knitted together by fungal hyphae, such as in the production of a fermented food, and this would be prevented by agitation. Beyond this, agitation can crush substrate particles if they do not have sufficient mechanical strength or can cause sticky particles to agglomerate, in either case producing a paste in which O$_2$ transfer is greatly hindered. Unfortunately, the balance between positive and negative effects of agitation has not been well characterized. It will be necessary to undertake your own studies at laboratory-scale in which the performance of agitated and non-agitated fermentations is compared, with both being forcefully aerated in order to minimize transport limitations, thereby isolating agitation as the factor responsible for any differences.

How fast does the organism grow and how sensitive is it, and product formation by it, to increases in temperature? Control of the temperature of the substrate bed is one of the key difficulties in large-scale SSF processes, especially in those processes that involve fast-growing microorganisms. At large scale, it may be difficult to prevent the temperature from reaching values that are quite deleterious to the microorganism. The various bioreactors differ in the efficiency of heat removal, with the temperatures reached depending on a complex interaction between the organism and the type of bioreactor and the way in which it is operated. These considerations may determine key decisions such as maximum bed depths.

What are the aeration requirements of the system? The majority of SSF processes involve aerobic growth. There are essentially two aeration options in SSF processes. One is to circulate air around the bed, but not to blow air forcefully through it. The other is to blow air forcefully through the bed. Agitation can influence the efficiency with which fresh air is delivered to the substrate particles. Note that in forcefully aerated beds the air phase plays an important role in heat removal. In fact aeration rates are typically governed by heat removal considerations since the air flow rates required for adequate heat removal are usually more than sufficient to avoid limitations in the supply of O_2 to the particle surface.

These considerations will be crucial in determining the agitation and aeration regimes that are appropriate. The bioreactors can then be compared on the basis of their ability to provide the desired regimes. More advice on how these various factors should be weighted in selecting an appropriate bioreactor type are considered in Sect. 3.4.

3.2.2 Other Questions to Consider

Once a bioreactor giving a certain agitation and aeration regime has been selected, various considerations will affect the details of its design:

- How important is it to have aseptic operation?
- To what degree is it necessary to contain the process organism?
- Is continuous operation desirable?
- How easy is loading and unloading and how much does labor cost?
- How much substrate is to be fermented?
- Will the bioreactor also be used for one or more of the downstream processing steps?

The degree to which sterile operation is required. Some SSF processes involve fast-growing organisms growing under conditions of low moisture that give the process organism a competitive advantage over contaminants. For example, in many fungal processes, the water activity is below that which is optimal for bacteria, so there are not serious problems with growth of bacterial contaminants, although fungal contaminants might cause problems. It may be possible to operate without strict asepsis: The process organism might be given sufficient advantage over any contaminants through cooking of the substrate, avoidance of gross contaminations, and the provision of a relatively pure and vigorous inoculum. However, in other cases the organism grows slowly and care must be taken to design the bioreactor for sterile operation and to operate it in such a manner as to prevent contamination. In this case it is necessary to sterilize the bioreactor before operation, to properly seal openings, to filter the inlet air and to add solutions to the bioreactor during the fermentation in an aseptic manner. The various bioreactors that have been used to date differ with respect to their ability to operate aseptically.

The degree to which containment of the process organism is required. In general, transgenic organisms are not used in SSF, and processes rarely involve dangerous pathogens (although some do involve opportunistic pathogens). However, many processes do involve fungi and workers can suffer from allergies or other health problems if spores are allowed to escape freely into the environment. The bioreactor may need to be enclosed, and filters may be required on the outlet air stream. Bioreactors that have been used to date differ with respect to the ease of containing the process organism.

The desirability of continuous operation. Continuous operation in a well-mixed bioreactor is not a useful option for SSF. In SLF the nutrients added to a continuous stirred tank reactor are distributed throughout the bioreactor, becoming available to all the microorganisms. In SSF, any solid particles added to the fermentation would need to be colonized, a process that would take a significant period of time. Even if the particles were inoculated at the time of addition, early growth might be expected to be slow, especially in a mixed bed, and an unduly high fraction of poorly colonized substrate particles would leave in the outflow. However, continuous operation of the "plug-flow type" certainly is an option.

The ease of loading and unloading and the cost of labor. Loading and unloading of the bioreactor are handling operations that are required for all SSF processes. Note that the type of operation can affect how loading and unloading must be done: In continuous bioreactors the loading and unloading operations must be continuous or at least semi-continuous, while in batch operation they are done at distinct times. These operations have received little attention. The general principle is that, depending on labor costs, it may be desirable to avoid bioreactor types that require manual handing in the loading and unloading steps.

The amount of substrate to be fermented. The dimensions of the bioreactor will be determined by the volume of substrate that it must hold at any one time. This will depend on the mass of substrate that it must hold and the bulk packing density of the bed. Note that the allowable height of the bed might be limited by the mechanical strength of the substrate particles.

Involvement of the bioreactor in downstream processing steps. At times, it might be desirable either to dry the substrate bed or to leach a product from it as one of the first downstream processing steps. It may be desirable to undertake such steps within the bioreactor itself. This may influence bioreactor design.

3.3 Overview of Bioreactor Types

Many different bioreactors have been used in SSF processes, and have been given different names by different authors. However, based on similarities in design and operation, SSF bioreactors can be divided into groups on the basis of how they are mixed and aerated (Fig. 3.2).

- Group I: Bioreactors in which the bed is static, or mixed only very infrequently (i.e., once or twice per day) and air is circulated around the bed, but not blown forcefully through it. These are often referred to as "tray bioreactors".
- Group II: Bioreactors in which the bed is static or mixed only very infrequently (i.e., once per day) and air is blown forcefully though the bed. These are typically referred to as "packed-bed bioreactors".
- Group III: Bioreactors in which the bed is continuously mixed or mixed intermittently with a frequency of minutes to hours, and air is circulated around the bed, but not blown forcefully through it. Two bioreactors that have this mode of operation, using different mechanisms to achieve the agitation, are "stirred-drum bioreactors" and "rotating drum bioreactors".
- Group IV: Bioreactors in which the bed is agitated and air is blown forcefully through the bed. This type of bioreactor can typically be operated in either of two modes, so it is useful to identify two subgroups. Group IVa bioreactors are mixed continuously while Group IVb bioreactors are mixed intermittently with intervals of minutes to hours between mixing events. Various designs fulfill these criteria, such as "gas-solid fluidized beds", the "rocking drum", and various "stirred-aerated bioreactors".

Note that this division is made on the basis of the manner in which the bioreactor is operated, and not on the outward appearance of the bioreactor. For example, there are bioreactors that are essentially identical with the "stirred drum", but in which the air is introduced within the substrate bed through the ends of the paddles. Such a bioreactor should then be classified as a "stirred-aerated bioreactor", although the bed will not be as efficiently aerated as when the bed receives an even aeration across its whole cross-section. Also note that the distinction is not always perfectly clear. It is an arbitrary decision as to what frequency of mixing is separates "static" and "agitated" operation. The advantage of grouping bioreactors on the basis of the manner in which they are operated is that principles derived on the basis of work with one member of a certain group of bioreactors can be applied to other bioreactors in the group.

3.3.1 Basic Design Features of the Various Bioreactor Types

This section presents basic design features of the various bioreactors types. More details are given in Chaps. 6 to 11, but sufficient information is presented here to allow a general comparison.

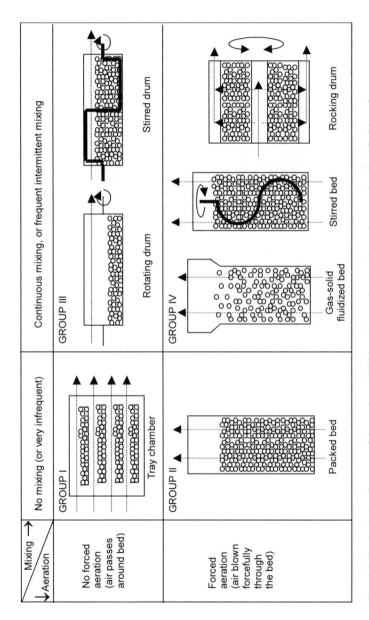

Fig. 3.2. Basic design features of the various SSF bioreactors, showing how they can be classified into four groups on the basis of how they are mixed and aerated. From Mitchell et al. (2000) with kind permission from Springer Science and Business Media

Group I bioreactors. These typically consist of a chamber containing a large number of individual trays, stacked one above the other with a gap in between (Fig. 3.2, upper left quadrant). Conditioned air (i.e., with control of humidity and temperature) is blown into the chamber and circulates around the trays. Agitation, if done, is very infrequent, and is typically done by hand. The trays themselves may be constructed of wood, bamboo, metal or plastic. They are typically open at the top and have perforated bottoms to increase the accessibility to O_2, but there are other possibilities. For example, micro-perforated plastic bags containing substrate fall within this category.

Group II bioreactors. A typical packed-bed bioreactor consists of a column of cylindrical or rectangular cross section, oriented vertically, with a perforated base plate on the bottom which supports a bed of substrate (Fig. 3.2, lower left quadrant). Air is blown up through the base plate.

Group III bioreactors. These typically consist of a drum of cylindrical cross section lying horizontally (Fig. 3.2, upper right quadrant). The drum is partially filled with a bed of substrate, and air is blown through the headspace. In rotating drums, the whole drum rotates around its central axis to mix the bed. In stirred drums, the bioreactor body remains stationary and paddles or scrapers mounted on a shaft running along the central axis of the bioreactor rotate within the drum.

Mixed and forcefully aerated bioreactors. There are several types of designs that fall into this group (Fig. 3.2, lower right quadrant). They can be operated with continuous or discontinuous mixing.

- Stirred-bed bioreactors are similar to the static packed bed in that a bed of substrate sits on a perforated base plate and air is forcefully blown through the bed, but rather than being static, an agitator is inserted and provides continuous or intermittent mixing. Such stirred beds are typically aerated from the bottom, and have the agitator inserted from the top.
- Rocking-drum bioreactors consist of three concentric cylinders - an inner perforated cylinder, an outer perforated cylinder, and an outer solid cylinder. The substrate sits in the space between the two perforated cylinders. Air is blown through into the central cylinder, passes through the substrate bed and then into the space between the outer perforated cylinder and the outer solid cylinder, before leaving through the air outlet. The two outer cylinders rotate in relation to the inner cylinder, thereby mixing the substrate bed, although not very effectively.
- Air-solid fluidized beds (ASFBs). In this bioreactor air is blown upwards through a perforated base plate at sufficient velocity to fluidize the substrate bed, which then behaves as though it were a fluid.

3.3.2 Overview of Operating Variables

Operating variables are variables that the operator can manipulate in an attempt to control the conditions within the bioreactor. The question of optimum operating strategies for the various bioreactor types is covered in the individual bioreactor chapters (Chaps. 6 to 11) and the modeling case studies (Chaps. 21 to 25). However, it is worthwhile to make some general comments here:

- Regardless of whether the air is blown forcefully through the bed or circulated around the bed, it is possible to control the flow rate, temperature, and humidity of the air supplied at the inlet to the bioreactor or chamber. The costs of supplying air will depend on the volumetric flow rate and the pressure drop in the bed, and the need to heat or refrigerate the air. Pressure drop will be discussed in Chap. 7, which deals with packed-bed bioreactors, since its importance is greatest for this type of bioreactor.
- The conditions in the surroundings of the bioreactor can be controlled. The bioreactor may be placed in a room or other location where the air temperature, humidity, and circulation are controlled. Alternatively, the bioreactor may be fitted with a water jacket. The flow rate and temperature of the cooling water at the inlet of the jacket can be controlled. Note that if the desired air or water temperatures are different from the temperatures at which they are available, either cooling or heating will be necessary, which entails extra costs.
- Additions can be made to beds that are mixed, even if only intermittently; for example, water can be sprayed onto the bed during mixing.
- In beds that are mixed, it is possible to control the frequency, duration, and intensity (i.e., revolutions per minute of the agitator) of the mixing.

Given the difficulties in controlling the conditions in SSF bioreactors, which were mentioned in Chap. 2 and are discussed in more detail in Chap. 5, it is not a simple matter to maintain the bed conditions at the optimum values for growth and product formation by manipulating these operating variables. The aim therefore is to select combinations of operating conditions that make the best balance in:

- minimizing deviations from the optimum temperature;
- minimizing damage to the organism;
- minimizing deviations of the bed water activity from the optimum value;
- maximizing the supply of O_2 to the particle surface.

Chapters 6 to 11 will give some idea of what we already know about how to do this for the various bioreactor types. It must be stressed that, although our knowledge is increasing, it is as yet far from complete.

3.4 A Guide for Bioreactor Selection

The answers to the questions and issues raised in Sect. 3.2 will determine which of the bioreactor types shown in Fig. 3.2 is most suitable. Figure 3.3 shows how the various considerations might be used to arrive at the decision to use a particular bioreactor. For example, if the microorganism is very sensitive to shear, then a bioreactor type with a static bed must be chosen. This might cause heat removal to be a problem. If some shear can be tolerated, it is not clear which of the agitated bioreactors is best, since shear effects during the mixing of solids in the various bioreactors are not well understood.

Figure 3.3 can give only general guidelines about bioreactor choice. The final decision comes down to bioreactor performance for a particular substrate-microorganism-product combination. However, it is not a simple matter, on the basis of laboratory-scale studies, to say which bioreactor design will perform best at large scale. Also, typically, neither large-scale nor even pilot-scale bioreactors of the various types will be available for comparative studies. Nor is the budget for the development process likely to be sufficient to build several pilot-scale bio-reactors. One of the main arguments of this book is that, in the face of these limitations, mathematical modeling of bioreactor performance is a very useful tool in such scale-up tasks. Scale-up should not be done solely on the basis of experimental studies; rather it should involve a combined experimental and modeling program. This issue will be returned to in Chap. 5, after a consideration of basic heat and mass transfer principles in Chap. 4.

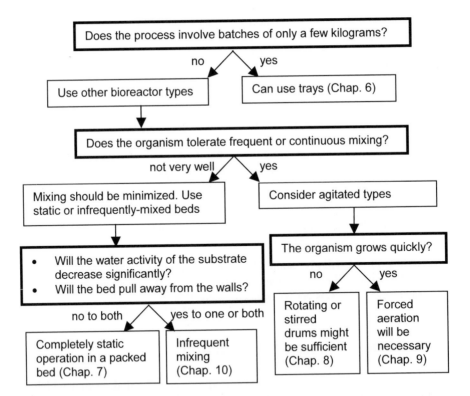

Fig. 3.3. A suggested key for SSF bioreactor selection

Further Reading

General considerations about bioreactor performance, which are relevant regard-less of whether the bioreactor is used for submerged liquid fermentation or solid-state fermentation
Lübbert A, Jørgensen SB (2001) Bioreactor performance: a more scientific approach for practice. J Biotechnol 85:187–212

Economic analysis of solid-state fermentation processes, economic performance being the most important criterion in bioreactor selection
Castilho LR, Polato CMS, Baruque EA, Sant'Anna Jr GL, Freire DMG (2000) Economic analysis of lipase production by *Penicillium restrictum* in solid-state and submerged fermentations. Biochem Eng J 4:239–247
Ghildyal NP, Lonsane BK, Sreekantiah KR, Sreenivasa Murthy V (1985) Economics of submerged and solid state fermentations for the production of amyloglucosidase. J Food Sci Technol 22:171–176

General reviews of bioreactor designs for SSF
Durand A (2003) Bioreactor designs for solid state fermentation. Biochem Eng J 13:113–125
Durand A, Renaud R, Maratray J, Almanza S, Diez M (1996) INRA-Dijon reactors for solid-state fermentation: Designs and applications. J Sci Ind Res 55:317–332
Fasidi IO, Isikhuemhen OS, Zadrazil F (1996). Bioreactors for solid state fermentation of lignocellulosics. J Sci Ind Res 55:450–456
Hardin MT, Mitchell DA (1998) Recent developments in the design, operation and model-ling of bioreactors for solid-state fermentation. In: Kaowai F, Sasaki K (eds) Recent research developments in fermentation and bioengineering, vol. 1. Research Signpost, Trivandrum, pp 205–222
Mitchell DA, Berovic M, Krieger N (2000) Biochemical engineering aspects of solid state bioprocessing. Adv Biochem Eng/Biotechnol 68:61–138
Robinson T, Nigam P (2003) Bioreactor design for protein enrichment of agricultural residues by solid state fermentation. Biochem Eng J 13:197–203

Description of bioreactor types used in the koji industry
Mudgett RE (1986) Solid-state fermentations. In: Demain AL, Solomon NA (eds) Manual of Industrial Microbiology and Biotechnology. ASM Press, Washington DC, pp 66-83
Sato K, Sudo S (1999) Small-scale solid-state fermentations. In: Demain AL, Davies JE (eds) Manual of Industrial Microbiology and Biotechnology, 2nd edn. ASM Press, Washington DC, pp 61–79

Recent experimental and modeling studies in which various bioreactor types have been compared
Couto SR, Moldes D, Liebanas A, Sanroman A (2003) Investigation of several bioreactor configurations for laccase production by *Trametes versicolor* operating in solid-state conditions. Biochem Eng J 15:21–26

Oostra J, Tramper J, Rinzema A (2000) Model-based bioreactor selection for large-scale solid-state cultivation of *Coniothyrium minitans* spores on oats. Enzyme Microbial Technol 27:652–663

4 Basics of Heat and Mass Transfer in Solid-State Fermentation Bioreactors

David A. Mitchell, Marin Berovič, Oscar F. von Meien, and Luiz F.L. Luz Jr

4.1 Introduction

Macroscale heat and mass transfer phenomena play important roles in determining the performance of SSF bioreactors. Therefore, in order for a mathematical model to describe bioreactor performance reasonably, it must describe these phenomena. The current chapter gives a qualitative overview of the various macroscale heat and mass transfer processes that occur within SSF bioreactors. These processes will be treated quantitatively in Chap. 18, where the various mathematical expressions that are used to describe them within bioreactor models will be presented.

Note that, in a particular SSF bioreactor, some of the heat and mass transfer mechanisms presented in this chapter may not be present, and the relative importance of the various mechanisms that are present may differ from bioreactor to bioreactor. Details specific to each bioreactor type will be covered in Chaps. 6 to 11 and 21 to 24. The current chapter focuses on the period of high heat generation, when it is necessary to remove energy from the bed, although early in the fermentation it might be necessary to transfer energy to the bed to maintain the temperature high enough to initiate growth.

4.2 An Overall Balance Over the Bioreactor

The bioreactor can be treated as a whole by drawing a system boundary around the outside of the bioreactor and only considering the exchanges of mass and energy between the bioreactor and its surroundings (Fig. 4.1). The air stream carries mass (N_2, CO_2, O_2, and water vapor) and energy into and out of the bioreactor, with the amount of energy carried depending on its humidity and temperature. The tendency is for the air to leave hotter and carrying more water than when it entered, and both the higher temperature and higher humidity contribute to the overall heat removal from the bioreactor. The composition of the air may be different at the air inlet and outlet. The outlet air is likely to have more CO_2, less O_2, and more water than the inlet air. Therefore, in terms of mass transfer, the effect of the airflow is not only to provide O_2 and remove CO_2, but also to dry the bioreactor. Note that it is almost impossible to prevent this drying effect if the air heats up as it passes through the bioreactor.

Energy can be exchanged by convective heat transfer between the bioreactor wall and surroundings, which could be air or could be cooling water in a water jacket. The convection of heat away from the outside of the bioreactor wall will occur by free convection if the outside wall of the bioreactor is in contact with the surrounding air and there is no forced flow of this air past the bioreactor. If the bioreactor is jacketed and water is pumped through the jacket, or if air is blown past the bioreactor surface, then the heat will be removed by forced convection. The significance of the contribution of this heat transfer to overall heat removal depends on the scale of the bioreactor. In the laboratory, small bioreactors have large surface-to-volume ratios, and this heat removal can make a large contribution. At large scale, the surface area-to-volume ratio will be smaller; therefore the contribution of this mechanism to overall heat removal may be very small or even negligible.

The change in energy of the bioreactor itself will manifest itself as a change in temperature ("sensible energy") or a change in phase of water between the liquid and vapor states within the bed ("latent energy").

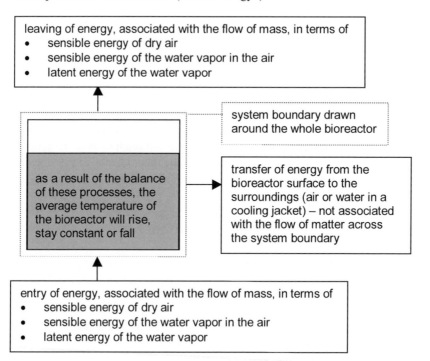

leaving of energy, associated with the flow of mass, in terms of
• sensible energy of dry air
• sensible energy of the water vapor in the air
• latent energy of the water vapor

system boundary drawn around the whole bioreactor

as a result of the balance of these processes, the average temperature of the bioreactor will rise, stay constant or fall

transfer of energy from the bioreactor surface to the surroundings (air or water in a cooling jacket) – not associated with the flow of matter across the system boundary

entry of energy, associated with the flow of mass, in terms of
• sensible energy of dry air
• sensible energy of the water vapor in the air
• latent energy of the water vapor

Fig. 4.1. How the transfer and storage of energy in bioreactors is treated when a global balance is undertaken by drawing the system boundary around the whole bioreactor

4.3 Looking Within the Bioreactor in More Detail

Each bioreactor type has three subsystems, the bioreactor wall, the substrate bed, and the headspace gases (Fig. 4.2). The substrate bed itself may be treated as two separate phases, the solid and air phases, or it may be treated as a single pseudo-homogeneous phase with the average properties of the solid and air phases. The arrangement of the subsystems and their relative importance vary with bioreactor type. A number of heat transfer and energy storage phenomena occur within and between these subsystems.

4.3.1 Phenomena Within Subsystems Within the Bioreactor

The average temperature of each subsystem may rise or fall, representing a change in the amount of energy stored in the subsystem. Other important phenomena that occur within each of the subsystems are discussed in the following subsections.

4.3.1.1 Phenomena Occurring Within the Substrate Bed

Several mass- and energy-related phenomena occur within the substrate bed. The phenomena listed here are for a static substrate bed treated as a single pseudo-homogeneous phase. The situation in which the air and solid phases in the bed are treated as separate phases is covered in Sect. 4.3.3.

Metabolic heat production. The bed is the site of microbial growth, and therefore the site of metabolic heat production.

Conduction. This occurs in response to temperature gradients, with energy flowing from warmer regions to cooler regions. Depending on the bioreactor, significant temperature gradients may exist in none, one, two, or three dimensions. This conduction occurs at different rates through the solid and air phases, so typically it is useful to consider the bed as though it were a single phase with the average properties of the air and the solid (a mass-weighted average). Conduction is usually of minor importance if the bed is forcefully aerated or mixed.

Diffusion. The gas phase components (O_2, CO_2, and water vapor) will diffuse within the inter-particle spaces in response to any concentration gradients. Typically the contribution of diffusion to the transfer of mass across the substrate bed is only important in Group I bioreactors (tray-type bioreactors).

Convective heat transfer. This occurs if the bed is forcefully aerated. As the air moves through the bed, energy is transferred to it from the solid phase, increasing the temperature and therefore the energy of the air. Since the air is moving through the bed, it carries the energy away from the site of production, and this represents a bulk flow of energy through the bed. Note that convective heat transfer in an unmixed bed leads to the establishment of axial temperature gradients, as explained in Fig. 4.3.

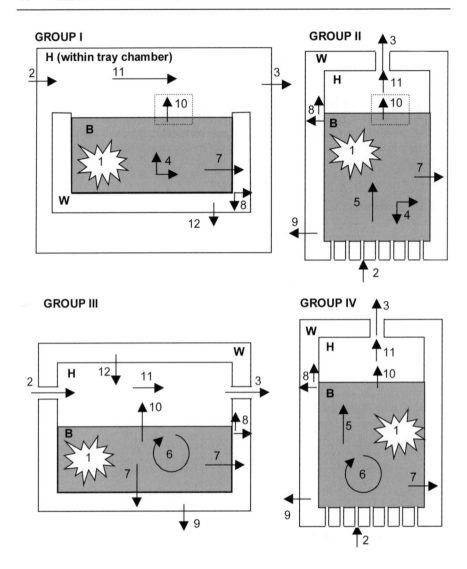

Fig. 4.2. Macroscale heat and mass transfer phenomena within and between the various subsystems in the bioreactor. Key: (H) headspace; (W) wall; (B) bed; (1) Liberation of waste metabolic heat during growth and maintenance; (2) Entry of mass and energy in the inlet air; (3) Exit of mass and energy in the outlet air; (4) Conduction and diffusion within the bed (makes a negligible contribution in mixed beds); (5) Convective flow of energy and mass within the bed due to aeration; (6) Solids flow due to mixing; (7) Heat transfer from bed to wall; (8) Heat conduction within bioreactor wall; (9) Heat transfer from wall to surroundings; (10) Mass and energy transfer from the bed to the headspace (see Fig. 4.4 for more details about the exchange that occurs in the boxed area); (11) Air flow within the headspace; (12) Heat transfer from wall to headspace. Note that in the case of Group I (tray-type) bioreactors, the focus is on an individual tray

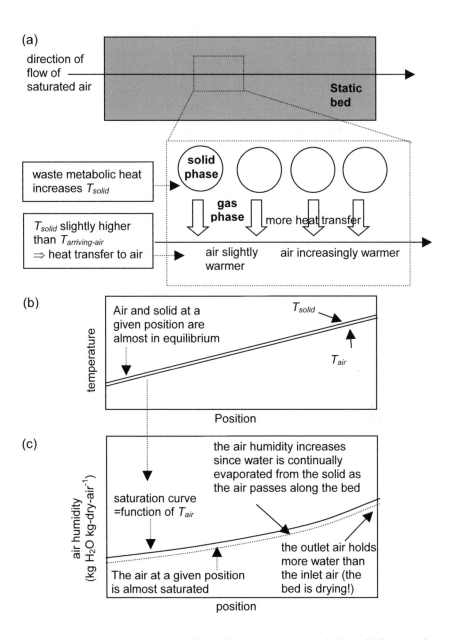

Fig. 4.3. Consequences of convective flow of air through a static bed in which an exother-mic reaction is occurring in the solid phase. It is assumed the column is fed with saturated air, at a relatively low superficial velocity. (a) Mechanism by which axial temperature gradients are established; (b) Axial temperature gradient (which may not be perfectly linear); (c) Consequences of the axial temperature gradient for evaporation

Evaporation. Water evaporates from the solid into the air phase, removing energy from the solid phase in the form of the enthalpy of vaporization. The degree of evaporation depends on the saturation of the air, but even if saturated air is used to aerate a bioreactor, if the air temperature increases while the air is within the bed, the water-carrying capacity of the air increases (Fig. 4.3(c)). When the bed is treated as a single pseudo-homogeneous phase, evaporation represents a change of phase within the subsystem and not transfer between subsystems.

Convective mass transfer. As the air flows through a forcefully aerated bed it carries water vapor, O_2, and CO_2 with it, representing bulk flows of these components. The importance of natural convection currents in contributing to heat and mass transfer within beds that are not forcefully aerated has not been investigated.

4.3.1.2 Phenomena Occurring Within the Headspace

Typically the headspace gases are flowing, since, even in those bioreactors in which the bed itself is not forcefully aerated, air is typically circulated through the bioreactor such that it moves transversely across the bed surface. In forcefully aerated bioreactors the flow is normal to the bed surface, that is, the air leaving the bed moves perpendicularly away from the bed surface (although in this type of bioreactor, once the air has left the bed, little attention is paid to it). In either case, this bulk flow carries not only energy with it, but also O_2, CO_2, and water vapor. With this bulk movement of air, conduction and diffusion will typically make negligible contributions to heat and mass transfer within the headspace phase.

4.3.1.3 Phenomena Occurring Within the Bioreactor Wall

Heat will be transferred across the bioreactor wall by conduction if there is a temperature gradient across it. Note that, depending on the temperature gradient, conduction does not necessarily occur directly from the inside to the outside. For example conduction may occur from a warmer region of the bioreactor wall in contact with the substrate bed to a cooler region of the bioreactor wall in contact with the headspace gases. The distribution of temperatures in SSF bioreactor walls and its influence on conduction has received almost no attention.

4.3.2 Transfer Between Subsystems When the Substrate Bed Is Treated as a Single Pseudo-Homogeneous Phase

Heat transfer can occur between any of the three phases, the substrate bed, the headspace gases, and the bioreactor wall (Fig. 4.2). In all bioreactor types, heat can be transferred by conduction from the substrate bed to the wall. Also, there will be heat transfer by convection between the headspace gases and the bioreactor wall, the direction of this heat transfer depending on the relative temperatures of these phases. The heat and mass transfer between the bed and headspace will depend on how the bioreactor is aerated (Fig. 4.4):

- In bioreactors that are forcefully aerated, the convective flow of the air leaving the bed and entering the headspace carries energy and mass (water vapor, O_2, and CO_2) across the subsystem boundary. In this case the majority of water vapor leaving the bed was already in the vapor phase.
- In bioreactors where air is only circulated past the bioreactor surface, the heat and mass transfer occur by conduction and diffusion across a static gas layer at the bed surface, to the air circulating past the bed. In this case most of the O_2 and CO_2 will be exchanged between the inter-particle spaces and the headspace, whereas a significant amount of water may evaporate from the exposed substrate particles.

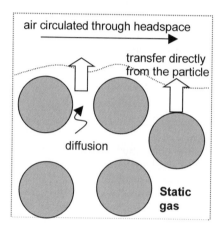 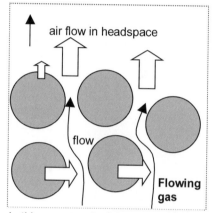

in this case much of the transfer from the solids occurs within the bed

Fig. 4.4. The difference in bed-to-headspace heat and mass transfer for unaerated and forcefully aerated beds. The regions shown here correspond to areas at the surface of the bed (see the dashed boxes within the Group I and Group II bioreactors in Fig. 4.2). In the left-hand diagram the dotted line represents the boundary between the static gas phase and the flowing gases within the headspace

4.3.3 Transfer Between Subsystems When the Substrate Bed Is Treated as Two Separate Phases

In some cases the substrate bed is not treated as a single pseudo-homogeneous phase, but rather as two separate phases. In fact, this is necessary in those cases in which it is not reasonable to assume that the substrate particles and inter-particle air are in thermal and moisture equilibrium.

Oxygen transfer between the solid and inter-particle gas phase has received some attention. Until recently, $k_L a$ was used as the transfer parameter, in analogy to SLF (Durand et al. 1988; Gowthaman et al. 1995). However, the two systems are different (Thibault et al. 2000a) (Fig. 4.5). In SLF, the major barrier to O_2 transport resides in a thin liquid film around each bubble, and there is no biomass and therefore no O_2 consumption within this film. Rather, the biomass is located within a well-mixed bulk phase. In SSF, the limiting step is diffusion within the static biofilm at the substrate surface, and simultaneous diffusion and consumption occur in this biofilm. Therefore k_La is not the appropriate parameter to character-ize O_2 transfer in SSF. Instead the biofilm conductance, k_Fa, should be used. It takes into account the diffusivity of O_2 within the biofilm and the thickness of the aerobic part of the biofilm, and therefore will very likely change during the fer-mentation (Thibault et al. 2000a). The biofilm conductance (k_Fa) might be able to be used to compare the influence of various operating parameters on the O_2 mass transfer in a given system, but it cannot be used to compare the performance of different microbe/substrate systems (Thibault et al. 2000a). This contrasts with SLF, in which k_La can be used to compare the efficiency of O_2 transfer in quite different systems. The question of O_2 transfer is further complicated by the fact that aerial hyphae, that is, hyphae exposed directly to the air within the inter-particle gas phase, can in some cases make a significant contribution to overall O_2 transfer (Rahardjo et al. 2002).

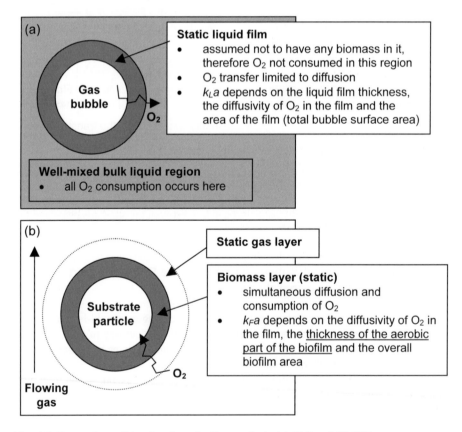

Fig. 4.5. Comparison of the situations for O_2 transfer in **(a)** SLF and **(b)** SSF

The substrate particle, the biofilm, and the static gas layer will contribute to the overall resistance to heat and water transfer. In the case of water transfer, note that the water changes phase as it leaves the solid, taking the energy of evaporation from the solid. This represents a combined heat-and-mass-transfer process. Heat and mass transfer from the particle to the inter-particle air has received little attention in SSF, although the literature about the drying of foods is relevant. At the high water activities typically encountered in SSF, there will typically be a film of liquid water at the surface and, for evaporation of this water, the major barrier is the static gas film that surrounds the particle.

4.3.4 Bulk Gas Flow Patterns and Pressure Drops

Relatively little attention has been given to gas flow patterns in SSF bioreactors. Those studies that have been done are discussed in Chaps. 6 to 11. Only general principles are given here. Basically, there are two extremes for gas flow patterns (Fig. 4.6): at one extreme the gas phase is well mixed and at the other it undergoes plug-flow. In the case of plug-flow, there is the question of axial dispersion: if a thin plug of colored gas molecules were introduced into the bioreactor, what would exit at the other end? A thin plug of the same thickness? Certainly diffusion makes the plug wider and more diffuse, and other phenomena, such as flow through torturous pathways, can increase the amount of dispersion. In this case the flow is referred to as "plug-flow with axial dispersion". In real bioreactors flow patterns can be more complicated, with the possibility of dead spaces, turbulence, and backflow.

The phenomenon of pressure drop arises due to the viscosity of air (Fig. 4.7). Air tends to stick to the surfaces in the bed, such as the particle surface, the surface of any biofilm growing at the particle surface and the surfaces of any hyphae growing into the inter-particle spaces. This retards the flow of air, due to the loss of energy by viscous friction between various layers of air. The air must still flow through the column at a steady rate, so this resistance to flow does not decrease the kinetic energy of the air, but rather decreases the pressure of the air. In other words, the pressure of the air falls as a gas flows through a column or bed.

In order to leave the outlet, the air leaving the bioreactor must be slightly above the barometric pressure (if the bioreactor is open to the air) or even at a higher pressure (if the outlet gas passes through a filter before entering the surroundings). The greater the resistance to flow, then the greater is the pressure gradient through the bed and the greater are the costs of pumping the air through the bed. Of course, going from an empty column to a column packed with a bed of particles, the area of solid surfaces increases dramatically (Fig. 4.7). It increases even further when a fungus on the surface of the solid particle begins to fill the voids with aerial hyphae. Therefore the pressure drop is greater in a packed-bed than in a hollow column and greater still when the inter-particle spaces in a bed are full of fungal hyphae.

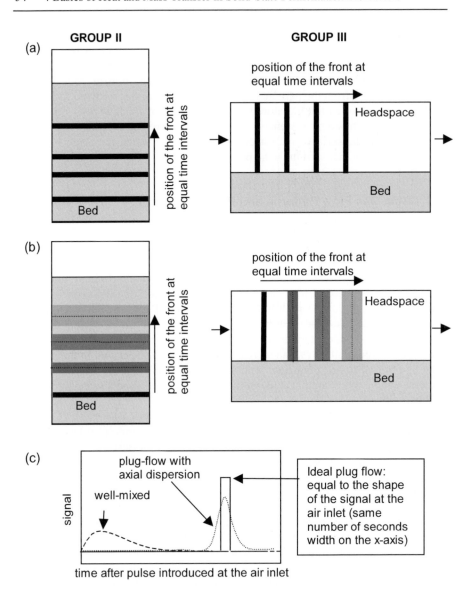

Fig 4.6. Gas flow patterns in bioreactors. **(a)** Ideal plug-flow **(b)** Plug-flow with axial dispersion. In both cases the movement of a front through the bioreactor is indicated (i.e., if it were possible at the air inlet to add a thin "plug" of tracer molecules across the whole cross-section). Examples are shown where air is forced through a static bed (Group II bioreactors) and through a headspace (Group III bioreactors). **(c)** Residence time distribution patterns for (—) ideal plug-flow, (· · ·) plug-flow with axial dispersion, and (- - -) a well-mixed system. Note that the inlet pulse is the same for all cases (the areas under the curves are equal)

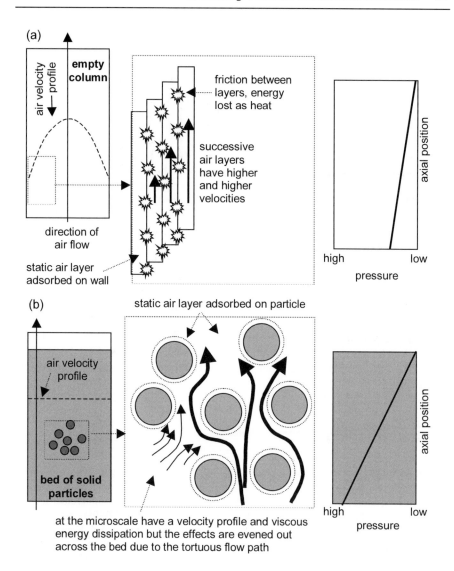

Fig. 4.7. The phenomenon of pressure drop. **(a)** Flow of air through an empty column. **(b)** Flow of air through a substrate bed constituted by small particles. In each case the diagram on the left is a schematic representation of the system and indicates the velocity profile (normal to the direction of air flow); the diagram in the middle shows a magnification of the microscale, highlighting the energy loss due to viscous interactions between successive air layers; the diagram on the right shows the pressure as a function of axial position within the bed (the diagram is reproduced in the same orientation as the bioreactor/column in the diagram on the left)

If for some reason the resistance is not uniform, then air can follow preferential paths: a majority of the air may flow through low resistance regions while air may hardly flow through high resistance regions, meaning that in these areas O_2 supply is limited to diffusion and heat removal is limited to conduction, the same situation as occurs in trays. This problem, called channeling, will be discussed in relation to packed-bed bioreactors in Chap. 7.

Due to these phenomena, one would expect laminar flow in a column without any filling, at least at the air flow rates typically used in packed beds, with each layer of air flowing at a different speed (Fig. 4.7(a)). There would be a parabolic velocity profile, with the flow rate being fastest at the center of the column and slowest near the wall. In fact the velocity is zero right at the wall, since a static boundary layer of gas molecules is absorbed to the wall. The situation is quite different when the column is packed with substrate particles. The air must pass through torturous pathways, with almost equal resistance across the whole column, which tends to even out the velocity profile across the bioreactor, such that the flow regime approaches plug flow.

4.3.5 Mixing Patterns in Agitated Beds of Solids

The mixing patterns that occur within beds of moist solids particles within SSF bioreactors have received relatively little attention. Chapters 9 and 10 report some work that has been done to characterize the effectiveness of mixing in Group III and Group IV bioreactors.

Further Reading

A discussion of the interactions between microbial growth kinetics and heat and mass transfer phenomena in SSF bioreactors
Mitchell DA, Stuart DM, Tanner RD (1999) Solid-state fermentation - microbial growth kinetics. In: Flickinger MC, Drew SW (eds) The Encyclopedia of Bioprocess Technology: Fermentation, Biocatalysis and Bioseparation, vol 5. John Wiley, New York, pp 2407–2429

A discussion of the use of k_La and k_Fa in SSF systems
Thibault J, Pouliot K, Agosin E, Perez-Correa R (2000) Reassessment of the estimation of dissolved oxygen concentration profile and k_La in solid-state fermentation. Process Biochem 36:9–18

The phenomenon of pressure drop in SSF bioreactors
Auria R, Ortiz I, Villegas E, Revah S (1995) Influence of growth and high mould concentration on the pressure drop in solid state fermentations. Process Biochem 30:751–756

5 The Scale-up Challenge for SSF Bioreactors

David A. Mitchell, Oscar F. von Meien, Luiz F.L. Luz Jr, and Marin Berovič

5.1 Introduction

Having now seen the various types of bioreactors used in SSF processes (Chap. 3) and the transport phenomena that occur within them (Chap. 4), we now return to the question of how the limitations on the efficiency of the transport phenomena within the bioreactor make it almost impossible to operate large-scale bioreactors in such a manner that the conditions within the substrate bed are maintained throughout the process at the optimum values for growth and product formation.

Is it really difficult to design an efficiently operating large-scale SSF bioreactor? In the case of SLF, there are examples of successfully operating bioreactors of hundreds of thousands of liters. Why cannot we do the same for SSF processes? Or can we? The answer is that the challenges in operating a bioreactor of several hundreds of thousands of liters are typically more difficult to overcome in SSF than in SLF, and it is no simple matter to develop efficient large-scale SSF bioreactors. This difficulty, often referred to as "the scale-up problem", is discussed in the following sections.

5.2 The Challenges Faced at Large Scale in SLF and SSF

The major challenge in the scale-up of aerobic submerged liquid fermentation processes is the transfer of O_2 into the liquid at a sufficient rate to obtain high cell densities. Scale-up strategies that address this transfer, which is characterized by the parameter $k_L a$, have long been available in the area of SLF (Kossen and Oosterhuis 1985). Although heat transfer calculations must be done, in order to provide sufficient cooling capacity, heat removal is typically not an overly challenging task. If the outer surface of the bioreactor does not provide a sufficiently large surface area to give the necessary rate of heat removal to the cooling water in a water jacket, then a cooling coil can be incorporated into the design without causing much complication in construction or operation.

On the other hand, in the case of SSF, heat removal is typically the major concern. It is more difficult to remove the waste metabolic heat from a bed of solids in which the inter-particle phase is occupied by air than it is to remove this heat from a continuous aqueous phase. There are two reasons for this:

- the thermal properties of a continuous aqueous phase, namely the thermal conductivity and heat capacity of liquid water, are superior to those of a bed of moist solids with inter-particle air;
- mixing greatly promotes heat removal by bringing the medium into contact with the cooling surfaces within the bioreactor. However, typically mixing must be minimized in SSF bioreactors, for several reasons: Firstly, it requires higher energy inputs to mix the bed of solid particles within an SSF bioreactor than to mix the liquid medium in an SLF bioreactor. Secondly, the presence of internal heat transfer surfaces such as plates or coils within the bioreactor will interfere much more with the mixing of a solid bed than it will with the mixing of a liquid medium. Finally, a liquid medium can be mixed reasonably well without causing undue shear forces, whereas in a bed of solids in an SSF process involving a fungus, even the slightest mixing action will cause significant physical damage to the mycelium growing at the particle surface.

The difficulty of heat removal from large-scale SSF bioreactors has two consequences for bioreactor design:

- evaporation may occur as a result of temperature rises in the bed (see Fig. 4.3.(c)), and in some cases it may in fact be promoted deliberately, given that it is one of the most effective heat removal mechanisms. However, continued evaporation can dry the bed out to water activities low enough to restrict growth. Therefore the maintenance of the water activity of the bed becomes a consideration that guides design and operation.
- given that in many SSF bioreactors the air phase plays a central role in heat removal and that the aeration rates needed in order to remove heat at a reasonable rate are more than sufficient to ensure a reasonable O_2 supply to the surface of the particles, O_2 supply is typically a minor consideration (except for Group I bioreactors, i.e., static beds without forced aeration).

The following discussion about the general scale-up problem therefore focuses on heat removal as the key scale-up criterion and maintenance of water activity as a related consideration. O_2 supply will not be covered in this general discussion, although something will be said about it in Chap. 6, which talks about Group I bioreactors.

5.3 The Reason Why Scale-up Is not Simple

Bioreactor design would be simple if all you needed to do was to obtain good performance in a laboratory-scale bioreactor and then simply construct a geometrically-identical larger version of this bioreactor. However, this is impossible to achieve. Recalling the argument presented in Sect. 2.3 (also see Fig. 2.3):

- the aim of the bioreactor is to control the conditions within the bed, such as the temperature and water activity, at the optimum values for growth and product formation;
- however, the growth of the organism causes deviations from the optimum conditions in its immediate surroundings, through the release of waste metabolic heat and the consumption of O_2, amongst other processes;
- in operating a bioreactor, we are limited to manipulating external operating variables;
- the effects of the operating variables on the conditions within the bioreactor, such as the bed temperature, are not direct. Between the manipulation that we make in the operating variable (for example, changing the temperature at which the air enters a forcefully aerated bioreactor) and any particular position in the bed, we have various transport phenomena. For example, to arrive at mid-height within a packed-bed bioreactor, the inlet air firstly has to pass through half of the bed, and the temperature of that air will have risen from the inlet value by the time it reaches the middle of the bed, due to the heat transfer that occurred over the intervening distance. This will decrease its ability to cool the middle of the bed (in fact, this phenomenon is the basis of the axial temperature profile shown in Fig. 4.3 for the forced aeration of static beds);
- the importance of these transport phenomena increases as the distance over which transport must occur increases. This distance typically increases as the size of the bioreactor increases.

So transport phenomena are of crucial importance in controlling how the bioreactor operates. Scale-up becomes a challenging task because the underlying physiology of the microorganism is independent of scale. The microorganism will respond in exactly the same way for a given set of conditions that it finds in its local environment, regardless of whether it is located within a bioreactor holding 10 g of substrate or a bioreactor holding many tons of substrate. In other words, in both bioreactors it will give the same rate of growth and heat release for a given combination of O_2 concentration, nutrient concentration, pH, temperature, and water activity.

The key question of the scale-up problem then becomes "Is it possible to keep the local environmental conditions at or very near optimal values as scale is increased?" Note that it is relatively easy to control the local environment within small-scale bioreactors. In fact, it is for this reason that thin columns are used for basic kinetic studies (which will be seen in Chap. 15).

It is important to understand that the conditions in the local environment depend on the balance between the changes caused by the microorganism and the transport phenomena that arise to counteract these changes. For example, the local temperature sensed by the organism (and which will affect its growth) depends on the balance between the rate of waste metabolic heat production and the rate of conduction of energy away to regions in which the temperature is lower (Fig. 5.1(a)). If the rate of waste heat production is higher than the rate of conduction, then the local temperature will rise, which of course occurs during the early periods of the fermentation when the growth rate is accelerating (Fig. 5.1(b)).

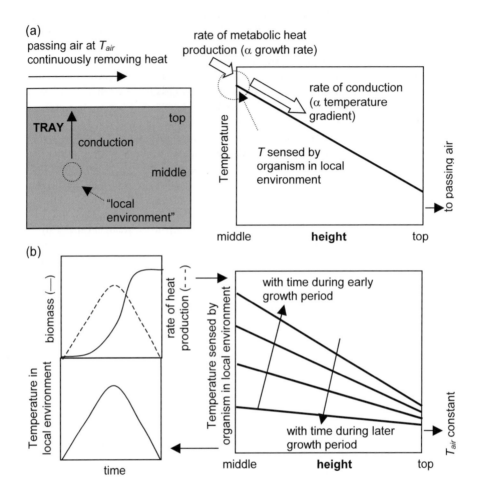

Fig. 5.1. The temperature in the local environment of the organism depends on the balance between heat generation and heat removal. This example is given in the context of a fermentation carried out within a tray, where the main heat removal mechanism in the bed is conduction. The "local environment" of interest is at mid-height in the bed. **(a)** Whether the temperature in the local environment remains constant, increases or decreases depends on the balance between the rate of metabolic heat production (which is proportional to the growth rate) and the rate of heat removal by conduction to the bed surface (which is proportional to the temperature gradient across the substrate bed). **(b)** Due to the change in the rate of production of waste metabolic heat as the growth rate changes, the temperature in the local environment changes over time. During early growth the rate of waste heat production increases. This causes the temperature to increase until the rate of heat removal once again equals the rate of heat production. However, since growth continues to accelerate, the rate of heat production continues to rise, so the local temperature must continue to rise in order to continue to increase heat removal. Later during growth, as the growth rate and therefore the rate of heat production decreases, the local temperature decreases

So the basic question that we need to answer in order to understand the scale-up problem has become: "What is the effect of scale on the ability of the transport processes to remove heat at a rate that is sufficient to prevent local temperatures from reaching values that limit growth?" The effect of scale on the effectiveness of transport phenomena will be discussed here in relation to convective and conductive heat removal in static beds. With respect to solids mixing phenomena, suffice to say that the effectiveness of mixing tends to decrease as scale increases.

Figure 5.2 illustrates the problem, using a packed-bed bioreactor as an example. As explained in Fig. 4.3, the convective flow of air through a static bed in which an exothermic reaction is occurring leads to an increase in the bed temperature between the air inlet and the air outlet. For a given organism, one of the major factors affecting the slope of the temperature gradient in the bed is the air flow rate. A laboratory-scale bioreactor may operate with the temperature exceeding the optimum temperature for growth by only a few degrees. However, as scale increases, the deviations from the optimum temperature will be much greater, especially if the same volumetric flow rate is used. It is possible to try to combat these deviations by changing key operating variables as scale increases. For example, it might appear reasonable to maintain the superficial air velocity constant (the superficial air velocity being the volumetric air flow rate divided by the overall cross-section of the bioreactor). In the simplest case, this will maintain the same temperature gradient in the bioreactor. However, due to the greater height, the temperature in the upper region of the bioreactor will reach much higher values than those that were reached at laboratory scale (Fig. 5.2). One strategy might be to increase the superficial velocity of the air (V_Z, m s^{-1}) in direct proportion to the height (H, m) of the bioreactor (that is, to maintain V_Z/H constant). This might in fact prevent the bed from ever exceeding the maximum temperature observed in the laboratory bioreactor, however, it might also lead to unacceptably high pressure drops, or the required air velocity might fluidize the bed.

The problem is more severe in the cases where significant amounts of heat are removed from the bed at small scale by conduction, such as in a tray bioreactor, or within a packed-bed bioreactor with a cooled surface. If geometric similarity is maintained, then the distance between the center of the bed and the surroundings or heat transfer surface increases with increase in scale. The effectiveness of conduction in removing heat decreases in proportion to the square of the distance over which conduction must occur. Therefore, maintaining geometric similarity will decrease the relative contribution of conductive heat removal. In fact, it is desirable to maintain the "conduction distances" constant as scale increases. For this reason tray bioreactors are scaled-up by increasing the number of trays, and not the thickness of the substrate layer within the tray. Likewise, as will be seen in Chaps. 7 and 23, it may be interesting for large-scale packed beds to have internal heat transfer plates arranged such that the large-scale version has the same "conduction distances" as a laboratory-scale bioreactor.

In general, as a bioreactor is scaled-up from the laboratory to production scale, it is not a simple matter to keep constant either V_Z/H or the distance over which conduction must occur. As a consequence, the local conditions, at least in some

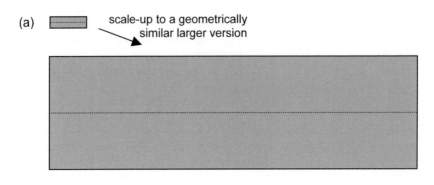

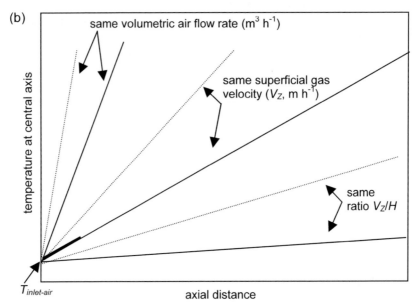

Fig. 5.2. Scale up on the principle of geometric similarity is not a simple matter. **(a)** Scale-up on the basis of geometric similarity. Both the radius and length have increased 10-fold. **(b)** Temperature profiles along the central axis that might be expected at the time of peak heat production. Key (—) Temperature profile in the small-scale bioreactor; (—) Temperature profiles that might be expected in the large-scale bioreactor for different strategies regarding the aeration rate, if the results with the small-scale bioreactor had been obtained under a condition where the side walls were insulated (i.e., with no heat removal by conduction though the side walls); (• • •) Temperature profiles that might be expected in the large-scale bioreactor for different strategies regarding the aeration rate, if the results with the small-scale bioreactor had been obtained under a condition where the side walls were not insulated and heat was removed by cooling water in a jacket or waterbath. The different strategies regarding the aeration rate are indicated directly on the figure

regions of the bioreactor, will be less favorable for growth than those that the organism experienced at laboratory scale. The average volumetric productivity of the large-scale bioreactor (kg of product produced per cubic meter of bioreactor volume per hour) will then be smaller than the volumetric productivity achieved with the laboratory-scale bioreactor. The scale-up problem becomes more difficult when we realize that this discussion has not explored all the potential problems and complications. Some further considerations are:

- in mixed beds, the efficiency of mixing is likely to decrease with scale;
- in some beds both convection and conduction play important roles in heat removal. The optimum combination of these two mechanisms may change with scale. For example, in some cases conduction plays an important role in removal at small scale, but its contribution decreases as scale increases as the surface area to volume ratio of the bioreactor decreases;
- bioreactor design will affect the ease of substrate handling, and ease of substrate handling may be an important consideration in the economics of the process, especially in relation to the need for manual labor;
- pressure drop and fluidization considerations may put a limit on possible air flow rates;
- the sensitivity of the microorganism to damage by mixing may put a limit on the frequency with which the bed can be mixed;
- increases in bed heights may have side effects, such as the deformation of particles at the bottom of the bed, affecting inter-particle void fractions, or even crushing the particles.

Given this complexity, we are only likely to achieve the maximum possible efficiency in large-scale bioreactors if we understand the phenomena that combine to control bioreactor performance and if we use quantitative approaches to the scale-up problem.

5.4 Approaches to Scale-up of SSF Bioreactors

Various quantitative approaches have been proposed for scale-up of SSF bioreactors, including the use of mathematical models and of various "simplified approaches" that have some similarity with the "rule-of-thumb" approaches to scaling-up SLF bioreactors. Given the complexity of SSF systems, models will be more powerful tools, and should be preferred where possible, especially since various fast-solving models are available in the literature, and can be adapted to new systems without requiring an onerous amount of work. Some of these mathematical models are presented in Chaps. 22 to 25, where their potential uses are demonstrated and discussed.

It is worthwhile remembering, as noted in Chap. 2, that the inter-particle phenomena themselves are independent of scale, since we will typically be using the same sized substrate particles at small scale and large scale. Significant intra-particle mass transfer limitations, of O_2 and nutrients, may occur even in particles of only 1 to 5 mm diameter. These limitations are intrinsic to SSF. The best that can be done in the manner in which the bioreactor is operated is to control the inter-particle conditions, for example, to maintain the O_2 concentration in the gas phase in contact with the particle surface at as high a concentration as possible.

The knowledge framework concerning scale up of SSF processes can be characterized as follows:

- **in relation to current large-scale bioreactors:** there is no evidence in the literature that anything other than "best-guess" or "trial-and-error" approaches have been used for the development of almost all current large-scale SSF bioreactors. It is likely that some engineering calculations have been done, even if they were not reported. This is most likely in the soy sauce industry, but the knowledge about scale-up, if it has been generated, has not been made widely available because it is important proprietary information;
- **in relation to the strategies themselves:** Since the work of Saucedo-Castaneda et al. (1990), mathematical modeling work has been done with the aim of developing rational scale-up strategies for SSF bioreactors. However, although such models are potentially very useful tools for guiding the selection and design of large-scale bioreactors, there are no reports describing a scale-up study in which this has actually been done. To date the investigations have been limited to the use of models to demonstrate, using simulations, how models might be used to guide scale up.

Finally, it is important to point out that although mathematical models of bioreactor behavior can be used to predict how a bioreactor will perform before it is built, this modeling work does not replace the need to do experimental work, rather, it is a tool for guiding the experimental program. As will be shown later, mathematical modeling can help to raise questions about bioreactor operation that can be answered through experimentation, it can also help to eliminate ideas which appear reasonable but are actually unfruitful, without wasting time and money to test the ideas experimentally.

Further Reading

Scale up in submerged liquid fermentation processes
Kossen NWF, Oosterhuis NMG (1985) Modelling and scaling up of bioreactors. In: Rehm HJ, Reed G (eds) Biotechnology, vol 2. Verlag Chemie, Weinheim, pp 571-605
Reisman HB (1993). Problems in scale-up of biotechnology production processes. Critical Reviews in Biotechnology 13:195–253

Scale-up of SSF bioreactors
Mitchell DA, Krieger N, Stuart DM, Pandey A (2000) New developments in solid-state fermentation II. Rational approaches to the design, operation and scale-up of bioreactors. Process Biochem 35:1211–1225

6 Group I Bioreactors: Unaerated and Unmixed

David A. Mitchell, Nadia Krieger, and Marin Berovič

6.1 Basic Features, Design, and Operating Variables for Tray-type Bioreactors

Group I bioreactors, or "tray bioreactors", represent the simplest technology for SSF. They have been used for many centuries in the production of traditional fermented foods such as tempe and in the production of soy sauce *koji*. However, this chapter does not review these applications. Readers interested in traditional fermented foods should consult the reading listed at the end of the chapter. The current chapter considers tray bioreactors as candidates in the selection of bioreactors for newly-developed SSF processes. Trays may be appropriate for a new process if the product is not produced in very large quantities, if a "produced-packet" of fermented product can be sold directly, or if labor is relatively cheap.

The basic design features of tray-type bioreactors have already been presented in Chap. 3.3.1. Figure 6.1 shows these features in more detail. Some possible variations in the design include:

- the tray chamber may be relatively small, such as an incubator, or it may be a room large enough for people to enter;
- the tray may be constructed of various different materials, such as wood, bamboo, wire or plastic. In fact, a plastic bag might be used instead of a rigid tray;
- the bottom and sides of the tray may be perforated or not;
- water-cooled heat exchange surfaces might be incorporated.

The available design features for tray-type bioreactors are:

- the dimensions of the tray, namely length, width, and height;
- the positioning of the trays within the bioreactor;
- the presence of cooling surfaces within the tray chamber.

The available operating variables are:

- the temperature, humidity, and flow rate of the air entering the tray chamber and the velocity of circulation past the tray surface;
- if cooling surfaces are present, then the temperature of the cooling water.

Note that, although this type of bioreactor is nominally static, the bed may be mixed infrequently. For example, it is typical for the tray contents to be turned by hand once or twice per day.

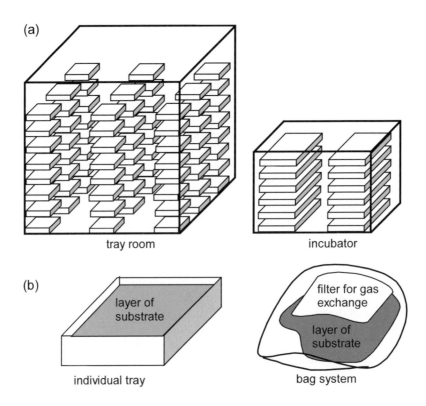

Fig. 6.1. Basic design features of tray-type bioreactors and possible design variations. **(a)** Different tray chambers, including tray rooms and incubators, in which the trays are arranged on shelves. **(b)** Different basic tray designs. The tray on the left could be made of wood, plastic, bamboo, or wire. The plastic bag on the right could be made entirely of a gas-permeable plastic, or could contain a filter insert that allows gas exchange

6.2 Use of Bag Systems in Modern Processes

Typically for newly-developed processes, plastic bags will be used, employing special plastics that allow the exchange of O_2 and CO_2, but do not allow the exchange of water, thus allowing the microorganism to respire but preventing the bed from drying out. The great advantage of this system over the traditional technology of open trays is that the plastic prevents contaminants from entering the bed. Either the whole bag may be made of this plastic, or the bag may be made of a gas-impermeable plastic and have a "window" made of a special plastic, paper, or fabric. This technology has been used for over 20 years to produce spore inoculum for the *koji* process (Lotong and Suwanarit 1983), although of course it is not an appropriate technology for the production of soy sauce *koji* itself in modern processes, where individual batches are of the order of several tons. However, it may be appropriate if smaller volumes are produced. For example, in Australia, a biopesticide based on spores of *Metarhizium anisopliae* is produced on rice grains within "self-aerating bags" (Milner 2000). In 1999, at the commercial trial stage, 9 tons of product was produced, which corresponds to a productivity of 25 kg per day, averaged over 365 days.

Cuero et al. (1985) used micro-porous plastic bags, consisting of polypropylene with 0.4 μm pores. The bags can be autoclaved. They allow gas and water vapor exchange, but neither the release of spores nor the entry of contaminants. Due to the fact that the bags allowed water vapor exchange, the bags were incubated in a high humidity environment (95% relative humidity).

6.3 Heat and Mass Transfer in Tray Bioreactors

Depending on the situation, it may be appropriate to consider either an individual tray or the whole-tray chamber as the bioreactor. For example, it would be appropriate to treat the whole tray chamber as the bioreactor when the trays are open to free gas and water exchange with their surroundings and the temperature and humidity of the air in the tray chamber are carefully controlled.

The question about optimum design of tray chambers has received little attention. For example, quantitative information is not available about the best way to position trays in the chamber. As a result, it is not possible to state what is the best spacing to leave between trays in order to maximize volumetric productivity (that is, the amount of product produced per unit volume of tray chamber). Most attention has been given to the individual trays.

As a generalization, within an individual tray, large O_2 and temperature gradients will arise in the substrate layer during the fermentation. The following subsections outline what is known about the limitations on O_2 and heat transfer within tray bioreactors.

6.3.1 Oxygen Profiles Within Trays

Since in a tray bioreactor air is not blown forcefully through the trays, O_2 and CO_2 can only move within the bed by diffusion. Potentially, due to the temperature gradients that arise, there could be natural convection within the void spaces within the bed, although this has not been studied. This discussion will focus on O_2. Similar considerations apply to CO_2, although it will typically be diffusing in a direction opposite to that of O_2.

The limitation of O_2 movement in the bed to diffusion through the void spaces, coupled with its simultaneous uptake by the microorganisms at the particle surfaces, leads to the establishment of O_2 concentration gradients within the void spaces (Fig. 6.2). Rathbun and Shuler (1983) noted O_2 gradients within the gas phase of a bed of tempe (which involves the cultivation of the fungus *Rhizopus oligosporus* on cooked soybeans) of the order of 2% (v/v) cm^{-1}. This represents a drop equal in magnitude to 10% of the gas phase O_2 concentration in air (~21% (v/v)) over 1 cm. Of course, the exact shape of the spatial O_2 concentration profile will depend on whether the bottom of the tray is perforated or not and the rate at which O_2 is being consumed by the organism.

Ragheva Rao et al. (1993) proposed an equation to estimate the maximum depth that a tray could be in order for the O_2 concentration not to fall to zero at any part in the tray during a fermentation. They referred to this depth as the critical depth (D_c, cm):

$$D_c = \sqrt{\frac{2 D C Y_{XO}}{R_{XM}}}. \tag{6.1}$$

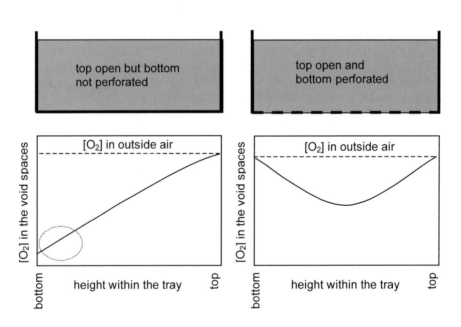

Fig. 6.2. O_2 concentration gradients within trays. Note that the spatial gradients will change over time, depending on the rate of growth of the microorganism. In the bioreactor with an unperforated bottom, O_2 limitation will be the factor that has the greatest influence on growth at the bottom of the bed (the area indicated by the dotted circle) since heat removal through the bottom of the bed will control the temperature in this region reasonably well

In Eq. (6.1) Y_{XO} is the yield coefficient of biomass from O_2 (g-dry-biomass g-O_2^{-1}), C is the O_2 concentration in the surrounding atmosphere (g cm^{-3}), D is the effective diffusivity of O_2 in the bed (cm^2 h^{-1}), and R_{XM} is the maximum growth rate (g-dry-biomass cm^{-3}-bed h^{-1}). Ragheva Rao et al. (1993) estimated D as 0.03 cm^2 s^{-1} and Y_{XO} as 1.07 g-dry-biomass g-O_2^{-1}. In dry air at 25°C and 1 atm pressure, C will be 2.7×10^{-4} g cm^{-3}. Using experimental estimates for R_{XM}, they concluded that the critical depth would be of the order of 2.4 cm. For a tray with a perforated bottom, oxygen can penetrate this distance from both the top and bottom surfaces, meaning that the bed depth in the tray can be twice the critical bed depth, namely 4.8 cm. This can be taken as a typical value for trays, although of course the exact value is influenced not only by the growth rate but also by the effective diffusivity of O_2 within the bed, which will decrease as the biomass grows into the inter-particle spaces during the fermentation.

6.3.2 Temperature Profiles Within Trays

Rathbun and Shuler (1983) found temperature gradients as high as 1.7°C cm^{-1} within a static bed of tempe, while Ikasari and Mitchell (1998) measured temperatures as high as 50°C at 5 cm depth during the cultivation of *Rhizopus oligosporus* on rice bran in a tray within a 37°C incubator.

Szewczyk (1993) derived a simplified equation that can be used to describe the temperature profile within a tray bioreactor, from the central plane ($z=0$) to the surface ($z=1$), when the top and bottom half of the tray are identical, that is, in the situation shown in Fig. 6.3(a):

$$T = \frac{(T_s + 273) + \Theta + N_{Bi}(T_a + 273)}{N_{Bi} + 1} + \frac{N_{Bi}}{N_{Bi} + 1}(T_s - T_a + \Theta)z - z^2\Theta , \qquad (6.2)$$

where T_s and T_a are the temperatures of the bed surface and surrounding air (°C), respectively. The spatial coordinate z is expressed as a dimensionless fraction of the total bed height (Z, m). N_{Bi} is the Biot number, given by $\alpha.Z/k$ where α is the heat transfer coefficient for bed-to-air heat transfer at the top of the bed (W m^{-2} °C^{-1}), and k is the thermal conductivity of the bed (W m^{-1} °C^{-1}). Finally, the symbol Θ represents the temperature difference that would occur between the bottom of the solid bed and the tray surface if there were no heat transfer through the bottom of the tray. It is given by the following equation:

$$\Theta = \frac{R_Q Z^2}{2k} , \qquad (6.3)$$

where R_Q is the volumetric heat production rate (W m^{-3}). It is not simple to apply Eq. (6.2), since the surface temperature of the tray needs to be known. The surface temperature depends on the value of the heat transfer coefficient α, but α also appears in Eq. (6.2), within N_{Bi}. A more complex modeling approach is needed to relate T_s and α. Szewczyk (1993) used such a model to derive the relationship

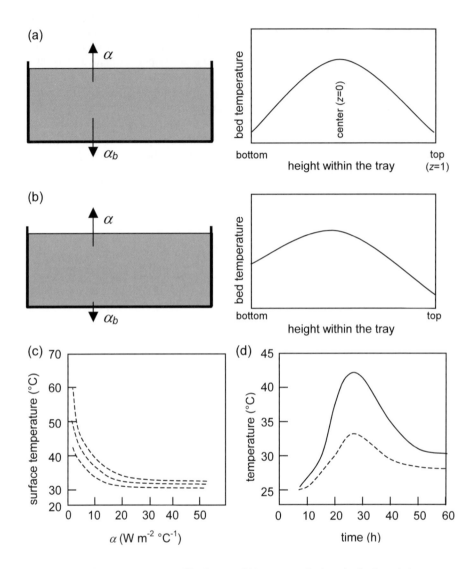

Fig. 6.3. (a) The temperature profile that would be expected when the bed-to-air heat transfer coefficients (α and α_b) at the top and bottom of the tray are identical. **(b)** The temperature profile that would be expected when the bed-to-air heat transfer coefficient at the bottom of the tray (α_b) is smaller than that at the top of the tray (α). **(c)** The effect of the bed-to-air heat transfer coefficient at the upper tray surface on the surface temperature, for three different heat generation rates (from bottom to top the curves represent 20, 40, and 60 W kg^{-1}, for a 6-cm-high bed of the type shown in Fig. 6.3(a) that is incubated in a 100% relative humidity atmosphere at 30°C. **(d)** How the temperatures at the center of the bed (——) and the bed surface (– – –) might typically be expected to vary over time, for the case where $\alpha = 10$ W m^{-2} °C^{-1}. This figure is based on data provided by Szewczyk (1993)

shown in Fig. 6.3(c). The temperature profile within the bed will depend on the relative values of the heat transfer coefficients at the top and bottom of the bed. If they are equal, then the profile will be symmetrical about the center plane of the bed (Fig. 6.3(a)). If not, then the profile will be asymmetrical (Fig. 6.3(b)). The surface temperature of the bed is greatly affected by the heat transfer coefficient α at values below 10 W m^{-2} °C^{-1}. Above this value, further increases in the heat transfer coefficient have little effect (Fig. 6.3(c)). The value of α will depend on the velocity at which air is circulated past the tray surface. Szewczyk (1993) simu-lated the growth of *Aspergillus niger* on wheat bran in a tray, with a value of α of 10 W m^{-2} °C^{-1}. At the time of peak heat generation, the center of the bed was 10°C hotter than the surface (Fig. 6.3(d)).

6.3.3 Insights from Dynamic Modeling of Trays

No modeling case study will be presented for trays in this book and therefore this section will discuss the insights that dynamic mathematical models of tray biore-actors have given into the relative importance of temperature and O$_2$ limitations in controlling the performance of trays.

Rajagopalan and Modak (1994) developed a model to describe heat and mass transfer in trays, which included the various processes shown in Fig. 6.4. They used their model to investigate the relative importance of high temperatures and low O$_2$ concentrations in determining the specific growth rate in a 6.4-cm-high tray. Since the tray was assumed to be symmetrical around the center plane it was only necessary to consider a depth of 3.2 cm from the surface to the central plane. Their results are shown in Fig. 6.5. In interpreting these results, it must be remem-bered that the overall growth rate (R_X, kg-dry-biomass m^{-3} h^{-1}) is a combination of the biomass density (X, kg-dry-biomass m^{-3}) and the specific growth rate accord-ing to the following equation:

$$R_X = \mu X \left(1 - \frac{X}{X_{max}} \right), \tag{6.4}$$

where X_{max} (kg-dry-biomass m^{-3}) is the maximum possible value of the biomass density. The specific growth rate constant μ (h^{-1}) is affected by both temperature and the biofilm O$_2$ concentration according to the relationship:

$$\mu = \mu_{max} \mu_{FT} \mu_{FO}. \tag{6.5}$$

In this equation, μ_{max} (h^{-1}) is the maximum value that the specific growth rate con-stant can have, that is, its value under optimal conditions for growth. On other hand, μ_{FT} and μ_{FO} are dimensionless fractions, that is, they vary between 0 and 1. The parameter μ_{FT} describes the limitation of growth by deviations from the opti-mum temperature of 38°C while μ_{FO} describes the limitation of growth at low O$_2$ concentrations.

bulk flow of O_2 and CO_2 with the headspace gases

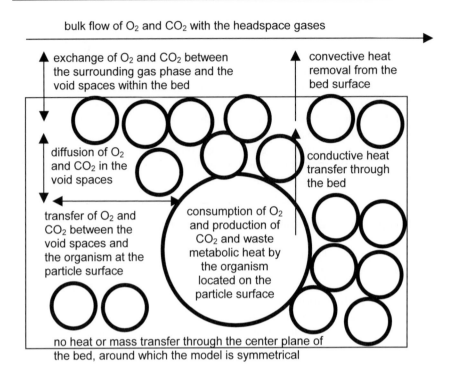

exchange of O_2 and CO_2 between the surrounding gas phase and the void spaces within the bed

convective heat removal from the bed surface

diffusion of O_2 and CO_2 in the void spaces

conductive heat transfer through the bed

transfer of O_2 and CO_2 between the void spaces and the organism at the particle surface

consumption of O_2 and production of CO_2 and waste metabolic heat by the organism located on the particle surface

no heat or mass transfer through the center plane of the bed, around which the model is symmetrical

Fig. 6.4. Heat and mass transfer processes described by the model of Rajagopalan and Modak (1994). Note that for simplicity, it was assumed that the whole biofilm was at the same O_2 concentration, although in reality there would be an O_2 concentration gradient due to the simultaneous diffusion and consumption of O_2

A key prediction of this modeling work of Rajagopalan and Modak (1994) is that limitation of growth due to lack of O_2 occurs even though the gas phase O_2 concentration never falls to very low values; in their simulations the gas phase O_2 concentration was always two-thirds or more of the O_2 concentration in the surrounding atmosphere, regardless of time or depth (Fig. 6.5(b)). Since the organism within the biofilm can consume O_2 much faster than the rate at which O_2 can transfer from the gas phase to the biofilm, biofilm O_2 concentrations can fall to low levels (Fig. 6.5(d)). This occurs at the top of the bed where, due to the fact that the temperature remains near to the optimum for growth since it is effectively cooled by the surrounding atmosphere (Fig. 6.5(a)), the organism grows rapidly and consumes the O_2 in the biofilm, reducing it to levels that significantly decrease the specific growth rate. In this case, the growth of the biomass is controlled by the rate at which O_2 is transferred to the biofilm ((Figs. 6.5(e) and (f)).

The highest temperatures occur at the central plane of the bed, and these are sufficiently high to decrease the specific growth rate significantly (Fig. 6.5(c)). Indeed, due to the fact that the high temperatures cause low growth rates in this

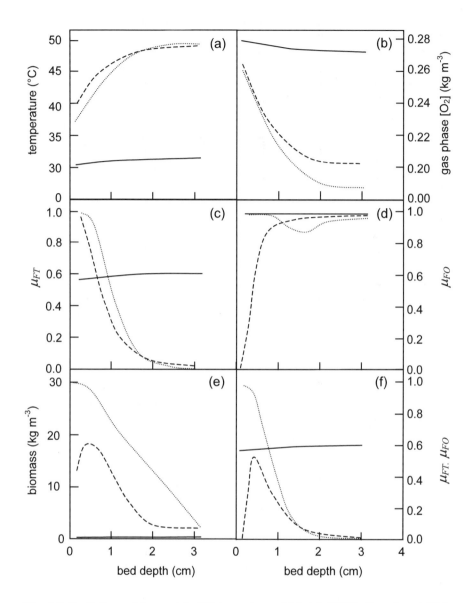

Fig. 6.5. Predictions of the model of Rajagopalan and Modak (1994). Key: (———) 20 h; (– – –) 60 h; (········) 100 h. Note that the top of the tray is represented by zero bed depth and the center plane corresponds to a bed depth of 3.2 cm. The fractional modifiers of the specific growth rate constant (μ_{FT} and μ_{FO}) are explained in the text (see Eq. (6.5)). Adapted from Rajagopalan and Modak (1994), with kind permission from Elsevier

region, the rate of O_2 consumption in the biofilms within this region is sufficiently low that the O_2 transfer from void space to biofilm can maintain a high O_2 concentration in the biofilm. In other words, in the center of the bed, temperature is the most important factor limiting growth. In fact, the temperature limitations are so severe in the middle of the bed that even after 100 h (when the organism would easily have reached its maximum concentration if it had been growing at the maximum possible specific growth rate), much of the bed has a biomass concentration significantly lower than the maximum biomass concentration (Fig. 6.5(e)).

The work of Smits et al. (1999) confirms that O_2 levels in the inter-particle spaces will generally not be a limiting factor. They used a heat and mass transfer model to investigate how the relative importance of O_2 limitation and temperature limitation depends on the thermal conductivity of the bed and the effective diffusivity of O_2 within the pores of the bed. For the growth of *Trichoderma reesei* in a tray, their model predicted that O_2 diffusion within the inter-particle spaces would only become limiting at a 10-cm bed depth if the effective diffusivity of O_2 in the bed was less than 4×10^{-6} m^2 s^{-1} and the thermal conductivity was greater than 0.45 W m^{-1} °C^{-1}. The effective diffusivity of O_2 in a bed with biomass at its maximum density is actually of the order of 4×10^{-6} m^2 s^{-1} (Auria et al. 1991) meaning that O_2 supply to a 10-cm bed depth can potentially become limiting, although this will only happen if there is a combination of high biomass concentration and high growth rate.

Smits et al. (1999) also modeled the diffusion of water vapor in the void spaces of the bed. When it was assumed that the air surrounding the tray was maintained at a high humidity, then the combination of metabolic water production with the relatively slow water vapor diffusion meant that the predicted water content of the substrate remained above the initial value. Under such conditions there will be no danger of the growth rate being limited by low water activities of the solid substrate. Of course, as Smits et al. (1999) point out, water could become limiting if the trays were incubated in an environment of low relative humidity. This would complicate operation since it would be necessary to periodically spray water onto the bed and to mix it in.

Rajagopalan and Modak (1994) used their model to investigate the effect of the height of the bed and the temperature of the surroundings on the average biomass concentration in the bed after 100 h of cultivation. For bed heights of 1.6 cm and less, the average biomass content reached its maximum possible value (i.e., 30 kg-dry-biomass m^{-3}) within 100 h only when incubated at temperatures near the optimum temperature of 38°C, namely at 35°C and 40°C (Fig. 6.6). This was because with these small bed thicknesses the bed temperature remained near the incubation temperature.

At a bed height of only 3.2 cm, the maximum biomass concentration was achieved by 100 h only when the bed was incubated at temperatures below the optimum temperature (i.e., between 30-35°C). Of course this lower outside temperature, combined with the metabolic heat production, combined to maintain the whole of the bed near the optimum temperature of 38°C.

For a bed height of 6.4 cm it was impossible to maintain the majority of the bed near the optimum temperature for growth, as evidenced by the fact that the highest value for the average biomass content at 100 h was only 16.6 kg-dry-biomass m^{-3}, obtained with incubation at 30°C. Incubation at lower temperatures controlled the temperature in the interior of the bed at values near the optimum, but cooled the surface to values at which growth was very slow. The problem of adequate temperature control became worse still at a bed height of 12.7 cm.

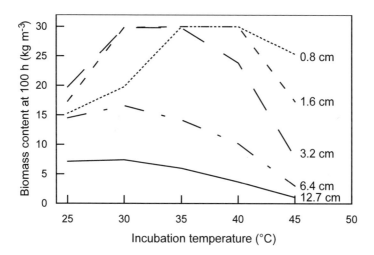

Fig. 6.6. Results obtained by Rajagopalan and Modak (1994) when they used their model to investigate the effect of the height of the bed and the temperature of the surroundings on the average biomass concentration in the bed after 100 h of cultivation. Adapted from a table presented by Rajagopalan and Modak (1994), with kind permission from Elsevier

6.4 Conclusions

The layer of substrate in trays is limited to a bed height of around 5 cm by considerations of heat and O$_2$ transfer within the bed. Therefore scale-up of the process cannot be achieved by increasing the bed height. The only manner to scale up a tray process to large scale is to increase the surface area of the trays, which is equivalent to saying that the large-scale process must use a large number of trays of the same size as those in which the laboratory studies were done. The use of large numbers of trays implies the necessity either for manual handling or highly sophisticated robotic systems, both of which can be inordinately expensive. However, in regions in which manual labor costs are low, such tray-type processes may find applications.

Further Reading

Traditional fermented food processes that involve tray technology
Hesseltine CW (1983) Microbiology of oriental fermented foods. Ann Rev Microbiol 37:575–601
Stanton WR, Wallbridge A (1969) Fermented food processes. Process Biochem April:45–51
Steinkraus KH (1986) Fermented foods, feeds, and beverages. Biotechnol Adv 4:219–243

An extension of the model of Rajagopalan and Modak (1994) that describes simultaneous diffusion and consumption in the expanding biofilm that develops on the particle surface
Rajagopalan S, Modak JM (1995) Modeling of heat and mass transfer for solid state fermentation process in tray bioreactor. Bioprocess Eng 13:161–169

Studies of heat and mass transfer in tray bioreactors
Ghildyal NP, Ramakrishna M, Lonsane BK, Karanth NG (1992) Gaseous concentration gradients in tray type solid state fermentors - Effect on yields and productivities Bioprocess Eng 8:67–72
Ghildyal NP, Ramakrishna M, Lonsane BK, Karanth NG, Krishnaiah MM (1993) Temperature variations and amyloglucosidase levels at different bed depths in a solid state fermentation system. Chem Eng J 51:B17–B23
Rathbun BL, Shuler ML (1983) Heat and mass transfer effects in static solid-substrate fermentations: Design of fermentation chambers. Biotechnol Bioeng 25: 929–938
Smits JP, van Sonsbeek HM, Tramper J, Knol W, Geelhoed W, Peeters M, Rinzema A (1999) Modelling fungal solid-state fermentation: the role of inactivation kinetics. Bioprocess Eng 20:391–404

7 Group II Bioreactors: Forcefully-Aerated Bioreactors Without Mixing

David A. Mitchell, Penjit Srinophakun, Nadia Krieger, and Oscar F. von Meien

7.1 Introduction

This chapter addresses the design and operation of SSF bioreactors under conditions where forced aeration is used but the substrate bed is not mixed. Typically these bioreactors are referred to as packed-bed bioreactors. This mode of operation is appropriate for those SSF processes in which it is not desirable to mix the substrate bed at all during the fermentation due to deleterious effects on either microbial growth or the physical structure of the final product.

The characteristics of this mode of operation also apply to the static phases of forcefully-aerated bioreactors that are mixed once every few hours. The operation of such bioreactors will be discussed in Chap. 10; suffice to say for the moment that during the static phase they will act as packed-bed bioreactors, and therefore the principles developed in the present chapter will apply to this static phase.

7.2 Basic Features, Design, and Operating Variables for Packed-Bed Bioreactors

The basic design features of a packed-bed bioreactor have been already presented in Sect. 3.3.1. Figures 7.1 and 7.2 show these features in more detail. Some possible variations in the design include:

- the column may have a cross section other than circular.
- the column may lie horizontally, or for that matter, at any angle. This alters the relative directions of the forces due to gravity and air pressure.
- the column may be aerated from either end. For a vertical column, the air may enter the bed from either the top or the bottom. Aerating from the top avoids the fluidization of particles at high air velocities, but will contribute to bed compaction since the air flow is in the same direction as gravity.

- the column may have a perforated tube inserted along its central axis, allowing an extra air supply in addition to the end-to-end aeration (Fig. 7.1(b)). However, this will only be effective for very small bioreactor diameters.
- the column may be water-jacketed or heat transfer plates may be inserted into the bed. In this chapter, packed-bed bioreactors with internal heat transfer plates will be referred to as "Zymotis packed-beds", using the name coined by Roussos et al. (1993), while those lacking such plates will be referred to as "traditional packed-beds" (Fig. 7.2).

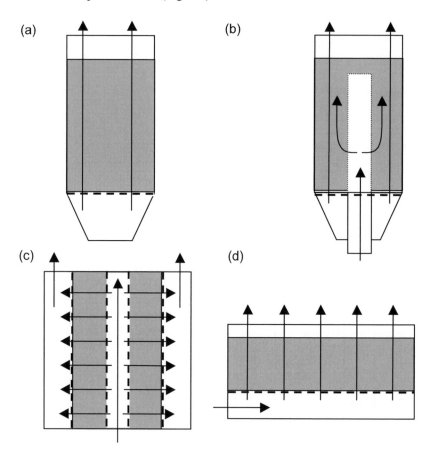

Fig. 7.1. Basic design features of packed-bed bioreactors and possible design variations. **(a)** A simple "traditional" packed-bed design. **(b)** A packed bed with a perforated tube inserted along its central axis: The benefits of this will only be apparent if the bed is relatively thin or, in a wide bed, if many perforated tubes are inserted. This is due to the fact that the forced aeration in the axial direction will tend to force the radial flow to follow the axial direction also. **(c)** Radial flow packed-bed: The advantage of this design is that, compared to a column of the same dimensions, the distance of flow through the bed is decreased. It is similar to the use of a wider "traditional" packed bed with a lower bed height. **(d)** A "short-wide" packed-bed

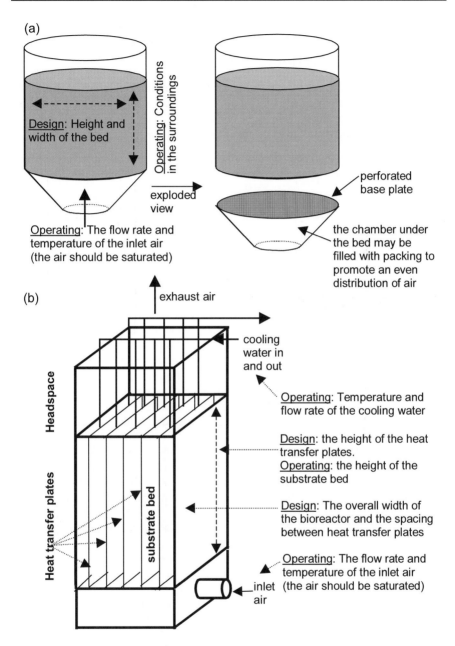

Fig. 7.2. Basic design and operating features of **(a)** traditional packed beds and **(b)** the Zymotis packed-bed with internal heat transfer plates of Roussos et al. (1993)

Taking the most common design, namely a vertical column in which the bed is aerated from the bottom and without any internal perforated tubes, the available design variables for a packed-bed are (Fig. 7.2):

- the presence or absence of a cooling jacket or internal heat transfer plates;
- the height and width of the bioreactor. The height to diameter ratio can vary over quite a wide range;
- if internal cooling plates are used, their height and the spacing between them.

The available operating variables are (Fig. 7.2):

- the aeration rate;
- the temperature of the inlet air;
- the temperature of the "surroundings" (which might be cooling water).

In a static bed, the relative humidity of the inlet air is not a useful operating variable. The problem is that it is not practical to add water into an unmixed bed in such a way as to distribute it evenly amongst the substrate particles; therefore evaporative water loss must be minimized. If the air entering the bed were not saturated with water, this unsaturated air would promote evaporation and dry out the bed, eventually decreasing the water activity to values unfavorable for growth and product formation. In order to minimize evaporation, saturated air must be supplied at the air inlet, which removes manipulation of the inlet air humidity as an available operating variable. Note that the use of saturated air does not prevent evaporation from occurring within the bed (see Fig. 4.3), but it does minimize evaporation compared to the use of unsaturated air. As will be discussed in Chap. 10, it is possible to replenish water during the mixing events of intermittently mixed beds, in which case unsaturated air can be used to aerate the bed.

At large scale, water-jacketing of the side walls of the bioreactor is not a good idea for the traditional design, since the water jacket will influence only the outer 20 cm or so of the bed. If cooling surfaces are to be used, then the internal cooling plates used in the Zymotis bioreactor will be more effective, as long as they are reasonably closely spaced. Optimum spacing of the plates will be discussed later. Another option for cooling surfaces is given by the "Prophyta" and "PlaFractor" designs (Fig. 7.3), two bioreactors that use a number of thin beds coupled with cooling plates oriented normal to the air flow (Lüth and Eiben 1999; Suryanarayan and Mazumdar 2000; Suryanarayan 2003). The difference between the two bioreactors is that in the Prophyta design the same air passes through each successive bed while in the PlaFractor design the air is introduced separately into each bed.

Important phenomena that are affected by the values chosen for the design and operating variables are:

- the axial and radial temperature gradients in the bed. In packed-beds it is impossible to prevent temperature gradients from arising within the bed, so the aim is generally to minimize the size of any temperature gradients.
- the evaporation of water from the bed. Efforts must be made to minimize evaporation in order to prevent the bed or parts of the bed from drying out.

- the pressure drop through the bed. This will depend on the bed height and the degree to which the organism fills the inter-particle spaces, with the resulting pressure drop affecting the design of the aeration system and its operating costs, and maybe placing a limit on the bed height that can be used.

In general, O_2 supply to the particle surface will not be considered in the selection of design and operating variables. The aeration rates that are chosen on the basis of heat removal considerations will typically be high enough that sufficient O_2 supply is ensured. However, note that problems such as channeling are possible, in which O_2 transport to large parts of the bed can be limiting (Fig. 7.4). Channeling is discussed in more detail in Sect. 7.3.3.5.

This chapter explains what is known, on the basis of experimental studies, about how these design and operating variables influence bioreactor operation. Later, Chap. 24 will show how mathematical models can be used to explore further the design and operation of packed-beds.

7.3 Experimental Insights into Packed-Bed Operation

This section presents and discusses the knowledge that experimental work has given firstly into the phenomena that occur within packed-bed bioreactors, and secondly into the operability of this type of bioreactor.

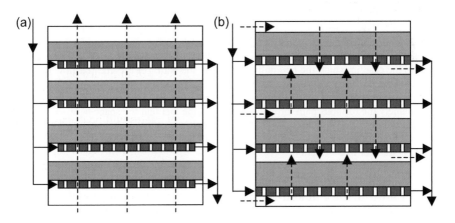

Fig. 7.3. The use of heat transfer surfaces normal to the air flow direction within packed beds **(a)** as used in the Prophyta bioreactor (Lüth and Eiben 1999) and **(b)** as used in the PlaFractor bioreactor (Suryanarayan and Mazumdar 2000; Suryanarayan 2003). In each case the substrate beds are in *light gray* and the heat transfer plates are in *dark gray*. The *white* regions represent empty spaces for air flow. *Solid arrows* represent the flow of cooling water and *dashed arrows* represent the flow of air

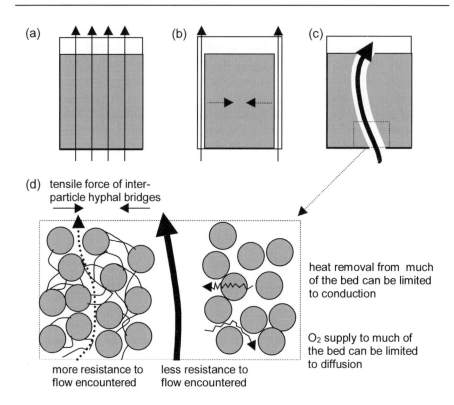

Fig. 7.4. The phenomenon of channeling. **(a)** The desirable situation, with uniform flow across the whole cross section of the bed. **(b)** Preferential flow between the bed and the wall in the case in which the bed pulls away from the wall. **(c)** Preferential flow through a crack in the bed. **(d)** Microscale view of a channel, showing how the preferential flow through the channel arises due to two sources of resistance to flow through the bed of particles, namely the tortuous path through the bed and the fact that the space between the particles is partially filled with biomass. Note that in extreme cases of channeling, there may be no bulk flow through the inter-particle spaces, with mass transfer being limited to diffusion and heat transfer to conduction

7.3.1 Large-Scale Packed-Beds

SSF bioreactors are only rarely operated at large scale as packed-beds throughout the entire cultivation period, although related intermittently stirred designs have been used quite successfully (see Chap. 10). Static packed-bed operation has been used at large scale in the production of *koji*, although details of the operation and performance of the bioreactors involved are not available. Only very brief and general descriptions are available. A simple design (Fig. 7.5) has a capacity for 1000 kg of *koji*, and has no special devices for substrate handling. Also, it is not designed for fully aseptic operation (Sato and Sudo 1999).

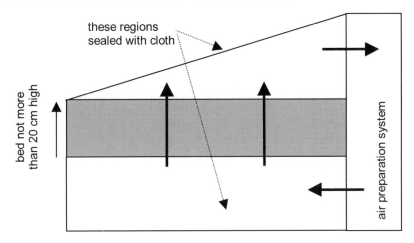

Fig. 7.5. Simple packed-bed of the type used in the *koji* industry for soy sauce production by Churitsu Industry Co. Ltd., Tokyo, Japan. Bioreactors of this type have capacities of up to one ton of substrate. This is a simplified version of a diagram presented by Sato and Sudo (1999)

7.3.2 Pilot-Scale Packed-Beds

Roussos et al. (1993) developed a pilot-scale packed-bed bioreactor with internal heat transfer plates, called the "Zymotis" bioreactor (Fig. 7.2(b)). The outer casing was acrylic, and it was 65 cm high, 50 cm wide, and 40 cm deep from front to back. This gave a total volume of 130 L, with a working capacity of 100 L. The aeration rate was varied from 0.1 to 0.2 L h^{-1} g-dry-substrate^{-1}.

The stainless steel heat transfer plates were 60 cm high, 38 cm wide (fitting within the 40 cm front to back depth of the outer casing), and 0.46 cm thick. There were 10 of these, and they occupied a volume of 9.44 L of the bioreactor vessel. Each heat exchanger plate contained serially placed tubes through which water was circulated. The bioreactor was designed to be flexible in that the number of heat transfer plates inserted and the spacing between them could be changed as desired. This bioreactor was emptied by raising the whole bioreactor and letting the substrate bed fall out of the bottom, although at large scale this might not be feasible.

Substrate loadings from 4 to 12 kg dry substrate matter (15 to 55 kg substrate on a wet basis) were tested, but detailed performance data was not provided, only the final enzyme levels. Heating of the circulating water was required during the first 10 h of fermentation, after which cooling was necessary. Thermistors placed in different locations were used to control the temperature of the cooling water. Uniformity of growth and absence of temperature gradients was claimed when the gap between plates was no larger than 5 cm, but experimental results showing this were not presented (Roussos et al. 1993). The water content also remained close to the original value, increasing by only 5%.

7.3.3 Laboratory-Scale Packed-Beds

The use of very small and thin packed-bed bioreactors in laboratory-scale studies of growth kinetics will be discussed in Chap. 15. A range of slightly larger bioreactors, typically up to 30 cm high and from 5 to 15 cm diameter, have been used to investigate how macroscale transport phenomena can influence bioreactor performance. The limitation of growth by transport phenomena is possible even at this small scale, as demonstrated in the following subsections.

7.3.3.1 Axial and Radial Temperature Gradients in Static Beds

The gas flow pattern within the bed of a packed-bed bioreactor that does not suffer from channeling problems is probably closest to plug flow with axial dispersion (see Fig. 4.6). However, studies have neither been done to confirm this nor to quantify the degree of axial dispersion. This plug-flow of the gas phase has implications for the operation of packed-beds. Firstly, the inlet end tends towards the inlet air temperature but, due to the lack of mixing and the unidirectional air flow, the temperature of the air increases as it flows along the bed towards the outlet end (Fig. 4.3(b)). One of the major challenges in designing and operating large-scale packed-beds will be to avoid excessive axial temperature gradients.

The increase in the temperature of the air as it flows through the bed increases the water-carrying capacity of the air and therefore evaporation will occur. Note that evaporation will occur even if saturated air is used at the air inlet (Fig. 4.3(c)).

In general, conduction along the axis in the direction of the air flow will be negligible compared to the convective and evaporative heat removal (Gutierrez-Rojas et al. 1996). The contribution of conduction normal to the direction of the air flow will depend on the design of the packed-bed. In traditional packed-beds that have diameters of the order of a few centimeters and in the Zymotis design, it can make a significant contribution, and there can be significant temperature gradients normal to the air flow. In contrast, in large-scale packed-beds, which might typically have diameters of the order of 1 m or more, the amount of energy removed from the bed by transfer through the side walls is likely to be small, even if the bed is water-jacketed. Various studies have been done that show how the appearance of axial and radial temperature gradients depends on the design and operation of the bioreactor. These are described below.

Temperature gradients in thin bioreactors. Saucedo-Castaneda et al. (1990) used a bioreactor of 6 cm diameter, containing a bed 35 cm high. Further, the column was immersed in a constant temperature waterbath at 35°C. They noted a steep temperature gradient in the first 5 cm along the axis of the bed, where the temperature increased by up to 12°C (Fig. 7.6). In contrast, in the upper 30 cm of the bed, the maximum increase in temperature along the axis was approximately 3°C. Note that, at some times and in some regions, the temperature actually decreased with axial distance, which might be related to evaporative cooling. However, Saucedo-Castaneda et al. (1990) did not measure water contents in the bed, so it is not possible to confirm this. The axial temperature in the upper 30 cm of the column did not remain constant; rather it increased with time over the period of 15 to 26 h. In contrast to the axial temperature gradients, the radial gradients were quite steep: at the time of peak heat production, there was an 11°C difference between the central axis and the bioreactor wall, which represents a distance of only 3 cm. These results suggest that in the case of thin bioreactors a significant amount of heat is removed through the side walls.

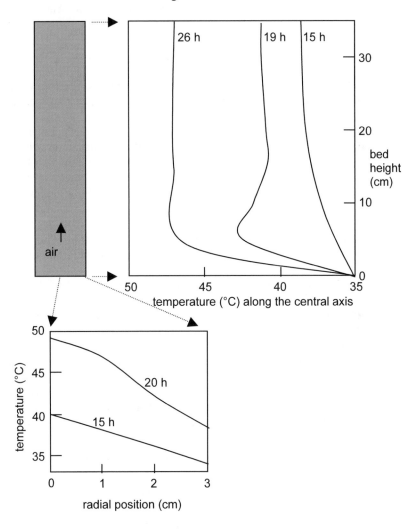

Fig. 7.6. Radial and axial temperature gradients at various times within a thin packed-bed bioreactor when *Aspergillus niger* was cultivated on cassava chips (Saucedo-Castaneda et al. 1990). The radial temperature gradient was determined at approximately mid-height in the bed. The superficial velocity of the air was 1 cm s^{-1}. Adapted from Saucedo-Castaneda et al. (1990) with kind permission from John Wiley & Sons, Inc.

Temperature gradients in wide or insulated bioreactors. The studies of Ghild-yal et al. (1994), Gowthaman et al. (1993a, 1993b), and Weber et al. (2002) allow insights into heat transfer in wider bioreactors. The bioreactor of Ghildyal et al. (1994) and Gowthaman et al. (1993a, 1993b) was 15 cm in diameter with a 34.5 cm bed height, while that of Weber et al. (2002) was 20 cm in diameter with a 50 cm bed height. Note that the bioreactor of Weber et al. (2002) was insulated on the sides, in order to mimic the situation at large scale where heat transfer through the walls makes a negligible contribution to heat removal. Note also that different organisms were used in the various studies, with quite different optimal temperatures for growth, so the actual temperatures involved are quite different.

Weber et al. (2002) measured the temperature as a function of time at various axial positions (Fig. 7.7(a)). At all heights, there was a temperature peak, whose maximum value occurred around day 4, with the height of the peak (that is, the maximum temperature reached) increasing with bed height.

Ghildyal et al. (1994) presented results that show the effect of the aeration rate on the axial temperature profile. Three different experiments were done, with air flow rates of 5 L min^{-1}, 15 L min^{-1}, and 25 L min^{-1}. The temperature was monitored at the central axis at mid-height in the bed. The height of the temperature peak increased as the air flow rate was decreased (Fig. 7.7(c)). They interpolated their data points to obtain a three-dimensional graph of the peak temperature obtained as a function of both bed height and air velocity (Fig 7.7(d)). In general terms, the temperature appears to increase linearly with increase in bed height and also to increase linearly with decrease in the air flow rate, except at the lowest bed height, where the peak temperature first increased only slowly as the air flow rate was decreased, but then shot up steeply at low air flow rates.

7.3.3.2 Oxygen Gradients in Static Beds

The convective flow of air will also lead to axial concentration gradients of O_2 and CO_2, although these are not likely to be important in controlling the performance of the bioreactor. Outlet gas concentrations remain reasonably high, above 18% (v/v), even at low airflow rates (Fig. 7.8) (Gowthaman et al. 1993a,b). Excessive temperatures are always likely to be a greater problem than the supply of O_2 to the particle surface within packed-bed bioreactors.

7.3.3.3 Evaporation and Water Gradients in Packed-Beds

As pointed out in Sect. 7.2, it is impossible to prevent evaporation from occurring in packed-bed operation, even if the air supplied to the bed is saturated. Ghildyal et al. (1994) and Gowthaman et al. (1993a,b) showed that the rate at which bed drying occurred depended on the position within the bioreactor and the air flow rate (Fig. 7.9).

Looking at the effect of air flow rate on the temporal variations in water content at each bed height, the key observations are (Fig. 7.9(a)):

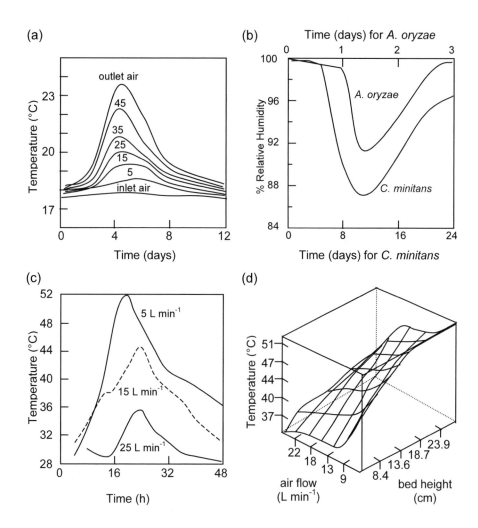

Fig. 7.7. Temporal and spatial temperature gradients in packed-bed bioreactors. **(a)** Temporal temperature profiles at different axial positions in the 50-cm-high packed-bed bioreactor of Weber at al. (2002), for growth of *Coniothyrium minitans* on hemp impregnated with nutrients. The numbers above each curve represent the cm height in the bed at which the temperature was measured. Adapted from Weber et al. (2002) with kind permission from John Wiley & Sons, Inc. **(b)** Off-gas relative humidity for fermentations undertaken with *Coniothyrium minitans* and *Aspergillus oryzae*. Adapted from Weber et al. (2002) with kind permission from John Wiley & Sons, Inc. **(c)** Temporal temperature profiles, at a bed height of 17 cm, in the 35-cm-high packed-bed bioreactor of Ghildyal et al. (1994), during the growth of *Aspergillus niger* on wheat bran. Adapted from Ghildyal et al. (1994) with kind permission of Elsevier. **(d)** Effect of air flow rate and bed height on the maximum temperature experienced during the cultivation. Adapted from Ghildyal et al. (1994), with kind permission of Elsevier

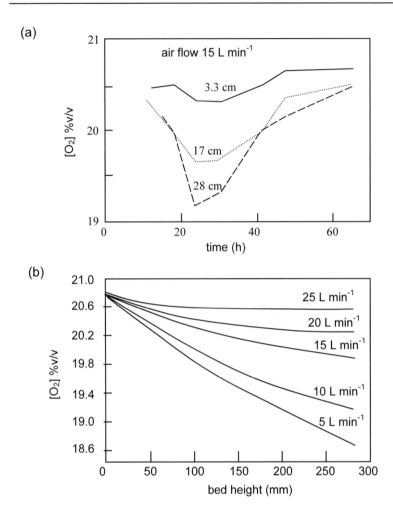

Fig. 7.8. Oxygen profiles during growth of *Aspergillus niger* on wheat bran in a 35-cm-high packed-bed bioreactor (Gowthaman et al. 1993b). **(a)** Temporal O_2 profiles at different axial positions, for an air flow rate of 15 L min⁻¹. **(b)** Axial O_2 profiles, at the time of peak growth rate (24 h), at various different air flow rates. Adapted from Gowthaman et al. (1993b) with kind permission of Elsevier

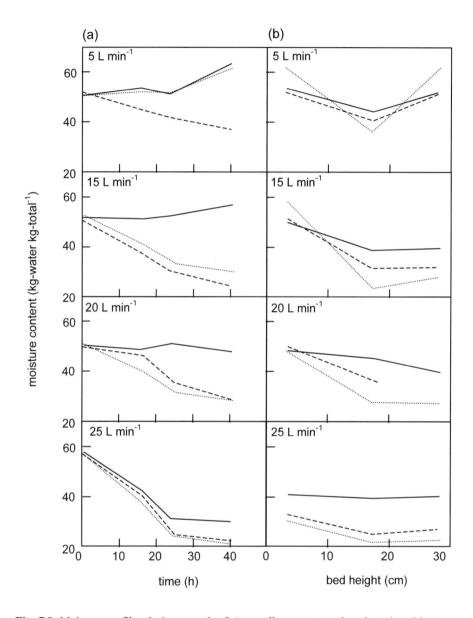

Fig. 7.9. Moisture profiles during growth of *Aspergillus niger* on wheat bran in a 35-cm-high packed-bed bioreactor (Gowthaman et al. 1993b; Ghildyal et al. 1994). **(a)** Temporal moisture profiles at different axial positions, for various different air flow rates. Key: bed heights of (———) 3.3 cm (- - -) 17 cm (· · ·) 28 cm. **(b)** Axial moisture profiles, at different times, for various different air flow rates. Key: Times of (———) 16 h (- - -) 24 h (· · ·) 40 h. Adapted from a table presented by Gowthaman et al. (1993b) with kind permission from Elsevier

- At 3.3 cm height, the water content of the bed increased over time or stayed constant when the air flow rate was 20 L min^{-1} or below, but decreased at 25 L min^{-1}, although even at 25 L min^{-1} this lower region did not dry out as quickly as the upper regions of the bed.
- The mid height of the bed (17 cm height) tended to dry out at all air flow rates.
- The water content of the bed at the upper position (28 cm height) increased during the fermentation for an air flow rate of 5 L min^{-1} but decreased with time for air flow rates of 15 L min^{-1} and above.

Inspecting the same results, but looking at the axial temperature gradients as a function of the air flow rate and how the axial temperature gradients varied over time, the key observations are (Fig. 7.9(b)):

- at 5 L min^{-1} the bed was driest at 17 cm, with the 3-cm and 280-cm heights remaining near and even exceeding the original water content of 51% (w/w). This observation holds for 16, 24, and 40 h.
- at 15 L min^{-1} the bed remained wet at 3 cm height, in fact, by 40 h it was significantly wetter than the original value, but became dry at 17-cm and 28-cm heights, with these two upper heights being reasonably close in water content, with values around 25-40 %(w/w)
- at 20 L min^{-1} the pattern was similar to that obtained for 15 L min^{-1} except that the water content at the 3-cm bed height remained close to the original value throughout.
- at 25 L min^{-1} all regions of the bed dried. By 16 h the water content had fallen to around 40% (w/w) at all bed heights. At 24 and 40 h the 3-cm height still had a water content around 40% w/w, but at both the upper bed heights the water content had fallen to around 22-26% (w/w).

These results suggest that drying patterns can be quite complex. Chapter 25 presents a model that can be used to explore these patterns.

Weber et al. (2002) monitored the off-gas relative humidities during packed-bed fermentations with two different fungi (Fig. 7.7(b)). At the time of the peak heat generation rate, the off-gas relative humidity fell to values around 90%. This could imply that either the transfer of water from the particle to the air is limiting or simply that the bed is drying out. It would be necessary to monitor the water activity of the bed contents in order to distinguish between these two possibilities.

7.3.3.4 Pressure Gradients in Packed-Beds

This phenomenon, introduced in Chap. 7.2.4, is of particular importance in packed-beds due to the combination of static operation with forced aeration. The static operation means that the hyphae that grow into the inter-particle spaces are not disrupted or squashed onto the particle surface, and therefore these hyphae represent an extra impediment to air flow, increasing the pressure drop. The maximum pressure drop expected during the fermentation is an important consideration because it will affect the pressure that the blower or compressor must be capable of supplying.

Excessive pressure drop tends not to be a problem in intermittently agitated packed-beds because the agitation prevents the hyphae from binding the substrate bed into one large mass and it also squashes the hyphae onto the particle surface. In fact, infrequent agitation events might be used in packed-beds with the major purpose of decreasing the pressure drop. Figure 7.10 illustrates this point.

Despite its potential importance at large scale, pressure drop has received most attention in small-scale packed-bed bioreactors, and in these experiments the interest was in using the pressure drop to quantify the growth.

Auria et al. (1993, 1995) used a column of 6.5-cm height and 2-cm internal diameter, and a superficial velocity (calculated as volumetric flowrate divided by the total cross-sectional area of the column) of 0.435 cm s^{-1}. The substrate was an artificial substrate based on an amberlite resin impregnated with nutrients. The maximum pressure drop observed during the fermentation ranged from 0.21 to 0.69 cm-H$_2$O cm-bed^{-1}, for various different initial nutrient concentrations. With bagasse as the substrate and a superficial velocity of 0.379 cm s^{-1} the maximum pressure drop obtained in the same column was 2.75 cm-H$_2$O cm-bed^{-1}. In this case the pressure drop was already 0.45 cm-H$_2$O cm-bed^{-1} at the beginning of the fermentation. In a larger column of 15-cm height and 4-cm diameter, they obtained a maximum pressure drop of 0.12 cm-H$_2$O cm-bed^{-1} with a wheat bran substrate and a superficial velocity of 0.675 cm s^{-1}. With a much higher superficial velocity of 11.2 cm s^{-1} Gumbira-Sa'id et al. (1993) obtained a pressure drop of 1.38 cm-H$_2$O cm-bed^{-1} with a substrate based on cooked sago-beads.

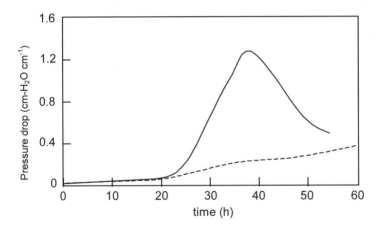

Fig. 7.10. Typical temporal profiles for the pressure gradient in the bed, based on the results of Gumbira-Sa'id et al. (1993) for the growth of *Rhizopus oligosporus* on a substrate based on sago beads. (——) Evolution of the pressure drop in a fermentation in which the bed was not disturbed by the removal of samples. Note that the decrease in the pressure drop after 40 h is due to the substrate bed shrinking and pulling away from the bioreactor wall; (- - -) Evolution of the pressure drop in a fermentation in which the bed was periodically disturbed by the removal of samples. Adapted from Gumbira-Sa'id et al. (1993) with kind permission of Elsevier

As yet there is insufficient information to predict the pressure drops that can be expected during a large-scale fermentation, although the experimental values reported here give some idea of the orders of magnitude that might be expected. The initial pressure drop will depend on the substrate and how it packs together, which in turn will depend on how the substrate is prepared. The maximum pressure drop achieved during the fermentation will depend on how the microorganism grows within the bed, although it is also a function of the superficial velocity.

Note that the pressure drop across the bed can decrease later in the fermentation. This can happen due to the bed pulling away from the walls, leaving a gap through which the air can pass (Gumbira-Sa'id et al. 1993; Weber et al. 2002). As described in the next subsection, this is undesirable because it will lead to heat and mass transfer limitations within the bed.

7.3.3.5 Channeling

Channeling is a potential problem in packed beds and the static phase of operation of intermittently-mixed beds. Channeling in intermittently-mixed beds will be discussed in Chap. 10. Channeling is problematic because air will flow preferentially through the cracks, such that in the regions of the bed where the particles are bound together, there will be no bulk flow, such that O_2 transfer will be limited to diffusion and heat transfer will be limited to conduction (see Fig. 7.4).

One of the major causes of channeling in packed-beds is the shrinkage of particles due to the consumption of the solid material of the particle, combined with the fact that, in many fungal fermentations, the substrate bed is bound together by "inter-particle hyphal bridges". These two phenomena mean that, as the bed volume reduces, the particles will not simply settle downwards but rather the bed is drawn inwards, pulling away from the walls or cracking in the middle. For fungi that do not produce these hyphal bridges, the substrate particles remain free flowing as the bed shrinks and the bed does not pull away from the wall or develop cracks, but simply reduces in height (Weber et al. 2002).

For fungal fermentations in which the fungus does bind the particles together, shrinkage problems can be minimized by the use of "inert" hemp impregnated with nutrients (Weber et al. 1999). However, there is not necessarily free choice of substrates in SSF processes.

7.3.3.6 Condensation on the Bioreactor Walls in the Headspace

Often the bioreactor wall in the headspace is cooler than the temperature of the gases leaving the bed, which can cause condensation of water on the inner surfaces of the bioreactor walls in the headspace. This can be problematic, since the water can run down and flood the top of the bed, greatly interfering with O_2 transfer in this region.

7.4 Conclusions on Packed-Bed Bioreactors

Packed-bed bioreactors are the natural choice when the microorganism does not tolerate mixing well. The major challenge in developing large-scale packed-bed bioreactors for new applications will be to minimize the axial temperature gradients. There are two main strategies by which this can be done:

- to use traditional packed-beds but use a low bed height
- to use a Zymotis-type bioreactor, with internal heat transfer plates.

If the organism can tolerate infrequent mixing events, of the order of once every few hours, or even as infrequent as once per day, then the traditional design should be chosen. These mixing events allow the pressure drop to be decreased, and also the addition of water to replenish the water lost in evaporation. This mode of operation is discussed in more depth in Chap. 10. Agitation is not a feasible option in the Zymotis packed-bed due to the presence of the heat transfer plates.

If a traditional packed-bed is chosen, then the bed height will need to be no more than say 20 cm to 1 m, in order to prevent high temperatures at the outlet end of the bed. The other possibility, of using tall water-jacketed columns of 15 cm or less in diameter, is unrealistic, since, to hold large amounts of substrate, the bioreactors will either need to be very tall or a large number of bioreactors will be needed.

If it is not desired to mix the bed at all, due to the sensitivity of either the organism or the substrate to damage by mixing, then the Zymotis design should be strongly considered, on the basis of considerations of the water balance. The contribution of conduction to heat removal will decrease the axial temperature gradient, and this will decrease the evaporation rate, as long as saturated air is used at the air inlet. Further, the greater the sensitivity of the process to high temperatures, the more the Zymotis bioreactor is indicated. For the same bioreactor height, the maximum temperature reached in a Zymotis bioreactor is lower than for the traditional bioreactor. This will be explored in the modeling case study in Chap. 24.

However, the Zymotis bioreactor does have some disadvantages in its operability compared to the traditional bioreactor. Both bioreactors have a potential problem with water condensing from the saturated outlet air onto the exposed bioreactor surfaces above the substrate bed, and this condensate can flood the top of the bed, causing O_2 limitations in this region. This problem will be greater with the Zymotis bioreactor than for traditional packed-beds if the cooling plates extend above the top of the bed.

Additionally, the traditional packed-bed will be easier to load and unload than the Zymotis packed-bed. For example, for the traditional packed-bed it will probably be possible to (1) have a hinged base plate, in which the substrate can be dropped into a screw conveyor or (2) open a side and use a backhoe or (3) insert a pneumatic conveying tube to suck the substrate out. These operations will not be so easy in the Zymotis packed-bed due to the presence of the heat transfer plates.

A more detailed comparison of these two designs will require more work than is currently in the literature. The Zymotis design has not received much experimental attention since the early 1990s. However, as Chap. 24 shows, mathematical models can be used in a preliminary evaluation.

Further Reading

Early studies to elucidate the importance of temperature and gas gradients in packed-bed bioreactors
Gowthaman MK, Ghildyal NP, Raghava Rao KSMS, Karanth NG (1993) Interaction of transport resistances with biochemical reaction in packed bed solid state fermenters: The effect of gaseous concentration gradients. J Chem Technol Biotechnol 56:233–239
Gowthaman MK, Raghava Rao KSMS, Ghildyal NP, Karanth NG (1993) Gas concentration and temperature gradients in a packed bed solid-state fermentor. Biotechnol Adv 11:611–620
Ghildyal NP, Gowthaman MK, Raghava Rao KSMS, Karanth NG (1994) Interaction between transport resistances with biochemical reaction in packed bed solid-state fermentors: Effect of temperature gradients. Enzyme Microb Technol 16:253–257
Gumbira-Sa'id E, Greenfield PF, Mitchell DA, Doelle HW (1993) Operational parameters for packed beds in solid-state cultivation. Biotechnol Adv 11:599–610

Selected recent examples of use of packed-bed bioreactors
Couto SR, Rivela I, Munoz MR, Sanroman A (2000) Ligninolytic enzyme production and the ability of decolourisation of Poly R-478 in packed-bed bioreactors by *Phanerochaete chrysosporium*. Bioprocess Eng 23:287–293
Jones EE, Weber FJ, Oostra J, Rinzema A, Mead A, Whipps JM (2004) Conidial quality of the biocontrol agent *Coniothyrium minitans* produced by solid-state cultivation in a packed-bed reactor. Enzyme Microbial Technol 34:196–207
Lu MY, Maddox IS, Brooks JD (1998) Application of a multi-layer packed-bed reactor to citric acid production in solid-state fermentation using *Aspergillus niger*. Process Biochem 33:117–123
Lu MY, Brooks JD, Maddox IS (1997) Citric acid production by solid-state fermentation in a packed-bed reactor using *Aspergillus niger*. Enzyme Microb Technol 21:392–397
Shojaosadati SA, Babaeipour V (2002) Citric acid production from apple pomace in multi-layer packed bed solid-state bioreactor. Process Biochem 37:909–914
Veenanadig NK, Gowthaman MK, Karanth NGK (2000) Scale-up studies for the production of biosurfactant in packed column bioreactor. Bioprocess Eng 22:95–99

8 Group III: Rotating-Drum and Stirred-Drum Bioreactors

David A. Mitchell, Deidre M. Stuart, Matthew T. Hardin, and Nadia Krieger

8.1 Introduction

This chapter addresses the design and operation of rotating-drum bioreactors and those stirred-drum bioreactors in which the air is blown into the headspace and not forcefully through the substrate bed itself. This type of bioreactor might be chosen for continuous processes, which will be discussed in Chap. 11. It can also be used for batch processes, which will be the focus of this chapter. Note that there are several bioreactors that are very similar in appearance to rotating- and stirred-drum bioreactors in which air is introduced directly into the bed. This forced aeration would tend to place them in the Group IVa bioreactors considered in Chap. 9 (continuously-agitated, forcefully-aerated bioreactors), however, whether the bioreactor performs more closely to this type of bioreactor or to a rotating-drum or stirred-drum bioreactor depends on the effectiveness of this forced aeration.

8.2 Basic Features, Design, and Operating Variables for Group III Bioreactors

The basic design features have already been presented in Sect. 3.3.1. Some possible design variations include (Fig. 8.1):

- the inclusion of baffles (or, more correctly, "lifters");
- periodic reversal of the direction of rotation;
- use of drum cross-sections that are not circular;
- inclination of the drum axis to the horizontal.

Design variables for both baffled rotating-drum and stirred-drum bioreactors include (Fig. 8.2):

- the length and diameter of the bioreactor. Note that the geometric proportions can vary over quite a wide range;
- the inclination of the central axis of the bioreactor to the horizontal;

- the size and shape of the mixing device within a stirred-drum and the number, size, and shape of baffles in a baffled rotating-drum;
- the design of the inlet and outlet of the aeration system, which will affect the gas flow patterns in the headspace;
- the presence or absence of an external water jacket. Note that for a rotating drum this will increase the weight to be rotated and also will require a rotating water seal on the inlet and outlet water lines;
- whether internal features such as baffles or paddles are designed to aid in cooling;
- the design of the system for the addition of water or other additives to the bed during the process;
- in continuous operation, the design of the substrate inlet and outlet.

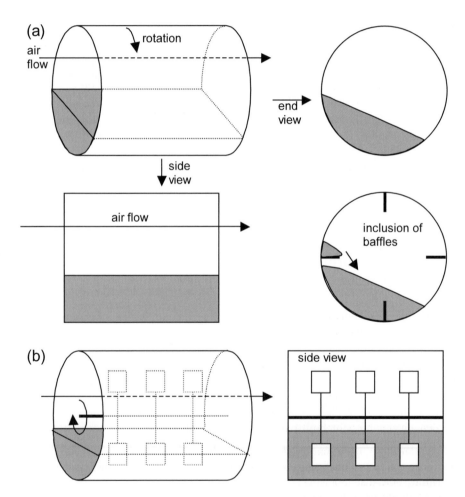

Fig. 8.1. Basic features of **(a)** rotating-drum bioreactors and **(b)** stirred-drum bioreactors

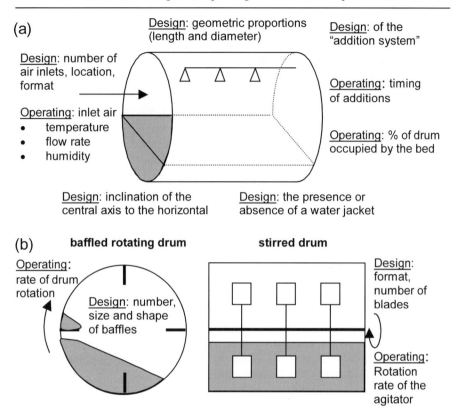

Fig. 8.2. Design and operating variables that are available with rotating-drum and stirred-drum bioreactors. **(a)** Design and operating variables that are the same for both types. **(b)** Design and operating variables that are specific for the bioreactor type

The operating variables that are available include (Fig. 8.2):

- the solids loading used;
- the rotational speed for a rotating-drum bioreactor, the stirring speed for a stirred-drum bioreactor. If rotation or stirring are done intermittently, then the frequency, duration, and speed of rotation or stirring events;
- the flow rate, temperature, and humidity of the air blown into the headspace;
- the timing of water additions;
- the temperature of the cooling water if a jacket is used; if a jacket is not used then whether air is forcefully blown past the drum wall or not.

Most of these operating variables can be changed at will during the fermentation. The solids loading is an operating variable that is fixed at the beginning of each run. Note that it does change during the fermentation as part of the solids is converted into CO_2, but once a fermentation has commenced, the solids loading cannot be freely controlled. The solids loading is related to the "fractional filling" of the drum, that is, the fraction of the whole drum volume occupied by the bed, here represented by the symbol ω. Typically, fractional fillings should be kept below 0.4, in order to enable reasonable mixing of the bed.

Since the bed is mixed, water can be replenished by spraying a fine mist onto the bed during mixing. Therefore evaporation can be promoted as part of the cooling strategy, meaning that unsaturated air can be used at the air inlet.

The values chosen for the design and operating variables will be affected by the following considerations:

- the heat production rate in the bed will strongly influence decisions about the loading of the bioreactor, the aeration rate, and the humidity of the inlet air.
- the rotation rate or stirring speed chosen will represent a balance between the promotion of heat and O_2 transfer within the bed and between the bed and the headspace and the minimization of shear damage to the microorganism.
- the strength of the particles may affect the maximum diameter and the substrate loading that can be used. Soft particles may be crushed by the weight of a large overlying bed.

This chapter explains what is known, on the basis of experimental studies, about how these design and operating variables influence the performance of rotating-drum bioreactors and stirred-drum bioreactors. Chapter 23 shows how mathematical models can be used to explore further the design and operation of rotating-drum and stirred-drum bioreactors.

8.3 Experimental Insights into the Operation of Group III Bioreactors

8.3.1 Large-Scale Applications

Takamine (1914) developed a process for the production of amylase by *Aspergillus oryzae* on wheat bran, first in tray bioreactors and then later in rotating-drum bioreactors. This work was later extended by Underkofler et al. (1939). Ziffer (1988) was involved in work during the early 1940s in which penicillin was produced at commercial scale by SSF of wheat bran, in a plant containing 40 rotating-drum bioreactors of 1.22 m diameter and 11.28 m length, meaning that each bioreactor had a total volume of 13 m^3 (Fig. 8.3(a)). He described how the system was operated, but not how it performed.

The bran was mixed with the nutrient solution externally and then added through the access hatches. These were closed and the bioreactor was sterilized by direct injection of steam, at 1 atm above ambient pressure, while being rotated at 24 rpm. Inoculum was added through spray nozzles, while the drum was rotating at 24 rpm. The aeration rate was maintained between 0.28 and 0.42 m^3 min^{-1}

until 30 h, then it was increased to 1.13 m^3 min^{-1}, which was maintained until the end of the fermentation. The rotation rate was maintained at 24 rpm during the first 6 h. It was reduced to 5 rpm between 6 and 30 h and then increased to 24 rpm, which was maintained until the end of the fermentation. Water was sprayed onto the external surface in order to aid temperature control. At the end of the fermentation (112 h), the fermented substrate was removed through the access hatches by a pneumatic vacuum system.

Rotating-drum bioreactors have also been used in the *koji* industry. Sato and Sudo (1999) report the use of a rotating-drum bioreactor of 1500 kg capacity, which is designed to rotate intermittently (Fig. 8.3(b)). They report that accurate temperature control is difficult in this type of bioreactor, but provide no details.

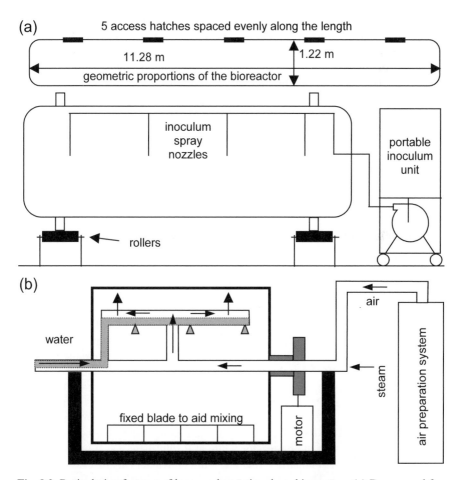

Fig. 8.3. Basic design features of large-scale rotating-drum bioreactors. **(a)** Drums used for penicillin production. This is a simplified version of a diagram presented by Ziffer (1988). **(b)** *Koji* bioreactor. This is a simplified version of a diagram presented by Sato and Sudo (1999)

8.3.2 Pilot-Scale Applications

Fung and Mitchell (1995) investigated the effect of the presence and absence of baffles on the performance of a 200-L rotating-drum bioreactor, in which *Rhizopus oligosporus* was grown on wheat bran. The drum had an internal diameter of 56 cm and an internal length of 85 cm. When baffles were used, four baffles of 17 cm width and 85 cm length were attached at right angles to the inner wall of the drum, with uniform spacing between them (i.e., in the manner indicated in Fig. 8.1). There was no external temperature control; the bioreactor operated within a room that varied from 17 to 26°C. Pre-humidified air was blown through the bioreactor at the optimum temperature for growth of the organism of 37°C.

The aeration at 37°C was not sufficient to maintain the bed temperature at a value suitable for initial growth. As a result, there was a long lag period, with bed temperatures below 30°C (Fig. 8.4(a)). This was especially true for the baffled drum, for which heat transfer between the bed and surroundings was more efficient. In the unbaffled drum the temperature was slightly higher during the lag phase and, as a result, the lag phase was slightly shorter.

The temperature then increased, over a period of 10 h, to values around 45°C. In the baffled drum the temperature then decreased quickly again. The bed was at temperatures of above 40°C for only 10 h. In the unbaffled bioreactor the temperature remained at values above 40°C for 30 h.

Peak O_2 consumption rates were higher in the baffled drum, as shown by the greater slope of the cumulative O_2 uptake profile in Fig. 8.4(a). However, due to the longer lag phase, the O_2 uptake was not significantly better in the baffled drum. Note that the baffled drum would have outperformed the unbaffled drum over the first 30 h, if the bed temperature in both drums had been maintained at 37°C during the first 10 h. This could be achieved with a water jacket or by placing the bioreactor in a 37°C room. Alternatively, it might be sufficient simply to insulate the outer surfaces of the bioreactor during the early stages of the fermentation, such that heating of the bed by the inlet air would be more efficient. Obviously, such insulation would need to be removed once rapid growth began.

These results, obtained at pilot scale, demonstrate a major challenge to be overcome in rotating-drum bioreactors, namely the adequate removal of the waste metabolic heat from the substrate. For example, in the fermentations described above that were undertaken with *R. oligosporus*, it was highly desirable to avoid temperatures above 40°C, but this value was exceeded for long periods.

Stuart (1996) grew *Aspergillus oryzae* on wheat bran in the same 200-L bioreactor, without baffles. Performance in terms of O_2 consumption was significantly better at 9 rpm than at 2 rpm (Fig. 8.4(b)). This is most likely due to the effect of the rotational speed on the effectiveness of the mixing within the bed. At 2 rpm the bed slumped within the bioreactor while at 9 rpm there was a tumbling flow regime (flow regimes in unbaffled rotating-drum bioreactors are discussed in more detail in Sect. 8.4.1). The maximum temperatures reached during the fermentations were about 43°C at 9 rpm and about 38°C at 2 rpm. The higher temperature occurred at the higher rotational rate due to better mixing, which allowed better O_2 transfer from the headspace into the bed, which in turn allowed faster growth.

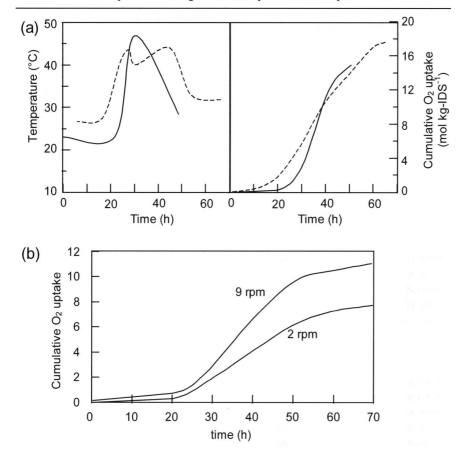

Fig. 8.4. (a) Effect of baffles on the performance of a 200-L rotating-drum bioreactor. Adapted from Fung and Mitchell (1995) with kind permission of Springer Science and Business Media. Key: (- - -) unbaffled drum; (—) baffled drum. **(b)** Effect of rotation rate on the performance of an unbaffled 200-L rotating-drum bioreactor (Stuart 1996). IDS is "Initial Dry Solids"

8.3.3 Small-Scale Applications

Stuart et al. (1999) undertook studies in a bioreactor of 85 cm length and 19 cm internal diameter, giving a volume of approximately 20 L. The bioreactor was operated with various substrate loadings and rotational speeds. *Aspergillus oryzae* was used, in some cases grown on wheat bran, in others on an artificial gel substrate. The temperature in the chamber was maintained at 30-32°C and air was supplied at this temperature. Relevant observations were:

- **Regarding the effect of rotational speed on growth on the gel substrate.** Between 0 and 10 rpm, there was a beneficial effect of rotation, with an increase in the maximum specific growth rate, while the amount of protein enrichment that occurred remained constant (Fig. 8.5). As the rotational speed used in the fermentation increased from 10 to 50 rpm, there were decreases in both the amount of protein enrichment obtained during the fermentation and the maximum specific growth rate observed.

- **Regarding the effect of the substrate used on the temperatures reached within the bed.** During fermentations with the gel substrate, the bed temperature did not exceed 35°C, while during fermentations with the wheat bran substrate the temperature peaked at values between 45 and 50°C. This effect is probably related to the availability of the carbon sources in the two substrates. The gel substrate contained slightly less than 5% starch by weight (on a wet basis) while the wheat bran substrate contained 15% starch by weight (on a wet basis) and in addition had protein and fat available. Also note that the majority of wheat bran particles were flat and smaller than 4 mm diameter while the gel substrate consisted of 6 mm cubes, such that the wheat bran substrate had a much larger surface area to volume ratio.

De Reu et al. (1993) built a rotating-drum bioreactor in which one end was fixed, in order to allow the insertion of sensors into the bed (Fig. 8.6(a)). This complicates the design, as it is necessary to have a seal between the rotating body and the fixed end-plate. The bioreactor had an inner diameter of 20 cm and a length of 15 cm, giving a total volume of 4.7 L. Although the bed volume was not specifically mentioned, fermentations were undertaken with 1 kg of cooked soybeans (inoculated with *Rhizopus oligosporus*), meaning that the bed volume was probably between 2 and 2.5 L. Air was introduced into the headspace through the central axis. The bioreactor was placed in a room with an air temperature of 30°C.

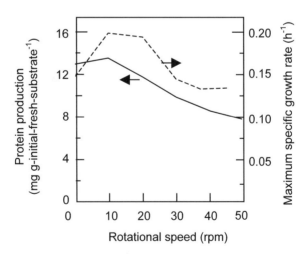

Fig. 8.5. Effect of rotational speed on the performance of a laboratory-scale rotating-drum bioreactor in which *Aspergillus oryzae* was grown on a gel-based artificial substrate. Key: (—) Amount of protein produced during the fermentation; (- - -) maximum value of the specific growth rate. Adapted from Stuart et al. (1999) with kind permission from John Wiley & Sons, Inc.

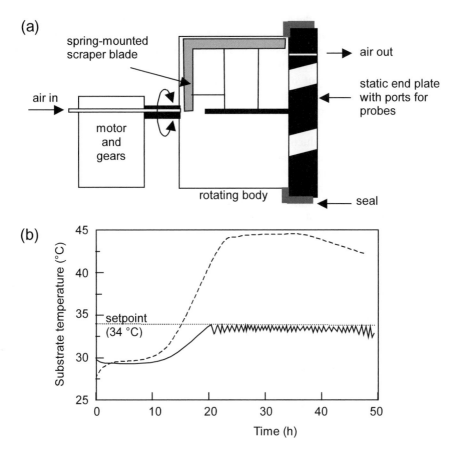

Fig. 8.6. Performance of a laboratory-scale rotating-drum bioreactor operated in a discontinuous agitation mode (de Reu et al. 1993). **(a)** Main features of the bioreactor used. **(b)** Control of bed temperature through the discontinuous rotation period. Key: (- - -) Temperature profile during a static fermentation; (—) Temperature profile during a discontinuously agitated fermentation. Adapted from de Reu et al. (1993) with kind permission of Springer Science and Business Media.

This bioreactor was used to investigate the use of discontinuous rotation for bed temperature control. Each time the bed temperature reached 34°C, a 60 s rotation period was triggered, with several clockwise and anticlockwise rotations, at rates of 4 to 6 rpm. Although it was possible to control the temperature of the 1-kg bed using this strategy (Fig. 8.6(b)), it is unlikely to be effective at large scale.

Kalogeris et al. (1999) developed a bioreactor that is a variation of a rotating-drum bioreactor (Fig. 8.7). In this bioreactor the substrate bed is held within a 10-L perforated cylinder that can be rotated. This perforated cylinder is inside a larger water-jacketed solid-walled cylinder through which air is passed. This bioreactor did work well for the cultivation of thermophilic organisms, but heat removal from the bed is unlikely to be sufficient for the cultivation of mesophiles, for two reasons. Firstly, the air blown into the headspace region will preferentially flow past the surface of the bed rather than through the bed itself (it is for this reason that this bioreactor is classified as a group III bioreactor). Secondly, there is no intimate contact between the bed and the water jacket; a layer of process air separates them. This type of operation was later adapted for an SSF process in which a nutrient medium was placed in the bottom of the bioreactor and nylon sponge cubes were regularly wetted with this nutrient medium as the inner perforated drum rotated at 3 rpm (Dominguez et al. 2001). The system was used for ligninolytic enzyme production by *Phanerochaete chrysosporium*.

Roller bottle systems are useful for testing, at laboratory scale, a number of different treatments for a process intended to be performed in a rotating-drum bioreactor. Figure 8.8 indicates one possible way in which a roller system can be constructed. Note that the direct introduction of air into the headspace of each individual bottle is complicated, although not impossible. In the majority of cases it would be more likely for each bottle simply to have a perforated lid, with a passive exchange of gases between the headspace and the surrounding air.

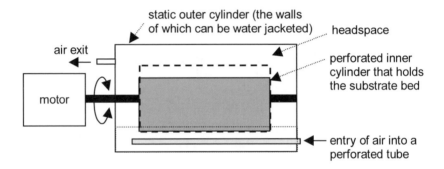

Fig. 8.7. General design principles of the Group III bioreactor used by Kalogeris et al. (1999) and Dominguez et al. (2001). The bottom of the bioreactor of Dominguez et al. (2001) was filled with a liquid nutrient medium up to the level shown by the dotted line. In the case of the bioreactor of Kalogeris et al. (1999) there was no liquid held by the outer cylinder. Adapted from Dominguez et al. (2001) with kind permission of Elsevier

8.4 Insights into Mixing and Transport Phenomena in Group III Bioreactors

The performance of rotating-drum and stirred-drum bioreactors will depend strongly on the effectiveness of the exchange of water and energy between the bed and the headspace gases. The effectiveness of this exchange will be affected by the flow patterns within the bed and headspace. It is unlikely that rotating- or stirred-drum bioreactors will be well mixed, unless specific attention is paid at the design stage to the promotion of mixing. Rather, air flow patterns and solids flow

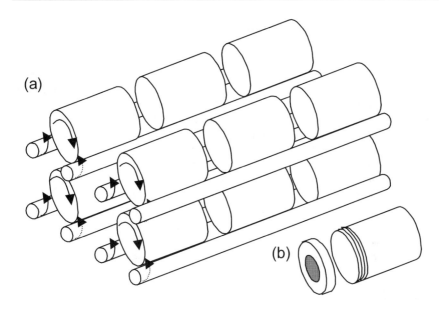

Fig. 8.8. (a) A simple roller bottle system, based on systems used for tissue culture. It is possible to have various layers of roller bars. The system would typically be placed in a temperature-controlled room. For each pair of roller bars that holds several roller bottles, one is a drive bar (*solid arrow*) and the other a slave bar (*dashed arrow*). **(b)** Typically each roller bottle would have a removable lid, with a mesh that allows the exchange of gases

patterns are likely to be complex. The flow patterns within the bed and the headspace of Group III bioreactors have only recently started to be explored. To date, the attention has been largely focused on rotating drum bioreactors. Note that quantitative approaches for determining bed-to-headspace transfer coefficients will be discussed in Sect. 20.5.

8.4.1 Solids Flow Regimes in Rotating Drums

Solids flow in both the radial and axial directions must be considered.

The radial flow regime within the solid bed is important because it affects the heat and mass transfer between the bed and the headspace and the homogeneity within the bed. Transfer of heat, water, and O_2 will be most effective when all substrate particles within the bed are regularly brought to the surface. However, this is not necessarily easy to achieve.

In non-SSF applications of unbaffled rotating drums, the various radial solids flow regimes that occur have been characterized (Fig 8.9). The flow regime depends on several factors, including the rotational rate and the percentage filling of the drum. It is convenient to relate the flow regimes to fractions of the critical rotational speed (N_C), which is defined as the speed at which the particles are held against the inside of the drum wall by centrifugal action (Ishikawa et al. 1980).

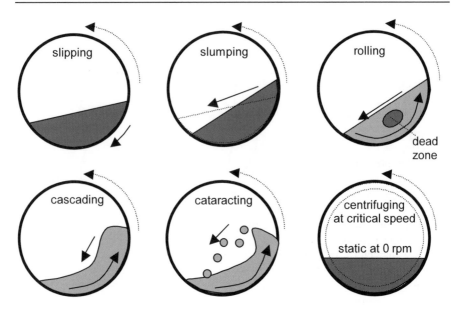

Fig. 8.9. Solids flow patterns within rotating drums without baffles. Relatively slow rotational rates are commonly used in SSF with rotating-drum bioreactors, giving the slumping flow regime. The *darker color* indicates poorly mixed parts of beds while the *lighter color* indicates better-mixed areas. *Solid arrows* indicate movement of the bed or particles within the bed. Adapted from Wightman and Muzzio (1998), with kind permission of Elsevier

This is a function of the drum diameter and for a horizontal drum is given by the following equation:

$$N_C = \frac{42.3}{\sqrt{D}} \tag{8.1}$$

where N_C is in rpm and D is the drum diameter in meters.

For both a static drum (0 rpm) and a drum rotating at the critical speed, there is no mixing action within the bed. For the slipping and slumping flow regimes, which occur when the rotational rate is less than 10% of the critical rotational speed, the bed moves essentially as a whole, meaning that the amount of mixing within the bed itself is negligible.

As the rotational speed increases through moderate rotational rates (from 10% to 60% of the critical rotational speed) the bed undergoes first rolling flow, characterized by a flat surface, and then cascading flow, characterized by a curved surface. There are no airborne particles. In both these flow regimes there is particle flow within the bed itself, although there may be dead zones. For rotational rates greater than 60% of the critical rotational speed, the flow changes to cataracting flow, in which particles are thrown into the air.

Most rotating-drum bioreactors are operated in conditions that give slumping flow, meaning that it is usually a good idea to attach baffles to the inner surface of the drum, in order to improve the mixing. However, it is also possible to operate unbaffled drums at high rotation rates. For example, the large-scale rotating-drum bioreactor reported by Ziffer (1988) had a diameter of 1.22 m, which gives a critical rotational speed of 38.3 rpm. During the period of peak growth rate the drum was rotated at 24 rpm, which represents 63% of the critical rotational speed, such that the bed must have been on the borderline between the cascading and cataracting flow regimes.

Schutyser et al. (2001) undertook studies of mixing in rotating drum bioreactors that give a greater insight into the radial mixing patterns that occur and how they are affected by baffles. They used a two-dimensional discrete-particle model, in which the predicted positions of a large number of individual particles are calculated by the model, with the change in the position of each individual particle during a time step depending on the sum of forces acting upon it as a result of collisions with other particles or with solid surfaces such as the bioreactor wall (Fig. 8.10(a)). They supported their modeling work with experimental validation in rotating drums containing cooked wheat grains.

They characterized the drum as being well mixed when the entropy of mixing was greater than 0.9 (see Fig. 8.10(b)) and compared the effectiveness of the mixing provided by a particular mode of drum design and operation on the basis of the number of drum rotations necessary to reach an entropy of mixing of 0.9.

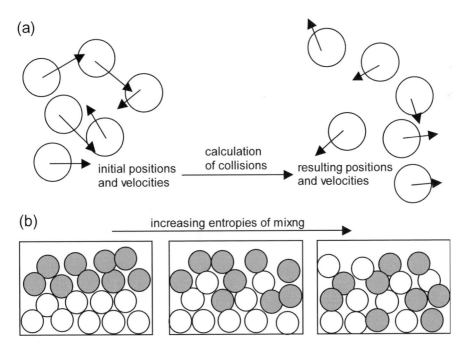

Fig. 8.10. (a) Basis of the discrete-particle modeling work done by Schutyser et al. (2001, 2002). The *arrow* originating from the center of each particle shows its velocity (magnitude and direction). **(b)** Concept of the entropy of mixing. The more random the distribution of particles in relation to their original position, the greater the entropy of mixing

The effects of drum rotational speed (0.5, 2, and 5 rpm), drum diameter (0.15, 0.3, and 1 m) and the fraction of the drum occupied by the bed (0.2, 0.33, and 0.4) were investigated. In the various experiments and simulations, between 1.5 and 10 rotations were necessary in order to reach the well-mixed state. The number of rotations required was essentially independent of the drum rotational speed, although of course for faster speeds the required number of rotations was completed in a shorter time. The effect of drum diameter and fractional filling of the drum were related to their effects on the ratio of the exposed surface area of the bed to the bed volume (R_B, m^{-1}), with the number of rotations required to achieve the well-mixed state initially falling quickly as this ratio increased, reaching a plateau of 1.5 rotations when this ratio had a value of 20 (Fig. 8.11(a)).

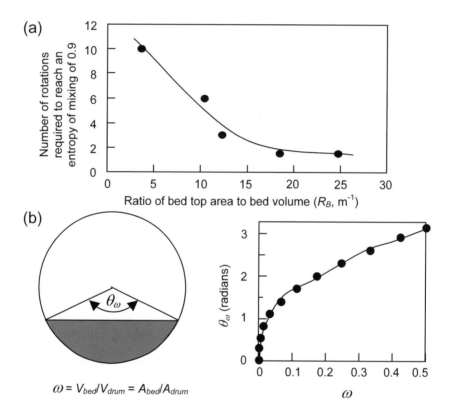

Fig. 8.11. Mixing in an unbaffled rotating drum. **(a)** Dependence of the number of rotations required to reach an entropy of mixing of 0.9 on the ratio of the bed top area to bed volume. The data are from a table presented by Schutyser et al. (2001), adapted with kind permission from John Wiley & Sons, Inc. **(b)** Equation (8.2) is expressed in terms of θ_ω, the angle subtended at the center of the drum by the bed surface, as shown on the left. Note that V and A represent volume and area, respectively. The graph on the right shows θ_ω as a function of the fraction of the drum occupied by the bed, according to Eq. (8.4)

Note that the ratio of exposed surface area to bed volume can be calculated as (Schutyser et al. 2001):

$$R_B = \frac{8}{D} \left(\frac{\sin(\theta_\omega/2)}{\theta_\omega - \sin(\theta_\omega)} \right) \tag{8.2}$$

where D is the drum diameter (m) and θ_ω is the angle in radians subtended at the center by the bed surface for a particular fractional filling ω (m^3-bed m^{-3}-total-bioreactor-volume). Note that θ_ω can be determined from the following relationship:

$$\omega = \left(\frac{\theta_\omega - \sin(\theta_\omega)}{2\pi} \right) \tag{8.3}$$

Unfortunately it is not possible to isolate θ_ω on the left hand side of this equation. However, it is possible to use this equation to plot θ_ω against ω and to fit a polynomial equation. Doing this for values of ω from 0 to 0.5 gives the following explicit equation for θ_ω in terms of ω (Fig. 8.11(b)):

$$\theta_\omega = -3412\omega^6 + 6461.3\omega^5 - 4738.7\omega^4 + 1697.5\omega^3 - 310.36\omega^2 + 31.567\omega + 0.326. \tag{8.4}$$

For unbaffled drums, for a particular fractional filling (ω), Eqs. (8.2) and (8.4) can be used to calculate R_B, which in turn can be compared against Fig. 8.11(a) in order to evaluate the effectiveness of radial mixing that can be expected.

Schutyser et al. (2001) did simulations to investigate the degree to which baffles affect mixing. They compared baffles of 5 cm and 10 cm width within a 30 cm diameter drum, fitting four straight baffles around the inner circumference of the drum (in the manner shown in Fig. 8.1). The smaller baffles had little effect in increasing mixing in the tumbling regime, although at low rotation rates they helped to prevent slumping flow. The larger baffles did improve the effectiveness of mixing.

Schutyser et al. (2002) extended the discrete-particle modeling approach to three dimensions and used it to analyze radial and axial mixing in three different drum designs: a drum without baffles, a drum with four straight baffles (each with a width of 66% of the drum radius) and curved baffles. Straight large baffles do increase axial mixing compared to that in an unbaffled drum, even though they are not designed specifically to push substrate along the axis of the drum. Schutyser et al. (2002) attributed this effect to the higher particle velocities that occur at the surface of the bed. The best design for good axial and radial mixing is a drum with curved baffles, in which the substrate is well mixed axially after three to four rotations. In the same drum without baffles, it can require of the order of 50 to 100 rotations for the bed to be well mixed in the axial direction. Schutyser et al. (2002) noted that with curved baffles it is interesting to incline the central axis of the drum (Fig. 8.12). It can be inclined up until the dynamic angle of repose of the solid, which in their case was 35°, although they suggested that 20° might be more appropriate.

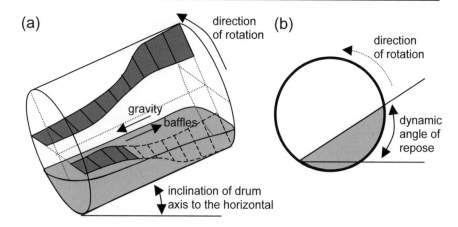

Fig. 8.12. (a) The use of angled baffles and an inclined axis in order to promote axial mixing within a rotating drum bioreactor (Schutyser et al. 2002). Only two baffles are shown, but more can be fitted. **(b)** The dynamic angle of repose of an agitated bed of solids, which represents an upper limit on the inclination of the drum axis that should be used. Adapted from Schutyser et al. (2002) with kind permission from John Wiley & Sons, Inc.

8.4.2 Gas Flow Regimes in the Headspaces of Rotating Drums

The necessity of knowing the headspace flow patterns in order to calculate bed-to-headspace exchange can be seen by using convective heat exchange as an example, although the argument also applies to exchange of O_2 and water. Convective heat removal to the headspace gases (R_{conv}, W) is described as follows:

$$R_{conv} = hA(T_{bed} - T_{head}) \qquad (8.5)$$

where h is the heat transfer coefficient (W m^{-2} °C^{-1}), A is the contact area between the bed and the headspace, T_{bed} is the bed temperature, and T_{head} is the headspace gas temperature. It is assumed that the bed is well mixed. As shown in Fig. 8.13, if the headspace is well mixed, then the driving force for heat transfer is constant, and the rate of heat transfer is the same at each location on the bed surface. On the other hand, if the flow through the headspace follows the plug-flow regime, then the driving force for heat transfer decreases as the gas heats up as it flows through the drum. In this case the rate of heat exchange between the bed and the headspace is greater near the inlet end of the drum than near the outlet end.

Some studies of headspace flow patterns have been undertaken. Stuart (1996) used a drum of 19 cm internal diameter by 85 cm length that was initially aerated with air and then a 5-minute pulse of pure N_2 was introduced. The outlet O_2 concentration was monitored with a paramagnetic O_2 analyzer and the shape of the response curve was compared with curves that would be expected for several theoretical flow regimes. She studied the effects of two flow rates (2.7 and 5.0 L min^{-1}) three substrate loadings (0, 1, and 2 kg of wheat bran substrate) and 4 rotational speeds (0, 5, 10, and 50 rpm).

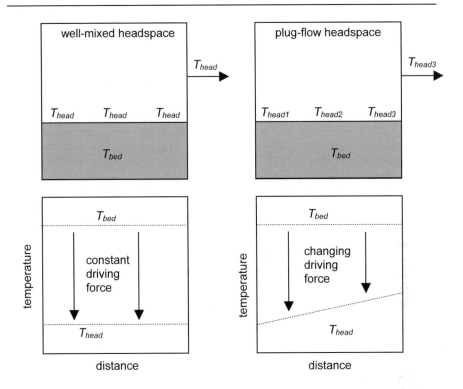

Fig. 8.13. The importance of the headspace flow patterns in affecting bed-to-headspace heat and mass transfer. Two extreme cases are shown. In both cases it is assumed that the bed is well mixed. **(a)** If the headspace is well mixed, then the driving force for heat transfer is equal at all axial positions **(b)** If the flow through the headspace occurs by plug flow, then the driving force for heat transfer decreases as the air flows past the bed surface

In some cases the curves were consistent with a flow regime consisting of several well-stirred regions in series (Fig. 8.14(a)). In other cases they were consistent with plug-flow with axial dispersion (Fig. 8.14(b)). The rotational speed did not affect the type of headspace flow regime. Drums without substrate gave patterns at both gas flow rates that were consistent with the presence of 1 to 2 well-mixed regions in series within the headspace. However, in the presence of substrate there was a difference between the flow patterns at the two different gas flow rates. At both substrate loadings, the response curves obtained with the gas flow rate of 2.7 L min^{-1} were consistent with the presence of 1 to 3 well-mixed regions in series within the headspace, whereas the response curves obtained with the gas flow rate of 2.7 L min^{-1} were consistent with plug-flow with axial dispersion.

Hardin et al. (2001) used CO as a tracer to study flow patterns in a 200-L drum. The patterns were consistent with those that would be expected for a central plug-flow region surrounded by a dead region (Fig. 8.15(a)). The dead region includes a part of the headspace gases and all of gas in the inter-particle spaces in the bed. The dead region is well mixed in the radial direction but there is no axial transport.

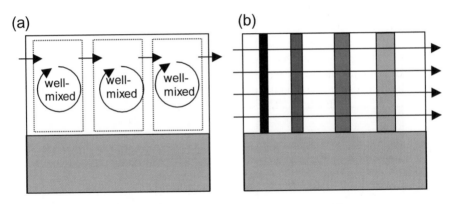

Fig. 8.14. In various different conditions, the residence time distribution patterns for gas flow in the headspace of a rotating drum followed either **(a)** a pattern consistent with several well-mixed regions in series or **(b)** plug flow with axial dispersion (Stuart 1996)

The presence or absence of baffles and the superficial velocity had the greatest effects on the fraction of the drum occupied by the dead region and the rate of transfer between the plug-flow and dead regions (Fig. 8.15(b)). With an increase in the superficial velocity of the air (defined as the volumetric air flow rate divided by the cross-sectional area of the empty drum) there was less mixing between the plug-flow and dead regions and the dead region occupied a greater proportion of the gas volume in the drum. Compared to the absence of lifters, the presence of lifters led to a greater degree of exchange between the plug-flow and dead regions and meant that the dead region represented a smaller proportion of the drum.

Unfortunately, it is not possible to make generalizations from these studies. Flow patterns within the headspace of rotating-drum bioreactors will be greatly influenced by the design and positioning of the air inlet and outlet. One thing is clear, however: If end-to-end aeration is used, it is not reasonable to assume that the headspace is well mixed.

8.5 Conclusions on Rotating-Drum and Stirred-Drum Bioreactors

The following conclusions can be made about the design and operation of rotating-drum and stirred-drum bioreactors on the basis of the experimental work reported above:

- If a rotating-drum bioreactor is used, a decision needs to be made about the rotational rate. If a rotational rate greater than 10% of the critical speed is to be used, then it may not be essential to include baffles within the drum. However, it will require large power inputs to maintain the high rotation rate. The other option is to use quite low rotation rates but to baffle the drum in order to promote mixing.

- End-to-end mixing should be promoted by using curved baffles and inclining the drum axis. Our knowledge is not sufficient to allow detailed advice on the best way of designing curved baffles but obviously the inclination of the central axis must not be greater than the dynamic angle of repose of the solids.
- Discontinuous rotation of the drum or agitator will probably be of little benefit at large scale. Discontinuous rotation brings the added disadvantage of having to overcome inertia, both when starting and when stopping rotation.

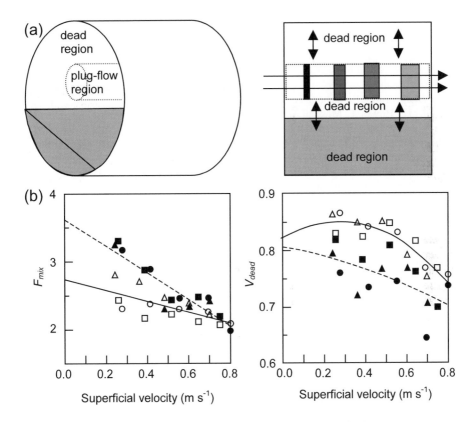

Fig. 8.15. Results of the residence time distribution studies of Hardin et al. (2001). **(a)** A descriptive model consistent with their results. **(b)** Effect of the superficial velocity, baffles, and fill depth on the exchange between the plug-flow and dead regions (characterized by the dimensionless variable F_{mix}) and the volume of the dead region (characterized by the dimensionless variable V_{dead}). F_{mix} is the volumetric exchange rate between the dead and plug-flow regions relative to the volume of the drum and the mean residence time, such that a F_{mix} of 1 would be equal to one volume of the drum exchanged per mean residence time. V_{dead} is the volume of the dead region relative to the total volume of the gas inside the drum. Key: *Hollow symbols* and *solid lines* represent an unbaffled drum. *Solid symbols* and *dashed lines* represent a baffled drum. The *circles* represent 26% filling, the *triangles* 19.5% filling and the *squares* 13% filling. Adapted from Hardin et al. (2001), with kind permission from John Wiley & Sons, Inc.

- Fractional fillings should not be more than 0.4 and may need to be less. In fact, the optimal fractional filling, that is, the filling that allows you to use as much of the drum volume as possible without compromising mixing too much, must be determined experimentally for each particular combination of substrate and microorganism.
- Our knowledge is not sufficient to allow detailed advice on the best design of mixers in the case of stirred-drum bioreactors.

Further Reading

Studies of heat and mass transfer in rotating-drum bioreactors
Fung CJ, Mitchell DA (1995) Baffles increase performance of solid-state fermentation in rotating drums. Biotechnol Techniques 9:295–298
Hardin MT, Howes T, Mitchell DA (2001). Residence time distribuition of gas flowing through rotating drum bioreactors. Biotechnol Bioeng 74:145–153
Hardin MT, Howes T, Mitchell DA (2002) Mass transfer correlations for rotating drum bioreactors. J Biotechnol 97:89–101
Schutyser MAI, Weber FJ, Briels WJ, Rinzema A, Boom RM (2003) Heat and water transfer in a rotating drum containing solid substrate particles. Biotechnol Bioeng 82:552–563

Selected examples of studies undertaken in rotating-drum bioreactors
Kalogeris E, Iniotaki F, Topakas E, Christakopoulos P, Kekos D, Macris BJ (2003) Performance of an intermittent agitation rotating drum type bioreactor for solid-state fermentation of wheat straw. Bioresource Technol 86:207–213
Kargi F, Curme JA (1985) Solid-state fermentation of sweet sorghum to ethanol in a rotary-drum fermenter. Biotechnol Bioeng 27:1122–1125
Marsh AJ, Mitchell DA, Stuart DM, Howes T (1998) O_2 uptake during solid-state fermentation in a rotating drum bioreactor. Biotechnol Letters 20:607–611
Mitchell DA, Tongta A, Stuart DM, Krieger N (2002) The potential for establishment of axial temperature profiles during solid-state fermentation in rotating drum bioreactors Biotechnol Bioeng 80:114–122
Stuart DM, Mitchell DA, Johns MR, Litster JD (1998) Solid-state fermentation in rotating drum bioreactors: Operating variables affect performance through their effects on transport phenomena. Biotechnol Bioeng 63:383–391

Simplified approach to making design and operating decisions for rotating drums
Hardin MT, Mitchell DA, Howes T (2000) Approach to designing rotating drum bioreactors for solid-state fermentation on the basis of dimensionless design factors Biotechnol Bioeng 67:274–282

9 Group IVa: Continuously-Mixed, Forcefully-Aerated Bioreactors

David A. Mitchell, Nadia Krieger, Marin Berovič, and Luiz F.L. Luz Jr

9.1 Introduction

This chapter addresses the design and operation of bioreactors that are forcefully aerated and are continuously mixed during the fermentation. Note that many of the bioreactors considered in this chapter can also be operated, if desired, with intermittent mixing, with intervals of minutes to hours between mixing events. This operation will be addressed separately, in Chap. 10. The choice between continuous and intermittent mixing will depend on both the sensitivity of the organism to shear effects during mixing and the properties of the substrate particles such as their mechanical strength and stickiness.

9.2 Basic Features, Design, and Operating Variables of Group IVa Bioreactors

There are various different ways in which bioreactors can be agitated, and therefore bioreactors in this group may have quite different appearances. The efficiency of mixing and aeration will vary significantly amongst the various designs. Figure 9.1 shows how these bioreactors can be divided into subgroups depending on how the agitation is achieved.

The general design variables associated with well-mixed bioreactors are:

- the geometrical shape and dimensions of the bioreactor;
- the design of the agitator;
- the presence of a water jacket;
- the presence of internal heat transfer surfaces. Note that, due to the fact that the bed is agitated, possibly the best manner to have internal heat transfer surfaces without interfering with mixing would be to have a hollow mixer and pass cooling water through it.

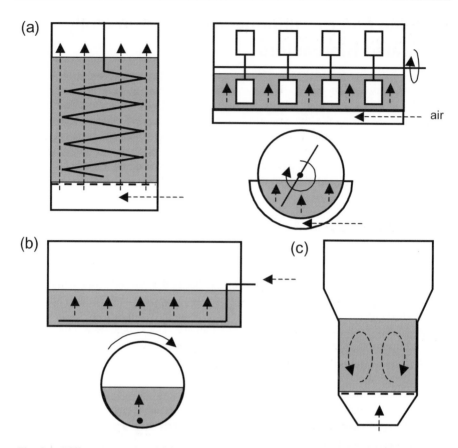

Fig. 9.1. Different ways in which agitation can be provided in continuously-mixed, force-fully-aerated bioreactors. **(a)** Mechanical agitation with an internal agitator, such as a bed with a vertical agitator or a drum with forced aeration and a central stirrer. **(b)** Agitation action caused by movement of the drum body. **(c)** Agitation caused by the movement of air

The general operating variables that can be manipulated are:

- the temperature, humidity, and flow rate of the inlet air;
- the intensity of mixing (rpm);
- the addition of water and other additives. Such additions create no difficulties, since the bed is already being continuously mixed.

This bioreactor type potentially has a significant advantage over packed beds. The inlet air for packed beds should always remain saturated, in order to minimize evaporation, given that it is not practical to add water to packed beds during the fermentation (Chap. 7). The use of dry air in packed beds would hasten the drying out of the bed to values that restrict growth. In contrast, since it is a relatively simple matter to add water uniformly to the bed in a continuously agitated bioreactor, it is possible to use unsaturated air in order to promote evaporation.

9.3 Where Continuously-Agitated, Forcefully-Aerated Bioreactors Have Been Used

Of course, bioreactors that are designed for continuous mixing can also be used in the intermittent-mixing mode. However, there is a difference: If a bed is to remain static for long periods, then it is important to design the aeration system to aerate the bed evenly during the static periods. Uniform distribution of the air may not be so crucial for a continuously mixed bioreactor, since the continuous mixing action should bring all parts of the bed to the well-aerated zone. Bioreactors that have been used in the intermittently mixed mode of operation will be mentioned here if they can also be operated effectively in the continuously mixed mode.

9.3.1 Stirred Beds with Mechanical Agitators

Some mechanically agitated bioreactors involve a substrate bed that sits on a perforated plate, such that air is blown through the whole cross-section of the bed. A mechanical agitator embedded in the bed mixes the bed. In the case of the 50-L bioreactor of Chamielec et al. (1994) and Bandelier et al. (1997) the bed is mixed with a planetary mixer, that is, the mixer blade rotates around its central axis while this central axis simultaneously rotates around the central axis of the bioreactor (Fig. 9.2(a)). This bioreactor has only been used for intermittently-stirred operation but can be used with continuous stirring. The modified solids-mixer of

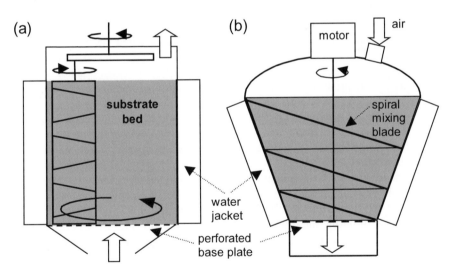

Fig. 9.2. Mechanically-agitated bioreactors that can readily be used in either the continuously-mixed or intermittently-mixed mode because they give good aeration of the bed when it is static. **(a)** A bioreactor with a planetary mixer (Chamielec et al. 1994; Bandelier et al. 1997); **(b)** A conical solids mixer aerated from the top (Schutyser et al. 2003b)

Schutyser et al. (2003b) has a helical blade that scrapes the inside wall of the bioreactor with a lifting action (Fig. 9.2(b)). It has a capacity for 20 kg of cooked wheat grain. No data is available for fermentations in this bioreactor. Schutyser et al. (2003b) used it to study mixing in the absence of the microorganism.

Other mechanically agitated bioreactors have been built in such a way that air only enters at specific points, and not over a wide cross-section of the bed. In this case, the efficiency of the aeration of the bed depends on the degree of mixing achieved by the agitation system, because it is the mixing action that brings the substrate particles into the aeration zone. Such bioreactors would not be particularly appropriate for operation in the intermittently mixed mode.

Nagel et al. (2001a) used a bioreactor that consisted of a 35-L horizontal drum, with paddles mounted on a central axis (Fig. 9.3(a)). The bed was aerated by forcing high-pressure air through holes in the ends of the paddles. They were able to control the temperature at 35°C in this continuously mixed bioreactor during the growth of *Aspergillus oryzae* on 8 kg of cooked wheat grains. In one experiment, they showed that temperature control could be achieved by heat removal through the wall to a cooling coil wrapped around the outside of the drum, with cooling water temperatures needing to be as low as 18°C during the time of peak heat production (Fig. 9.4(a)). In another experiment they promoted evaporative cooling by using high flow rates of dry air. For adequate temperature control at the time of peak heat production, the air flow rate needed to be about 75 L min^{-1} (Fig. 9.4(b)), or two volumes per volume per minute (vvm, m^3-air m^{-3}-total-bioreactor-volume min^{-1}). Despite this success at small scale, it is not clear how such a bioreactor would perform at large scale, the most important question being the efficiency of aeration of the bed.

Ellis et al. (1994) adapted a Z-blade mixer with an internal volume of 28 L as an SSF bioreactor (Fig. 9.3(b)). However, they only studied the mixing behavior, in the absence of microbial growth. The degree of mixing achieved did not depend on the rotational speed, but rather on the number of revolutions. It is not clear how effective the distribution of air will be in such a bioreactor, with air being introduced through four relatively small holes in the bottom of the bed.

Berovič and Ostroveršnik (1997) designed a stirred bed bioreactor in which a horizontal cylindrical drum was filled to two-thirds depth with substrate and air was introduced through a perforated central shaft embedded in the substrate bed and upon which mixer blades were mounted (Fig. 9.3(c)). If desired, the reactor could be rotated 90° to a vertical position to aid in loading or unloading operations and could even be operated in this orientation. This bioreactor was used for optimization of inoculation, sterilization, mixing, aeration, and temperature and humidity control during the production of pectinolytic enzymes by *Aspergillus niger* Berovič and Ostroveršnik (1997) and later for the production of fungal polysaccharides by *Ganoderma lucidum* (Habijanic and Berovič 2000).

The bioreactors of Nagel et al. (2001a), Berovič and Ostroveršnik (1997), and Ellis et al. (1994) had water jackets. However, if such designs were to be used at larger scale with geometrically similar proportions, the effectiveness of the water jacket would decrease, due to the decrease in the ratio of the surface area for heat transfer to the volume of the substrate bed.

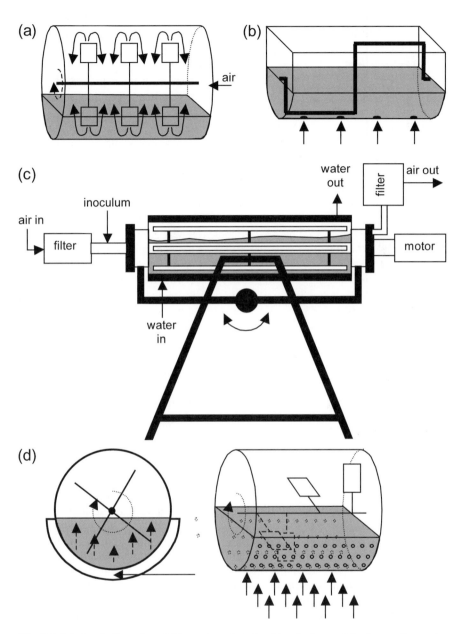

Fig. 9.3. Mechanically-agitated bioreactors that should only be used in the continuously-mixed mode because they do not give good aeration of the bed when it is static. **(a)** The stirred drum (Nagel et al. 2001a); **(b)** the Z-blade mixer of Ellis et al. (1994); **(c)** the horizontal stirred tank bioreactor of Berovič and Ostroveršnik (1997); **(d)** end and side views of a stirred drum with a perforated bottom ("stirred perforated-drum")

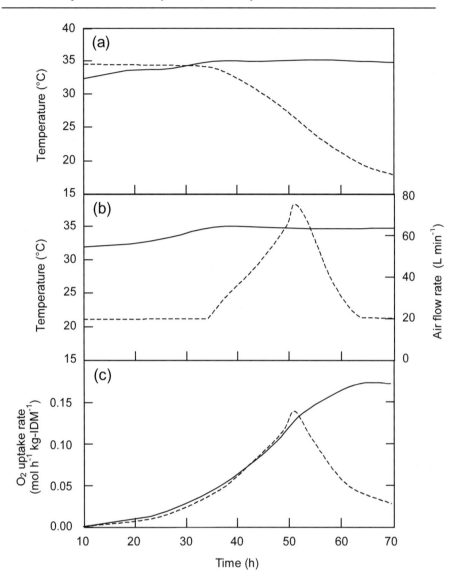

Fig. 9.4. Performance of the 35-L bioreactor of Nagel et al. (2001a), which is of the type shown in Fig. 9.3(a), when *Aspergillus oryzae* was grown on 8 kg of cooked wheat bran. **(a)** With wall cooling. Key: (—) bed temperature, (- - -) wall temperature; **(b)** with evaporative cooling. As the aeration rate increased the relative humidity of the inlet air decreased since the rate of water vapor addition to the inlet air stream was maintained constant. Key (—) bed temperature, (- - -) air flow rate; **(c)** O_2 uptake rate with (—) wall cooling, (- - -) evaporative cooling. The sudden decrease in the O_2 uptake rate at 50 h with evaporative cooling is due to the drying out of the bed to water activities low enough to restrict growth. IDM = initial dry matter. Adapted from Nagel et al. (2001a), with kind permission from John Wiley & Sons, Inc.

These various bioreactor designs suggest that possibility of a further design, which might be referred to as a "stirred perforated-drum" (Fig. 9.3(d)). This is a stirred drum, similar to that of Nagel et al. (2001a), but rather than blowing air through the ends of the paddles, it can be blown through the base of the drum, in the manner of the Z-blade bioreactor of Ellis et al. (1994). To improve the aeration, air can be introduced across a broad cross-section of the bed. Note that ideally such a bioreactor should be continuously mixed and not intermittently mixed because the different bed heights at different positions mean that during static operation there will be a preferential flow of air through the thinner part of the bed.

The design and the operation of the agitator are crucial for bioreactors with mechanical agitators, since they determine the effectiveness of the mixing. However, it is not a simple matter to establish general principles. Optimal design and operation of agitators will be affected by the properties of the substrate bed, which can vary widely between different substrates. It appears that many agitators have been designed on a best-guess approach, since there are no studies that relate the comparison of various different mixer types in order to select the best design. The work of Schutyser et al. (2003b) shows that the use of discrete-particle models that predict the movement of particles in agitated bioreactors is a powerful tool not only for selecting a particular agitator design amongst the various possibilities but also for optimizing the design and operation of the selected agitator.

9.3.2 Gas-Solid Fluidized Beds

Gas-solid fluidized beds consist of a vertical chamber with a perforated base plate through which air, or some other gas, is blown with sufficient velocity to fluidize the substrate particles (Fig. 9.5(a)). It is necessary to design the bioreactor with sufficient height to allow for expansion of the bed upon fluidization. Also, in order to facilitate separation of the solids, the upper regions of the bioreactor need to be somewhat wider than the fluidization region. Due to the greater cross-sectional area for flow, the superficial velocity of the air falls below the minimum fluidization velocity and the particles in this region therefore settle. It may be necessary to incorporate a mechanical mixer slightly above the base plate to help to break up any unfluidized agglomerates that may deposit there. In this type of bioreactor it is a relatively simple matter to make additions to the substrate bed. Water, or nutrient or pH correcting solutions can be sprayed onto the top of the substrate bed.

It may be interesting to recycle the process air, in order to reduce the air preparation costs involved with heating and humidification, although in an aerobic process care must be taken not to allow the O_2 level to fall too low and the CO_2 level to rise too high. Such bioreactors can be used for anaerobic processes if N_2 is used for fluidization, but recycling is then essential in order to minimize process costs.

The ability to use this type of bioreactor depends on the substrate properties. There are two potential difficulties. Firstly, large agglomerates will form if sticky particles are used and these agglomerates will not fluidize. Secondly, if the substrate particles have different sizes, then some particles might fluidize while others

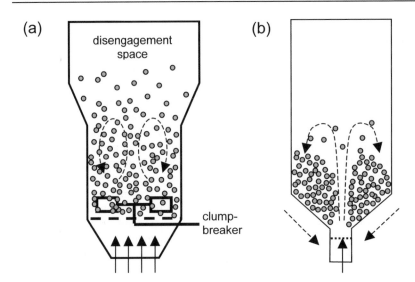

Fig. 9.5. Bioreactors in which the mixing action is provided by the gas stream. **(a)** Gas-solid fluidized bioreactors, in which the whole bed is fluidized. **(b)** A spouted bed in which only particles that fall into the central region are fluidized (Silva and Yang 1998)

might not. Even with a non-sticky substrate of uniform size, fluidized-bed operation would also be expected to face challenges given that the properties of the substrate particle can change markedly during a fermentation due to the consumption of nutrients within the particle by the microorganism and consequent loss of particle mass in the form of CO_2.

Matsuno et al. (1993) describe the use of two air-solid fluidized beds by the soy sauce company Kikkoman in the 1970s:

- a 16-L bioreactor 2 m high, with a diameter of 20 cm in the lower fluidization region and a diameter of 28 cm in the upper disengagement region;
- an 8000-L bioreactor 8 m high, with a diameter of 1.5 m in the lower fluidization region and a diameter of 2 m in the upper disengagement region. This bioreactor had a capacity for 833 kg of wheat bran at 40% moisture content.

According to Matsuno et al. (1993), Kikkoman claimed that the air-solid fluidized bed gave a higher productivity for the production of proteases and amylases by *Aspergillus sojae* on wheat bran powder than either static-bed SSF systems or submerged liquid culture. However, detailed information about this bioreactor and its operation is not available in the literature.

Gas-solid fluidized beds also received interest in the 1980s for the production of ethanol. A pilot-scale bioreactor of 55 cm diameter was built by Rottenbacher et al. (1987). In this case the system had some differences from "typical" SSF processes. Given that ethanol production requires anaerobic conditions, N_2 was used as the fluidizing gas. It was recycled through the bioreactor, with ethanol being condensed from the gas before it was returned to the bioreactor. The idea was that, in continuously stripping ethanol from the system, this strategy would minimize the inhibitory effects of ethanol and maximize ethanol yields. Another difference was that the solid phase was not a nutrient phase but rather consisted of pellets of compressed yeast. The bioreactor had a capacity for 20 kg of yeast pellets. A glucose solution was sprayed onto the bed surface, therefore each pellet received fresh nutrients as it circulated through the bed.

None of the workers who have used fluidized beds have mentioned any problems with temperature control. This is not unexpected, since the high flow rates required for fluidization should provide sufficient convective cooling capacity. In fact, due to the ease of temperature control, mathematical models that have been developed for fluidized bed operation (Rottenbacher et al. 1987; Bahr and Menner 1995) do not include energy balances. Further, due to the good homogeneity of the bed, they tend to be concerned with intra-particle phenomena.

A variant of the fluidized bed is the "spouted bed" (Fig. 9.5(b)). The major difference is that air is blown upwards only along the central axis of the bed, such that only part of the bed is fluidized at any one time. There is a continuous cycling of particles as the solids slip down the sloped sides at the bottom of the bioreactor.

Silva and Yang (1998) built a spouted-bed bioreactor of 7.6 cm diameter and 73 cm height. However, in their experiments, in which they grew *Aspergillus oryzae* on rice, the bed height was only 9 cm, which means that ratio of overall bioreactor volume to bed volume was quite large. Further, they used an aeration rate of around 250 L min^{-1}. This represents an aeration rate of 625 vvm (volumes of air per volume of bed per minute), which is unlikely to be practical to maintain at large scale. In this particular fermentation, continual spouting led to poorer growth and lower enzyme levels than in beds that were operated as packed beds (at an aeration rate of around 50 L min^{-1}) for most of the time and only spouted intermittently at 1 or 4 h intervals (at an aeration rate of around 250 L min^{-1}), presumably due to the shear damage caused by the continuous motion (Silva and Yang 1998). Note that in intermittently spouted beds, the aeration of the bed will not be uniform during the periods of packed-bed operation, since air is not introduced evenly across the whole section of the bed. It is not clear how suitable spouted operation will be for large-scale bioreactors.

9.3.3 Bioreactors Mixed by the Motion of the Bioreactor Body

It is also possible to obtain some mixing within the substrate bed through movement of the whole bioreactor body (Fig. 9.6). A rocking-drum bioreactor of 1.3 L holding volume was used by Barstow et al. (1988), Ryoo et al. (1991), and Sargantanis et al. (1993) in studies of bioreactor control strategies. The bioreactor consists of three concentric drums, an inner, a middle, and an outer drum (Fig. 9.6(a)). The inner drum and the middle drum are perforated, and the substrate bed is held, loosely packed, between these two drums. Air is introduced inside the inner drum. It passes through the perforations into the bed, through the bed and through the perforations in the middle drum to the space between the middle drum and the outer drum where it then moves to the air outlet. Water can be dripped

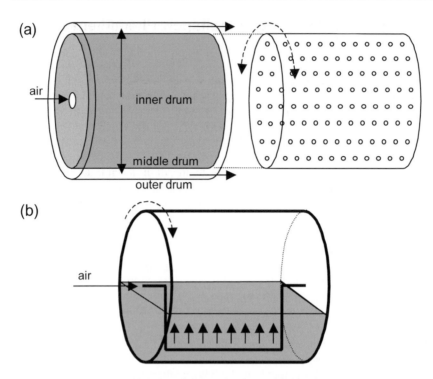

Fig. 9.6. Bioreactors in which the mixing action is provided by the motion of the bioreactor body. **(a)** The rocking-drum bioreactor. The substrate bed is held between the inner and middle drums, which are both perforated in the manner shown on the right in the exploded view of the middle drum. The diagram on the right also shows how the mixing action is provided by the forwards and backwards turning of the middle drum. **(b)** A drum in which a stationary pipe remains within the bed as the drum body rotates

through the perforations in the inner drum, moving by gravity through the bed. The two outer drums rotate in relation to the inner drum, this causing a mixing action within the bed. The name "rocking drum" arises because the rotation occurs with three-quarter turns in a clockwise-counterclockwise sequence, at a rate of 1 revolution every 5 minutes. At the scale of 1.3 L, good control is achieved, with the substrate bed temperature being controlled within ± 1°C of the set point of 37°C. However, it is questionable whether this bioreactor will be effective at large scale. Certainly, the mixing action generated by the relative motion of the inner and middle drums will be inefficient at large scale.

Schutyser et al. (2003a) used a 28-L drum bioreactor (30 cm internal diameter and 40 cm length) in which the air line entered at the central axis but then passed through the bed in a U-shaped tube, which had several small holes in the horizontal section that passed through the center of the bed (Fig. 9.6(b)). Mixing action was provided by the rotation of the drum. They used discontinuous mixing. However, from the point of view of good aeration, continuous mixing would be better.

9.4 Insights into Mixing and Transport Phenomena in Group IVa Bioreactors

Relatively little work is available that allows insights into mixing and transport phenomena in continuously-mixed, forcefully-aerated bioreactors. More work has been done on intermittently-mixed, forcefully-aerated bioreactors (see Chap. 10). This is not altogether surprising: Although some microorganisms can tolerate continuous mixing, the majority performs better when the mixing is intermittent.

Wall cooling can be effective in heat removal at small scale but it will not be sufficient to maintain the bed temperature at the desired value as scale is increased, if the bioreactor is scaled up on the basis of geometric similarity. This is demonstrated in Fig. 9.7, which is the result of a case study undertaken by Nagel et al. (2001a). If the length-to-diameter ratio is maintained constant, then the surface area of the wall per unit volume of bed decreases. Therefore, as scale is increased, eventually a "critical bioreactor volume" is reached, above which wall cooling alone cannot control the bed temperature. In Fig. 9.7 this critical volume corresponds to the bioreactor volume at which the curve intersects the horizontal line. Of course the critical bioreactor volume will depend on the maximum growth rate of the organism, the temperature difference between the bed and the cooling water and the length to diameter ratio of the bioreactor.

One strategy for heat removal at volumes above the critical bioreactor volume might be to include internal heat transfer surfaces. This could possibly be done by incorporating baffles, or by circulating cooling water through the mixing paddles. However, a situation will quickly be reached in which a further increase in internal heat transfer surfaces will interfere with the ability to mix the bed. The other strategy is to promote evaporation by using dry air. In this case, it will be essential to make periodic water additions in order to prevent growth from being limited by low water contents. Note that the sudden decrease in the O_2 uptake rate in Fig. 9.4(c) was due to the decrease in the water activity of the bed.

Flow patterns within mixed solid beds in SSF bioreactors have received little attention. It is often assumed that the bed is well mixed in such bioreactors. However, this might not necessarily be the case. Rather, there might be defined circulation patterns and the effectiveness of mixing may be different in different regions of the bed. This is best illustrated by the study of mixing within a conical solids mixer that was undertaken by Schutyser et al. (2003b). Positron emission particle tracking was used to follow the circulation of individual particles within the bed. Figure 9.8 shows how such studies can give information about the circulation patterns of particles within the bed. However, note that such studies require access to quite sophisticated equipment. Schutyser et al. (2003b) compared the experimental results obtained with positron emission particle tracking with predictions made using a discrete-particle mixing model (see Fig. 8.10(a) for a simple explanation of the basis of discrete-particle modeling). Once such a model has been validated, it can then be used to predict flow patterns.

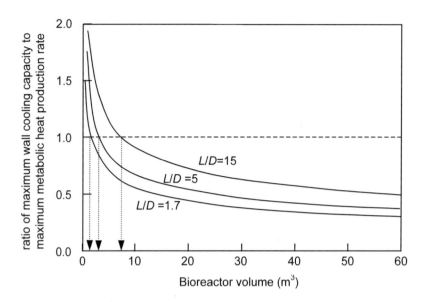

Fig. 9.7. Effect of scale on the ability to remove the waste metabolic heat by wall cooling in a continuously-mixed, forcefully-aerated bioreactor, of the type shown in Fig. 9.3(a), as calculated by Nagel et al. (2001a). The Y-axis represents the ratio of the maximum wall cooling capacity to the maximum metabolic heat production rate. The calculations are done for three length-to-diameter ratios in a situation in which the maximum O_2 uptake rate is 0.0191 mol s^{-1} m^{-3} bed, the overall heat transfer coefficient for heat transfer across the wall to the cooling water is 100 W m^{-2} $°C^{-1}$ and the temperature difference between the wall and the bed is 20°C. Where the curve is above the dashed horizontal line, wall cooling is sufficient to maintain the bed temperature at the desired value. Conversely, where the curve is below the dashed horizontal line wall cooling is not sufficient to maintain the bed temperature. Adapted from Nagel et al. (2001a) with kind permission from John Wiley & Sons, Inc.

Pressure drop has been little studied in continuously-mixed, forcefully-aerated bioreactors. It would not be expected to be a problem, since the mixing action should squash hyphae onto the particle surface, preventing them from growing into the inter-particle spaces. Likewise, the appearance of cracks in the bed should not be a problem since the particles will not be bound together. However, preferential flow could occur due to differences in the bed height caused by:

- the mixing action. For example, in the conical bioreactor shown in Fig. 9.2(b), the rotation of the helical mixing blade causes the sides of the bed to be higher than the middle, as indicated by the vertical section of the bed shown in Fig. 9.8(b). Air will flow preferentially through the center of the bed.
- the design of the bioreactor itself. For example, in the bioreactor shown in Fig. 9.3(d), the curvature of the base of the bioreactor means that the bed height varies as a function of position, even if the top of the bed is horizontal. In this case air will flow preferentially through the sides where the bed is thinnest.

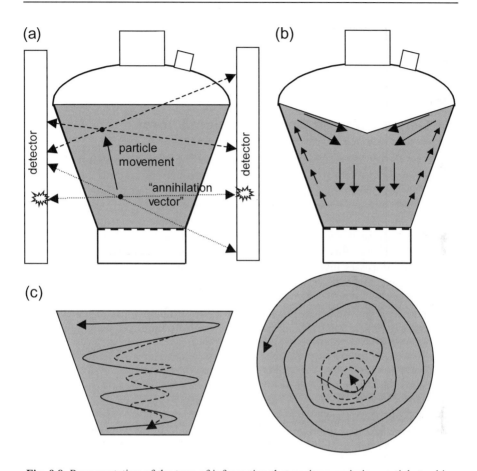

Fig. 9.8. Representation of the type of information that positron emission particle tracking studies can provide, which was demonstrated with a conical solids mixer adapted for use as a bioreactor (Schutyser et al. 2003b). As shown in Fig. 9.2(b), the bioreactor contains a spiral mixing blade that follows the inside of the wall and rotates counterclockwise, although this detail is not shown here for the sake of clarity. **(a)** The principle of positron emission particle tracking. The particle contains a radioactive isotope that emits positrons. Emitted positrons immediately annihilate with electrons and two 0.51 MeV gamma rays leave the annihilation site in diametrically opposed directions. These gamma rays are registered by detectors that are placed around the bioreactor. The position of the particle at any particular instant can be determined simply by finding the intersection of the "annihilation vectors" resulting from the various positrons emitted at that instant; **(b)** The data can be analyzed to give particle velocity vectors within any plane. In this particular case, the velocity vectors are shown in a vertical plane that passes through the central axis of the bioreactor. The longer the arrow the greater the velocity; **(c)** The trajectory of individual particles can be plotted. Shown here are smoothed trajectories of two particles, in side and overhead views. Key: (—) A particle being pushed up the wall of the bioreactor by the mixing blade; (- - -) A particle descending in the middle of the bioreactor. Adapted from Schutyser et al. (2003b) with kind permission from John Wiley & Sons, Inc.

In addition, the agitator may cause transient open channels as it mixes the bed. In other words, as the agitator moves, it may leave a gap behind it. Even though the solid bed may later collapse to fill the gap, air will flow preferentially through the gap while it is open.

9.5 Conclusions on Group IVa Bioreactors

There are in fact few examples of use at large scale of continuously-mixed, force-fully-aerated bioreactors. Perhaps this is not surprising. The majority of SSF processes involve filamentous fungi, and many of these will not tolerate continuous mixing well. However, there are exceptions: In the work of Nagel et al. (2001a), in which *Aspergillus oryzae* was cultivated on wheat grains, it appears that the fungus grew underneath the seed coat and was thereby protected from shear forces, although of course this would have meant restricted availability of O_2. On the other hand, processes involving bacteria should tolerate mixing well and therefore such processes might be expected to be suitable for the continuous-mixing mode of operation.

There is not yet sufficient knowledge to allow a judgment as to whether mechanical agitation or gas-based agitation (i.e., gas-solid fluidized beds) will be better for the continuously-mixed mode of operation.

Further Reading

General considerations about the effect of mixing in large-scale submerged liquid bioreactors on microbial physiology (the effects of mixing in SSF are much less understood)
Enfors SO, Jahic M, Rozkov A, Xu B, Hecker M, Jürgen B, Krüger E, Schweder T, Hamer G, O'Beirne D, Noisommit-Rizzi N, Reuss M, Boone L, Hewitt C, McFarlane C, Nienow A, Kovacs T, Trägardh C, Fuchs L, Revstedt J, Friberg PC, Hjertager B, Blomsten G, Skogman H, Hjort S, Hoeks F, Lin HY, Neubauer P, van der Lans R, Luyben K, Vrabel P, Manelius A (2001) Physiological responses to mixing in large-scale bioreactors. J Biotechnol 85:175–185

A recent study involving a continuously mixed forcefully aerated bioreactor
Nagel FJI, Tramper J, Bakker MSN, Rinzema A (2001) Temperature control in a continuously mixed bioreactor for solid state fermentation. Biotechnol Bioeng 72:219–230

Positron emission particle tracking method for studying mixing in SSF bioreactors
Schutyser MAI, Briels WJ, Rinzema A, Boom RM (2003) Numerical simulation and PEPT measurements of a 3D conical helical-blade mixer: A high potential solids mixer for solid-state fermentation. Biotechnol Bioeng 84:29–39

10 Group IVb: Intermittently-Mixed Forcefully-Aerated Bioreactors

David A. Mitchell, Oscar F. von Meien, Luiz F.L. Luz Jr, Nadia Krieger, J. Ricardo Pérez-Correa, and Eduardo Agosin

10.1 Introduction

This chapter concerns the design and operation of SSF bioreactors under conditions where forced aeration is used and the substrate bed is mixed intermittently. These will be referred to as intermittently-mixed, forcefully-aerated bioreactors or, more simply, as intermittently-mixed bioreactors. As explained in Chap. 3, this mode of operation is appropriate for those SSF processes in which continuous mixing is not tolerated well by the microorganism, but intermittent mixing events do not have unduly deleterious effects. For much of the fermentation the bioreactor operates as a packed-bed bioreactor. The advantage is that the mixing event prevents the pressure drop from becoming too high within the bed and that water can be added to the bed, in a reasonably uniform manner, during the mixing event.

10.2 Basic Features of Group IVb Bioreactors

The basic design features of intermittently mixed bioreactors are similar to those of the various continuously mixed designs (Chap. 9), the difference being in the mode of operation. Since the mixing is only intermittent and the bioreactor spends periods in the static mode of operation, designs should be preferred that give a uniform aeration of the bed when it is static. Forced aeration may or may not be applied during the mixing period, depending on the design. Figure 10.1 shows possible basic designs for intermittently mixed bioreactors.

Intermittently mixed bioreactors have the same design and operating variables as packed-bed bioreactors (Sect. 7.2), which affect the performance during the periods of static operation. In addition to this, the type of agitation is an extra design variable for intermittently mixed bioreactors. The bed may be mixed by a mechanical stirrer, by rotation of the whole bioreactor or by the air flow.

In addition to having the operating variables for packed-bed bioreactors, intermittently mixed bioreactors have several extra operating variables available.

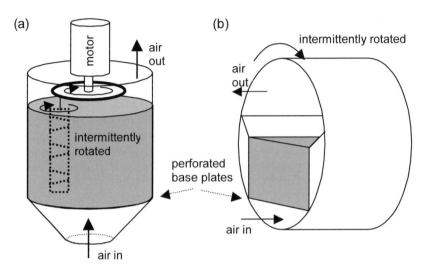

Fig. 10.1. Two basic options for mixing of intermittently mixed bioreactors. **(a)** The bed may be agitated by a mechanical agitator inserted into the substrate bed. In this case forced aeration can be applied during the mixing period. The agitator may simply rotate around its axis, in which case it will need to be almost as wide as the bioreactor, it may rotate with a planetary action (as shown) or it may travel from side to side across the bioreactor. **(b)** The bed may be agitated through rotation of the bioreactor around its central axis (Toyama 1976). There is no mechanical structure within the bed itself. In this case it is not practical to aerate the bed forcefully during the mixing period

Firstly, there is the strategy for initiating mixing events, which will determine the frequency of the mixing events. Secondly, the duration and intensity of mixing events can be varied. Thirdly, unlike packed beds, the relative humidity of the inlet air is potentially available as an operating variable. Since water can be added to the bed in a reasonably uniform manner during the mixing events, unsaturated air can be used to aerate the bed in order to promote evaporative cooling.

The values selected for these extra design and operating variables will be most affected by:

- the temperatures reached in the bed during static operation (e.g., mixing could be triggered by high temperatures at the outlet end of the bed);
- the water activities reached in the bed during static operation (e.g., mixing could be triggered when the outlet-air relative humidity falls below a set point);
- the pressure drop through the bed (e.g., mixing could be triggered when the pressure drop reaches unacceptably high values);
- the sensitivity of the organism to damage during mixing events, which will affect the frequency, intensity, and duration of mixing events.

Considerations affecting the selection of appropriate values of other operating variables, such as the air flow rate and temperature, will be similar to those for packed-beds (Sect. 7.2).

Channeling should be less of a problem for intermittently mixed beds than for static packed beds. The mixing events will tend to break up the bed so that the particles remain separate and these will tend to settle as the bed shrinks, rather than being knitted together and pulled away from the wall as happens with packed beds. However, channeling might be caused by an imperfect bed structure at the end of the mixing event. For example, in the case that the agitator stays in the bed during the static periods, it may leave a hole behind or around it as it comes to a stop. Alternatively, in the case that it is withdrawn from the bed, it may leave a hole as it leaves. In either case, if nothing is done to close the hole, the air will flow preferentially through it during the period of packed-bed operation.

This chapter explains what is known, on the basis of experimental studies, about how these design and operating variables influence bioreactor operation. Later, Chap. 25 will show how mathematical models can be used to explore further the design and operation of intermittently mixed bioreactors.

10.3 Experimental Insights into the Performance of Group IVb Bioreactors

This section presents and discusses the knowledge that experimental work has given into the phenomena that occur within intermittently mixed bioreactors and into the operability of this type of bioreactor.

10.3.1 Large-Scale Intermittently-Mixed Bioreactors

10.3.1.1 The Koji Industry

Intermittently agitated designs have been used in the *koji* industry. Sato and Sudo (1999) report a bioreactor with a capacity of 15 tons of rice *koji* on a 12-m diameter disk (Fig. 10.2). The inoculated substrate is placed in the upper chamber, where it remains for one day. After this period the screw mixer is used to transfer the substrate to the bottom chamber, where it is mixed intermittently. The bioreactor is computer controlled. However, Sato and Sudo (1999) give no further details. For example, it is not clear exactly how often the mixing is carried out.

Interestingly, Sato and Sudo (1999) note that, even for an industry with much experience, the maximum height of the substrate bed is of the order of 20 cm. This means that large-scale bioreactors will occupy a large area. The 15-ton capacity bioreactor has disks of 12 m diameter. In comparison, an SLF bioreactor would have a diameter of about 5 m to hold the same working volume, assuming that the solid bed has a packing density of 400 kg m^{-3} and therefore a volume of 37.5 m^3 and that the SLF bioreactor has a height to diameter ratio of 2:1.

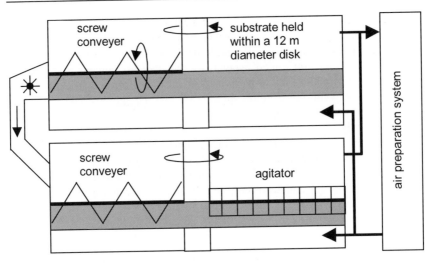

Fig. 10.2. Intermittently-mixed bioreactor of the type used in the *koji* industry for soy sauce production by Nagata Brewing Industry Co Ltd., Takarazuka, Japan. It has a 12 m diameter bed and a capacity for 15 tons of substrate. After one day the upper disk is rotated, with the upper screw conveyer transferring the substrate to the lower disk, where it can be agitated intermittently. Note that the mixers rotate in place and the whole circular bed moves to bring the substrate to the mixing point. This is a simplified version of a diagram presented by Sato and Sudo (1999)

10.3.1.2 The Bioreactor of INRA-Dijon

Durand and Chereau (1988) developed an intermittently mixed bioreactor at INRA, in Dijon, France. It is 2 m long, 0.8 m wide and has an overall height of 2.3 m, with a working bed height of 1 m. This gives a working volume of approximately 1.8 m^3, sufficient to hold a bed of approximately 1 ton of moist material. The mixing is provided by a number of screw augers (i.e., designed to lift the substrate as they turn) that are mounted on a carriage on top of the bioreactor (Fig. 10.3). This carriage travels from one end to the other at a top speed of 6.5 cm min^{-1}, meaning that it takes 35 minutes to traverse the bioreactor for one end to the other. The screws rotate at a top speed 22 rpm. The agitation regime, in terms of the frequency and duration of mixing events, can be varied according to the needs of the process, as determined by the particular microorganism and substrate used. Further, if necessary, different agitators such as hollow screws or helicoid screws, can be fitted, depending on the mixing behavior of the solid medium to be used in the fermentation. The carriage has spray nozzles fitted onto its underside, allowing the addition of inoculum, water, nutrient or pH-correcting solutions during mixing. The aeration system has a maximum capacity of 1500 m^3 h^{-1}, which means that it is possible to aerate with over 13 volumes of air per bed volume per minute (i.e., over 13 vvm).

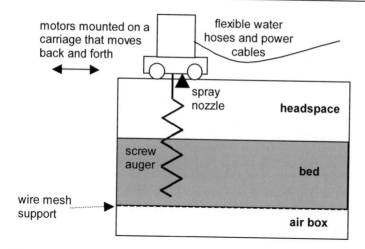

Fig. 10.3. Intermittently-mixed bioreactor of the type used by Durand and Chereau (1988). There are three motors/agitators mounted across the width of the bioreactor (behind the one shown in the side view given here). The bioreactor dimensions are given in the text. Adapted from Durand and Chereau (1988) with kind permission of John Wiley & Sons.

Durand and Chereau (1988) cultivated *Trichoderma*, which grows optimally at temperatures around 28°C, on a sugar beet pulp medium for the production of single cell protein. It was possible to maintain the temperature and water content of the bed within acceptable limits by maintaining the relative humidity of the inlet air constant at 90%, and by manipulating the flow rate and temperature of the air supplied to the bioreactor. They describe the operating regime as follows:

- three to four "turnings" (i.e., mixing events) during the 48 h cultivation;
- for the first 10 h, an air flow rate of 750 m^3 h^{-1} (i.e., about 7 vvm) at 29°C;
- as the growth rate accelerates, an increase of the air flow rate to 1000 m^3 h^{-1} (i.e., about 9 vvm) at 26°C.

With this operating regime, during exponential growth the temperatures at 85-cm depth in the bed and at 20 cm depth in the bed ranged from 26.5 to 29.0°C.

Since it was first reported, the use of this bioreactor has been extended successfully to the production of enzymes and biopesticides (Durand 2003).

Xue et al. (1992) adapted the bioreactor of Durand and Chereau (1988) to build a much larger scale process for the production of microbial protein from sugar beet pulp by *Aspergillus tamarii*. The bioreactor is built from concrete, having a length of 17.6 m, a breadth of 3.6 m, and an overall height of 2.0 m. A perforated stainless steel plate, designed to support the bed, is fixed at a height of 0.6 m. The actual bed height used is 0.7 m, leaving a headspace of 0.7 m, and giving a bed volume of 44 m^3. This corresponds to 25 tons of moist substrate, which, given a water content of 80% (wet basis), gives 5 tons of dry substrate. The carriage holding the screw mixers has a linear speed of 30 cm min^{-1} and rotates the screws at 13.3 rpm. The facility has two such bioreactors. The air system has a maximum capacity of 60,000 m^3 h^{-1}. Divided over two bioreactors, this is 11 vvm.

Very little performance data was provided. They used two turning cycles during the first 48 h of the 72 h process. The aeration rate used was 220 m^3-wet-air min^{-1}, at a relative humidity of 88% and at a temperature that ranged from 32 to 34°C. However, it was not specified whether this aeration rate was for one or both bioreactors. The outlet air was reported to have "operational parameters" of 100% relative humidity and 33°C, although it is not clear whether this was achieved.

10.3.1.3 The Bioreactor of PUC-Chile

Pérez-Correa and Agosin (1999) built a bioreactor with a capacity for a bed of 200 kg. The bioreactor has three sections (Fig. 10.4). The bottom section, which remains stationary, is simply the air box. The 150 cm diameter bed is held by the second section, which is rotated in its entirety by a motor. The top of the bioreactor represents a third section, which is stationary, and on which the agitators are mounted. The thermocouples can be withdrawn from the bed into the headspace during the mixing event. The bioreactor was designed to enable a bed height of 80 cm, although in practice the bed height was kept at 60 cm or below.

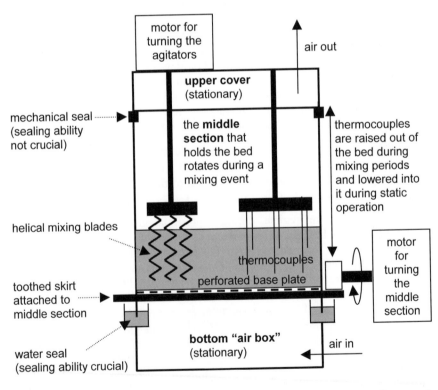

Fig. 10.4. The 200-kg capacity intermittently-mixed bioreactor used by Pérez-Correa and Agosin (1999). The upper cover and air box are maintained stationary by an outer frame while the middle section is rotated by the motor

The advantage of holding the agitators stationary and rotating the bed is that this simplifies the design of the agitator device. However, it also brings a disadvantage: The seal between the bottom and middle sections of the bioreactor, which move relative to one another, must not allow air to escape, otherwise air will leave the bioreactor without passing through the bed. The bioreactor shown in Fig. 10.4 has a water seal. However, since the depth of water in the seal is only 10 cm, this means that the pressure drop across the bed cannot be more than the equivalent of 10 cm of water; otherwise the air will simply bubble through the water in the seal and leave the bioreactor without passing through the bed. This limits the height of the bed that can be used and means that often mixing events are necessary simply to prevent the pressure drop from becoming too high, rather than being prompted by a need for water replenishment or temperature control.

Figure 10.5 shows typical data obtained from this bioreactor, for the growth of *Gibberella fujikuroi* on extruded wheat bran granules for the production of gibberellic acid. A bed height of 40 cm was used. It was possible to control most of the bed within the range of 25 to 30°C most of the time. However, there were hotspots formed, in which the bed temperature exceeded the temperature of the outlet air (compare Fig. 10.5(b) with Fig. 10.5(c)); these hotspots most likely represent regions that were receiving poor aeration due to channeling. The CO_2 production rate peaked at around 40 h (Fig. 10.5(c)). In an attempt to control the temperature in the bed at 28°C, the temperature (Fig. 10.5(c)), humidity (Fig. 10.5(e)), and flow rate (Fig. 10.5(f)) of the inlet air were manipulated. The pressure drop was kept well below 10 cm of water by the mixing events (Fig. 10.5(g)). These 30-min-long mixing events occurred, on average during a fermentation, once every 6 to 10 h, although during periods of high heat production they were as frequent as once every 4 h. Water needed to be replenished to replace evaporated water (Fig. 10.5(h)). This was done over the 30-min period of the mixing event, with the amount of water necessary being calculated from a set of mass balance equations.

10.3.2 Pilot-Scale Intermittently-Mixed Bioreactors

Pérez-Correa and Agosin (1999) also developed a bioreactor with a capacity for 50 kg of moist substrate (Fig. 10.6). This bioreactor used a similar strategy to that used in their larger scale bioreactor in the sense that mixing was achieved through movement of the bed past a number of fixed mixing blades. The bed was held within a perforated basket, 1.15 m in diameter and 0.28 m high, that was rotated when mixing was desired. It was necessary to have a seal between the basket and the body of the bioreactor to make sure that the air flowed through the basket and not around its sides. This bioreactor was capable of being operated aseptically. The whole lid could be raised to give access to the interior, but was hermetically sealed during the fermentation. This bioreactor was used for the production of gibberellic acid by *Gibberella fujikoroi* (Pérez-Correa and Agosin 1999) and for the production of *Trichoderma* (Agosin and Aguilera 1998).

In the 50-L bioreactor of Chamielec et al. (1994) and Bandelier et al. (1997), which is designed for sterile operation, the substrate bed is supported on a wire

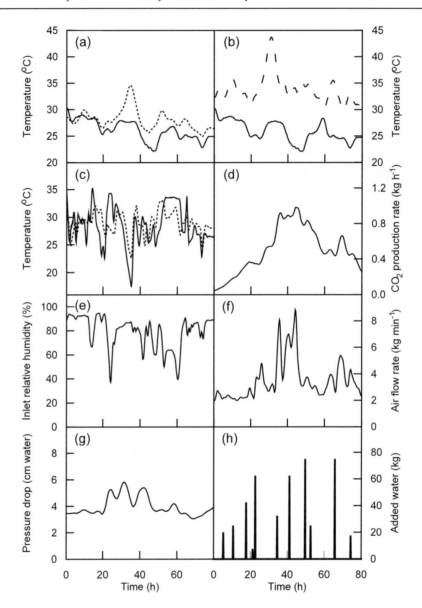

Fig. 10.5. Typical results from the 200-kg capacity bioreactor of Pérez-Correa and Agosin (1999). **(a)** Temperatures 10 cm from the wall at (——) 5 cm bed height and (- - -) 20 cm bed height; **(b)** Temperatures 10 cm from the center at (——) 5 cm bed height and (- - -) 20 cm bed height; **(c)** Air temperature at the (——) inlet and (- - -) outlet; **(d)** Rate of CO_2 production from the bioreactor, as an indicator of the overall growth rate; **(e)** Relative humidity of the inlet gas (controlled by the controller); **(f)** Flow rate of air into the bioreactor (controlled by the controller); **(g)** Pressure drop through the bed; **(h)** Addition of water during the fermentation (added over a 30 minute period during mixing events)

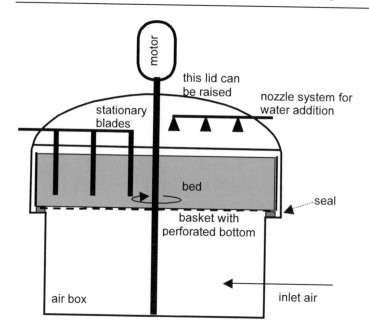

Fig. 10.6. The 50-kg capacity bioreactor of Pérez-Correa and Agosin (1999). Mixing is achieved by rotating the basket that holds the bed while maintaining the mixing blades stationary. The bed dimensions are given in the text

mesh, and is mixed by an agitator that undergoes a "planetary motion", that is, the agitator rotates around its central axis while this central axis simultaneously rotates around the central axis of the bioreactor (as in Fig. 10.1(a)). The bioreactor is fitted with a water jacket. It was used successfully in the production of gibberellic acid by *Gibberella fujikoroi*. The bioreactor contained 12 kg of moist wheat bran at a moisture content of 50% (wet basis). The bed was mixed for 10 s every 2 h, but this can be adapted as necessary according to process requirements. The air flow rate was 15 L min^{-1} kg-dry-matter^{-1}, which corresponds to a flow rate of 90 L min^{-1}. Since *Gibberella fujikoroi* is a relatively slow-growing organism, with the process taking 11 days, the major challenge was aseptic operation of the bioreactor. Given the low heat generation rate, temperature control was not difficult. The bed temperature was maintained within 1.3°C of the desired temperature of 28.5°C by maintaining the inlet air temperature at 28°C until 50 h and then reducing it progressively to 22°C at the end of the process (250 h).

The 50-L solids mixer of Schutyser et al. (2003b) presented in Sect. 9.3.1 could be used with intermittent mixing although use of this bioreactor in such fermentations has not yet been reported. Note that if the bioreactor were to be operated in the intermittent mixing mode, then aeration should be from top to bottom in order to ensure that the sides of the bed would be well aerated during the periods of static operation. Introducing air at the bottom of the bioreactor would tend to aerate only the central axis.

10.3.3 Laboratory-Scale Intermittently-Mixed Bioreactors

There are in fact very few reports about the use of intermittent mixing in force-fully-aerated bioreactors at pilot scale. The 7.6 cm diameter spouted-bed of Silva and Yang (1998) (see Fig. 9.5(b)) could be operated in either continuous- or intermittently-spouted mode. As noted in Sect. 9.3.2, intermittent spouting gave better results, presumably due to the lesser shear damage caused to the organism when compared to continuous spouting.

10.4 Insights into Mixing and Transport Phenomena in Group IVb Bioreactors

Intermittently-mixed bioreactors are typically static for most of the fermentation and therefore the principles of heat and mass transfer in them have many similarities to those of packed-bed bioreactors, or namely, axial and possibly radial temperature gradients will be established, the magnitude of which will depend on the combination of bed height, superficial air velocity and microbial growth rate (see Sect. 7.3). As pointed out in Sect. 10.2, the operating variables that intermittently mixed bioreactors have in addition to those of packed-bed bioreactors include the humidity of the inlet air, the strategy for initiating mixing events (which affects their frequency), and the duration and intensity of mixing events.

Little work has been done to characterize quantitatively the damage that intermittent mixing causes to the microorganism and the speed of recuperation, or not, of the microorganism after mixing. Schutyser et al. (2003a) reported a decrease of about 10% in the O_2 consumption rate immediately after mixing events in their intermittently agitated bioreactor, although they did not actually show the results.

Schutyser et al. (2003a) also investigated the timing of the first agitation event, concluding, at least in the case for fungi that produce significant amounts of aerial hyphae, that an early mixing event should be scheduled to prevent the formation of bound aggregates of substrate particles. If such aggregates are allowed to form, then they will be difficult to break apart in subsequent mixing events and O_2 supply to the particle surfaces within the aggregates will be greatly restricted. They showed that for *Aspergillus oryzae* growing on wheat, this "hyphae-disrupting" mixing event will be needed before it is necessary to make the first water addition, even if evaporation is the sole cooling mechanism.

There has been little effort to characterize experimentally the heat and mass transfer phenomena associated with the intermittent mixing mode of operation, although the modeling study of Ashley et al. (1999) suggests that this mode of operation can potentially lead to temperatures being reached that are higher than those that would be obtained in completely static (i.e., packed bed) operation (Fig. 10.7(a)). Immediately before a mixing event, the temperature profile in the bioreactor is identical to that which would be expected for packed-bed operation. In this situation the rate of heat removal is uniform at the different heights within the bed.

Immediately after a mixing event, due to the absence of an axial temperature gradient, the cooling effect is concentrated at the bottom of the bioreactor. As a result there is significant heat transfer to the air, warming it up to such a degree that it is ineffective in cooling the top of the bed. This allows the top of the bioreactor to heat up since in this region the metabolic heat is not being removed as fast as it is produced. The cooling effect travels up the bioreactor like a "wave-front" (indicated by the region within the dotted ellipse in Fig. 10.7(b)). Under the conditions simulated, it takes around 20 min for the cooling effect to reach the top of the bioreactor, during which time the temperature has risen to a value over 2 °C higher than the value for packed-bed operation. Once this cooling "wave-front" arrives, the temperature returns to the value for packed-bed operation.

Pressure drops will typically not be a crucial problem in intermittently-mixed, forcefully-aerated bioreactors, since the intermittent mixing will tend to disrupt the inter-particle hyphae that develop during static periods and squash aerial hyphae onto the surface of the substrate particles. After a mixing event the pressure drop through the bed will typically be significantly smaller than the pressure drop before the mixing event. In some cases the mixing event has been triggered exactly for this reason, that is, to reduce the magnitude of the pressure drop across the bed.

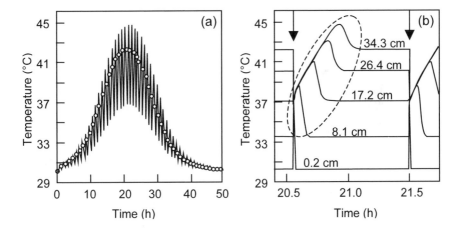

Fig. 10.7. Predictions of the modeling study of Ashley et al. (1999) about the temperatures reached in the intermittent-mixing mode of operation in a bioreactor or 34.5 cm height with a superficial air velocity of 0.0236 m s^{-1}. **(a)** Temperature profile predicted for a bioreactor mixed approximately every hour. At each mixing time the sensible energy in the bed is distributed evenly amongst the bed contents. The *hollow symbols* (o) represent the temperature profile expected for the absence of mixing events, that is, for simple packed-bed operation. **(b)** More detail of the temperature profiles at different heights in the bed, showing why the maximum bed temperature exceeds the value expected for packed-bed operation, which in this case is the value of 40.7 °C at the top of the bed immediately before the mixing event. The arrows mark the timings of the mixing events. The *dashed oval* shows how the "cooling wave-front" moves up the bed after a mixing event. Adapted from Ashley et al. (1999), with kind permission of Elsevier

10.5 Conclusions on Group IVb Bioreactors

Intermittently-mixed, forcefully-aerated bioreactors appear to have some potential, judging by the fact that several processes involving bioreactors that operate in this mode have been demonstrated at a reasonably large scale. They appear to offer some benefits in control of the conditions within the bed, while minimizing the deleterious effects that continuous mixing can have, at least for fungal processes.

Based on what is known to date, it would seem that the best strategy is not to try to use mixing of the bed directly as a temperature control strategy. For fungal fermentations such a strategy would lead to intolerably frequent mixing events. Rather, the mixing events should be used to:

- prevent undue aggregation of substrate particles, unduly high pressure drops, and the appearance of cracks and channels in the bed;
- replenish water in the bed in order to prevent low water activities in the bed from being one of the factors that limit growth.

Attempts to control the temperature in such bioreactors therefore should be focused on manipulation of the temperature, humidity, and flow rate of the inlet air. These have not been explored to any great extent, but Chap. 25 will present a mathematical model of an intermittently-mixed, forcefully-aerated bioreactor that can be used to explore the question of how best to operate such bioreactors in order to control the temperature.

Further Reading

Studies regarding the effect of intermittent mixing on aggregation of the substrate by fungal hyphae
Schutyser MAI, de Pagter P, Weber FJ, Briels WJ, Boom RM, Rinzema A (2003) Substrate aggregation due to aerial hyphae during discontinuously mixed solid-state fermentation with *Aspergillus oryzae*: Experiments and modeling. Biotechnol Bioeng 83:503–513

More detailed descriptions of the INRA Dijon and PUC Chile bioreactors
Agosin E, Perez-Correa R, Fernandez M, Solar I, Chiang L (1997) An aseptic pilot bioreactor for solid substrate cultivation processes. In: Wise DL (ed) Global environmental biotechnology. Kluwer Academic Publishers, Dordrecht, pp 233–243
Durand A, Chereau D (1988) A new pilot reactor for solid-state fermentation: Application to the protein enrichment of sugar beet pulp. Biotechnol Bioeng 31:476–486
Durand A, Renaud R, Maratray J, Almanza S, Diez M (1996) INRA-Dijon reactors for solid-state fermentation: Designs and applications. J Sci Ind Res 55:317–332
Fernandez M, Perez-Correa JR, Solar I, Agosin E (1996) Automation of a solid substrate cultivation pilot reactor. Bioprocess Eng 16:1–4

11 Continuous Solid-State Fermentation Bioreactors

Luis B. R. Sánchez, Morteza Khanahmadi, and David A. Mitchell

11.1 Introduction

The previous chapters have presented solid-state fermentation (SSF) bioreactors that operate in batch mode. Although batch operation is the most common type of operation in SSF processes to date, it is also possible to design and operate continuous SSF bioreactors. However, there are challenges faced in the operation of continuous SSF bioreactors that are not faced in classical continuous submerged liquid fermentation (SLF) processes and, consequently, true continuous-flow SSF bioreactors (CSSFBs) are currently scarce in industry. Improved design procedures and sensors promise a better future for these bioreactors.

This chapter deals with the design and operation of continuous SSF bioreactors and discusses the potential advantages that continuous operation can bring to SSF processes and also the various considerations that need to be addressed in order to arrive at a well-performing continuous process.

11.2 Basic Features of Continuous SSF Bioreactors

11.2.1 Equipment

In general, continuous chemical reactors can be classified into one of three groups: stirred tank reactors, tubular flow reactors, and designs that are between these two types (i.e., which combine some characteristics of both stirred and tubular flow reactors). This is also true for CSSFBs. Readers interested in exploring possible designs further are encouraged to consult references that deal with equipment used for mixing of solids (Sastry et al. 1999) and for feeding of solids (Bell et al. 2003), many of which could be adapted to act as CSSFBs.

In this chapter we will discuss three possible CSSFB designs: the Continuous Stirred Tank Bioreactor (CSTB), the Continuous Rotating Drum Bioreactor (CRDB), and the Continuous Tubular Flow Bioreactor (CTFB). The principles of operation of screw conveyors and belt conveyors are the same as those of CTFBs, so in this chapter these conveyor bioreactors will be used as examples of this group.

11.2.1.1 Continuous Stirred Tank Bioreactors (CSTBs) for SSF

Continuous Stirred Tank Bioreactors are designed to mix the whole content of the bioreactor thoroughly. In the ideal case, mixing is said to be *perfect* which means all the properties are identical everywhere inside the vessel at a given time.

In the case of SSF processes, it is impractical to *mix perfectly* due to two limitations imposed by the solid nature of the system. Firstly, wet solids have limited capacity for flowing and this makes mixing difficult. Secondly, wet solids tend to show a significant degree of flow segregation. The term *flow segregation* refers to the tendency of particles that have been in the vessel for different periods of time to remain segregated in different groups (Fogler 1999). As a result, any CSTB that is used for a continuous SSF process will behave to some degree as an intermittently mixed bioreactor. Despite these problems, perfect mixing behavior remains as an ideal model that serves as a paradigm for the analysis and design of these systems as we will see later in this chapter.

Note that there is a further limitation on *perfect mixing*. In *perfectly mixed* CSTBs for SLF, mixing is perfect even at the molecular level. However, in SSF, the bed of solids cannot be mixed at the molecular level unless the solid substrate particles are completely destroyed. If the solid particles are to remain intact, then *perfect mixing* can only occur at the "supra-particle" scale, with no mixing at the intra-particle scale. Further, transfer of liquid or biomass between particles will typically be quite limited. As a result, in SSF, even for a *perfectly-mixed* CSTB, each particle essentially acts as a "batch micro-bioreactor".

The main design variables for CSTBs are:

- The geometry of the vessel. Figure 11.1 shows a conical geometry that could favor both the mixing of the solids within the bioreactor and their flow through the bioreactor. The height to diameter ratio of the vessel and the way it is positioned (i.e., vertical, inclined or horizontal) will influence the agitation devices that should be used and also the portions of the flow that will be moved as plug-flow and as perfectly mixed flow.
- The availability of heat transfer devices. Temperature control is easier in this bioreactor because of mixing, so different approaches could be explored, such as the use of water jackets or water-cooled impellers.
- The design of the aeration system. The air can be circulated through the headspace or blown forcefully through the bed. If blown through the bed, the air flow can be in the same direction, in the opposite direction or normal to the solids flow. Of course the aeration system can be designed to allow changes in the direction of air flow during the process.
- The type and number of impellers. The solids mixing efficiency depends strongly upon the type of impeller used (Sastry et al. 1999). A careful study should be conducted to select the appropriate design and positioning.

- The features of the solids addition and removal devices. They may need to be designed to prevent the entry of contaminants into the bioreactor. The solids inlet and outlet should be designed and positioned in order to minimize the possibility of short-circuiting. That is, added solids should be mixed into the bed and should not simply flow directly from the solids inlet to the solids outlet. In the case of external recycling of part of the solids that exit the vessel, the design of the recycling system must prevent contamination and mix the recycled solids well into the fresh solids stream.
- The features of equipment for addition of water and nutrients. A large part of the metabolic heat may be removed via evaporation, in such cases continuous or semi-continuous water replenishment will be required. The equipment for makeup water distribution should be designed to allow an even distribution. Minerals and soluble carbon sources can be added by the same system.

The operating variables include:

- The dilution rate. This is defined as the ratio of the mass flow rate into the bioreactor to the total mass of solids within the bioreactor. It is a key factor in optimizing the productivity of the process and maximizing the concentration of products. Theoretically *washout* flow could occur, in a similar manner to that which occurs in continuous CSTB processes in SLF.
- Impeller velocity and frequency of stirring. These factors will influence the quality of mixing and will be very important in determining the distribution of temperatures and concentrations within the vessel.

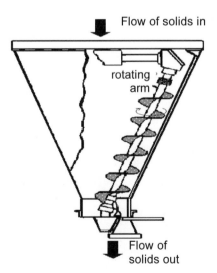

Fig. 11.1. Schematic representation of a Continuous Stirred Tank Bioreactor (CSTB) that could be used for SSF processes. Note that air could be blown into this bioreactor either at the top or at the bottom. Alternatively, it may even be possible to blow air into a hollow mixing device, with appropriately positioned holes allowing the air to pass into the bed

11.2.1.2 Continuous Rotating Drum Bioreactor (CRDB)

This bioreactor is similar to those of the stirred tank group but differs in the manner in which mixing is achieved: the CRDB consists of a cylinder that rotates horizontally around its axis. Bioreactors of this kind fall between perfectly mixed bioreactors and plug-flow bioreactors and hence might be referred to as mixed-flow bioreactors. Indeed, as in solid-drying equipment of this shape (see Moyers et al. 1999), they can have internal devices that promote forward and backward mixing. These devices could be static mixers, like the baffles in Fig. 11.2, or dynamic mixers, which stir and transport the solid internally within the vessel.

The design and operating variables of CRDBs are similar to those of CSTBs. Nevertheless the fact that the drum rotates without the motion of an internal agitator produces particular features in the stirring mechanisms. The number, shape, and position of the baffles are important factors that affect the flow through the drum and consequently the performance of this bioreactor.

In addition to heat removal by convection to the air flowing through the headspace, different strategies can be tried for removal of waste metabolic heat from the bed of fermenting solids. For example, the lower part of the external wall of the vessel could be immersed in a water bath.

The speed of rotation of the drum and the angle of inclination of the body of the bioreactor to the horizontal are very important factors affecting solids mixing and transportation. Rotational speeds as low as 2 to 3 rpm are commonly used in batch systems (Hesseltine 1977; Pandey 1991), although higher speeds have also been reported. The substrate normally occupies 10% to 40% of the volume of the bioreactor (Stuart 1996).

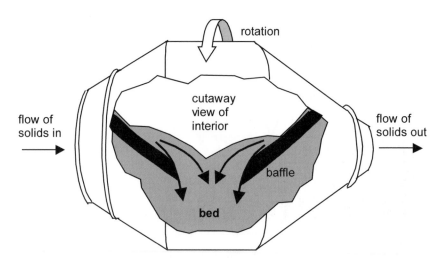

Fig. 11.2. Continuous Rotating Drum Bioreactor, which, in terms of solids-flow regimes, is placed between perfectly-mixed bioreactors and plug-flow bioreactors

Van de Lagemaat and Pyle (2001) used a 1-m-long CRDB with a diameter of 8 cm. By adjusting the baffle arrangement and the inclination of the central axis of the bioreactor to the horizontal, they achieved near-perfect mixing of un-inoculated solid substrate particles. The main goal of this design was to achieve sufficient back mixing so that the sterile feed could be inoculated by the fermented particles within the bioreactor, in such a manner as to remove the need for an external inoculation system. However, the efficiency of such "back-inoculation" has not yet been directly investigated. Further, as will be explained later, there may be problems with product uniformity in back-mixed CSSFBs.

A special kind of CRDB was tested for fermentation of a mixture of feedlot waste and coarsely cracked corn (Hrubant et al. 1989). The 91.5-cm-long bioreactor had a diameter of 22.8 cm and consisted of three chambers aligned axially and separated by bulkheads. Each bulkhead had a centrally located hole to permit unidirectional passage of fermenting substrate sequentially through the chambers. Each chamber had several baffles to ensure perfect mixing of the fermenting solids within the chamber and therefore this bioreactor acted like three perfectly-mixed continuous bioreactors in series. Fermentation runs as long as two months were conducted with this bioreactor. A pilot-scale bioreactor of this kind having three 468-liter chambers was also used at Illinois University.

11.2.1.3 Screw and Belt Conveyor Bioreactors

Screw conveyors and belt conveyors, which are examples of continuous tubular flow bioreactors (CTFBs), can move solids with almost zero mixing in the direction of flow (Fig. 11.3). When mixing is desired, which is often the case, static or dynamic mixers can mix the bed in the radial direction and, if desired, also in the axial direction. The current subsection focuses on the situation without axial mixing. Due to lack of back mixing, internal back-inoculation is not possible, however, external inoculation might be done by recycling a part of the fermented product, avoiding the need for a separate process for inoculum production.

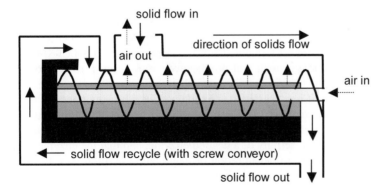

Fig 11.3. Screw bioreactor with recycling. The central axis is hollow and perforated, to allow the flow of air into the bed. The screw blade is mounted on this axis, which rotates

Gibbons et al. (1984, 1986) investigated a continuous screw-type bioreactor for farm-scale fuel ethanol production from various solid substrates such as fodder beets and sweet sorghum. Their system was not aerated and only anaerobic or microaerophilic fermentations could be carried out. Moreover, the void spaces between the solid particles contained significant quantities of liquid, meaning that the process actually represented a borderline case between SSF and a "slurry fermentation". In any case, their bioreactor could be adapted for true SSF processes, although an aeration system would need to be incorporated for the cultivation of aerobic organisms.

Some of the large-scale *koji* production bioreactors can work in this mode. For example, the rotary disk bioreactor shown in Fig. 10.2 can be operated in a manner in which the rotating disk acts as a circular conveyor belt. As the disks slowly rotate, particles are transferred from the upper disk to the lower disk. The empty space on the upper disk is then filled with freshly inoculated particles. Each particle entering the upper disk spends the same time before being transferred to the lower disk. Each particle entering the lower disk then spends the same time before being harvested. Production rates as high as 4150 kg h^{-1} have been reported (Yokotsuka 1985; Chisti 1999). Tower-type CSSFBs used in certain composting processes operate in a similar manner, with a semi-continuous flow of substrate from one chamber to the next.

11.2.2 Flow Patterns: Real-Flow Models

As for any other continuous chemical reactor, the flow of materials from the inlet to the outlet of a CSSFB could potentially fall anywhere between the ideal plug-flow and perfect-mixing regimes. In plug flow, all of the particles have the same residence time within the bioreactor as they move along parallel paths at the same speed. On the other hand, if the particles are mixed parallel to the direction of flow, they may spend different lengths of time in the reactor. In other words, different particles may have different residence times. In a completely mixed reactor, the residence time distribution for the population of particles is wide, some particles may exit almost immediately after they enter, while some other may remain within the reactor for longer times.

Theoretically, in order for all fermented substrate particles exiting a CSSFB to have the maximum possible growth and product formation, each particle should spend the same amount of time in the bioreactor between when it is inoculated at the solids inlet and when it is harvested at the solids outlet. The importance of this can be seen in a simple example. Let us assume that it requires 24 h, measured from the time of inoculation, for the microorganism on a particular substrate particle to produce the maximum activity of a desired enzyme, and that the enzyme activity falls off after 24 h due to denaturation or degradation by proteases. In this case, any particles exiting a CSSFB with residence times lesser than or greater than 24 h will have an enzyme activity less than the maximum possible value. In such a case it would be desirable for the residence time distribution to be as narrow as possible, with a mean of 24 h.

This is the consideration of "uniformity amongst harvested substrate particles", which is desirable, but may be difficult or impossible to achieve in practice. Note that "true" continuous operation does guarantee a uniform product regardless of whether all substrate particles have the same residence time or not, but in this case the concept of uniformity is applied differently: if a CSSFB does manage to establish a steady state, then the exiting product will have a uniform composition, averaged over the population of exiting substrate particles. In other words, the proportions of "young" substrate particles and "old" substrate particles in the harvested product will remain constant over time for true continuous operation, regardless of the flow regime and residence time distribution of the particles. However, in terms of bioreactor productivity, the exiting of a mixture of younger and older particles is disadvantageous when compared to the exiting of a uniform population of "fully-fermented" particles. There is a further consideration: heterogeneity of the inlet raw material and the presence of non-ideal flow patterns, dead volumes, air channeling, and solids short-circuiting may all contribute to fluctuations in the quality of the product exiting a CSSFB. These issues have received very little attention in SSF.

The wideness of the residence time distribution depends on the direction and extent of mixing (Fig. 11.4). Mixing of fermenting solid particles perpendicular to their flow direction in a bioreactor would typically be desirable, especially if the bed were forcefully-aerated with the air flow being perpendicular to the solids flow direction. In the absence of vertical mixing in the bioreactor shown in Fig. 11.4, undesirable temperature and moisture gradients, similar to those noted for packed-bed bioreactors (Chap. 7), would arise along the direction of the air flow. If it were possible to mix the solids in this bioreactor vertically (i.e., perpendicular to the solids flow direction) without any horizontal movement of the solids (i.e., parallel to the solids flow direction), then such mixing would have no influence on the residence time of the solid particles. However, this is an ideal that is impossible to achieve in practice: Mixing perpendicularly to the solids flow direction will also cause some mixing parallel to the solids flow direction. Mixing parallel to the solids flow direction, often called flow dispersion, leads to a broadening of the residence time distribution pattern.

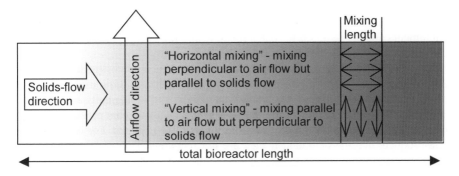

Fig. 11.4. Two main mixing directions in continuous SSF bioreactors

The amount of flow dispersion will depend on the design and operation of the bioreactor and the mixer and on the number of mixers used. It may be assumed to be roughly proportional to the sum of mixing lengths divided by total bioreactor length (see Fig. 11.4). For example, in a CSSFB composed of a single long belt carrying fermenting solids from inlet to outlet, the bed might not be mixed at all or it may be mixed occasionally so that sum of mixing lengths is negligible compared with total belt length. On the other hand, in a CSSFB composed of a rotating drum with curved baffles, the mixing length may be equal to total bioreactor length, leading to a wide residence time distribution.

11.3 Continuous Versus Batch Mode of Operation

11.3.1 Reduction of Upstream and Downstream Investment

In the batch mode of operation, a quantity of feed equal to the bioreactor capacity must be ready for loading at the start of each cycle. Chemical changes cannot be prevented when moist solid substrates are stored for long times, and there is always the danger of the growth of contaminants, so it is not feasible to prepare and cook the substrate gradually. Neither is it feasible to inoculate the substrate gradually if the bioreactor is to be operated in batch mode. Hence, the upstream equipment for substrate preparation and inoculation must be large enough to be able to prepare the whole batch of required substrate within a few hours before the start of each cycle. In contrast, in the continuous operation mode, smaller equipment can be used to process, on an hourly basis, the smaller substrate quantities that are fed into the bioreactor. In this manner, continuous operation can reduce the investment in upstream processing equipment. The degree of reduction becomes greater as the cycle time is increased. For example, suppose that a bioreactor with a capacity of 1000 kg is used in batch mode in a SSF process that has a 50-hour fermentation time. If the substrate is required to be prepared no sooner than 10 h before the start of each fermentation, then the capacity of the upstream equipment must be 100 kg h^{-1}. On the other hand, if the same bioreactor were used in continuous mode, then the required capacity of the upstream processing equipment would be 20 kg h^{-1}. Moreover, if the fermentation time were 100 h, the upstream equipment capacity required for batch mode would not change, while that for continuous mode would be reduced to 10 kg h^{-1}.

In the same manner, continuous operation will reduce downstream equipment costs. Continuous operation of the bioreactor will require continuous downstream processing since the fermented solids are chemically and biologically active and if stored for long times before a large batch is sent for downstream processing, then the fermentation may continue, leading to undesirable changes. For example, labile products may be degraded if they are not recovered from the solid medium soon after the fermentation. Consequently, fermented solids leaving the fermenter must be processed as soon as possible to the final product or stabilized via means such as drying or freezing. In a continuous system in which the fermented solids exit the bioreactor gradually, relatively small equipment could process them into the final product or a stabilized product. This contrasts with batch operation, in which a large amount of fermented solids is discharged from the fermenter during a very short time period, requiring equipment with a large capacity in order to minimize storage time.

Replacing a single large batch fermenter with several smaller ones having staggered start times would reduce the required capacity and cost of upstream and downstream equipments. However, this would be accompanied by an increase in the investment required in the bioreactor section of the process.

11.3.2 Uniformity of the Product from Batch and Continuous Bioreactors

Continuous operation permits a more uniform product than batch operation, especially in cases in which the solid bed is mixed only intermittently. This can be seen by comparing an intermittently-mixed, forcefully-aerated bioreactor operated in batch mode with the same bioreactor type operated in continuous mode. Note that, as shown in Fig. 11.5, in both bioreactors the air flow is perpendicular to the flow of solids. As described in Chap. 10, this design has been successfully proven in batch bioreactors at pilot scale. Note that the intermittent mixing is necessary in order to break up aggregates of solid particles and also to allow the replenishment of evaporated water.

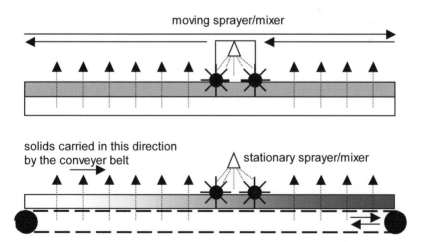

Fig. 11.5. Mixing schemes in intermittently-mixed, forcefully-aerated bioreactors operated in **(a)** batch mode, in which the mixer moves back and forth along the whole length of the bioreactor; **(b)** plug flow continuous mode, in which the mixer stays in place and the bed moves past it

In batch bioreactors of this type, mixing is often performed by a moving mixer, although in some cylindrical *koji* bioreactors the mixing system is stationary and the bed is moved past it via rotation of the base plate. In both cases, simultaneous mixing of the whole bed is difficult to achieve. In a system such as that shown in Fig. 11.5(a), parts of the bed located to the left of the mixer are mixed soon after the start of mixing while parts located to the right of the mixer are mixed only after a lag time that depends on the bed length and the speed with which the mixer travels back and forth along the bed. For large-scale bioreactors the lag time may become considerable, especially in cases in which the mixer travels slowly in order to enable homogeneous distribution of added water. Such lag times could have adverse effects on the product uniformity. Use of several mixers could reduce time lags in the batch mode but will imply a more expensive mixing system. On the other hand, in the continuous system shown in Fig. 11.5(b), the mixer stays in place as the substrate is moved past it and all of the fermenting solids are mixed or wetted at the same time interval after their entrance into the bioreactor, leading to more uniform product.

11.3.3 Enhanced Production Rates

Changing of the mode of bioreactor operation from batch to continuous saves the time required for loading, discharging, and cleaning of the bioreactor, since in continuous operation loading and discharging proceed simultaneously with the fermentation whereas in batch processes the bioreactor is not producing product while these "turnaround" operations are taking place. Assuming the same fermentation time for both batch and continuous operations, the saved time means that the volumetric productivity of the continuous plug-flow bioreactor is higher.

The extent of the increase in productivity depends on ratio of the turnaround time to the fermentation time. For example, assuming that the turnaround and fermentation times are 10 and 40 h, respectively, the volumetric productivity will increase by 25% upon changing from batch to continuous mode. The productivity increase is smaller for higher ratios of fermentation time to turnaround time. For example, with the same turnaround time of 10 h but a fermentation time of 70 h, the increase in productivity gained by changing from batch to continuous operation is only 14%.

As mentioned previously, replacing a single large batch bioreactor with several small ones operating in a staggered manner will reduce the difference in productivities. However, once again, it must be highlighted that this implies higher investment costs.

11.3.4 Contamination

The risk of contamination seems to be the major barrier to be overcome in the development of continuous SSF bioreactors.

In the batch mode of operation, the air flow and added water are the only streams that need to enter the bioreactor during the fermentation. It is typically not difficult to perform these operations aseptically. On the other hand, continuous operation involves a constant flow of a feed stream into the bioreactor and a product stream out of the bioreactor. It is more difficult to ensure that these operations are done aseptically, so the risk of contamination is higher in continuous operation than in batch operation. Moreover, the consequences of occasional contamination are more severe in continuous mode. Typically, the initial concentration of the contaminant is much less than that of the inoculated process organism. Hence, if the growth rate of the contaminant is not significantly higher than that of the process organism, it is not able to reach high concentrations before the fermentation terminates when the bioreactor is operated in batch mode. In the continuous plug-flow mode the situation may or may not be different. If the contaminant is simply carried along with the flow, it poses no greater a problem than it poses for batch operation. However, any back-mixing that occurs allows some particles to remain in the bioreactor for longer times and also it is possible for some contaminated particles to attach to stationary surfaces in a particular part of the bioreactor. This may allow sufficient time for the contaminant to reach high levels on some particles, which would act as seed for inoculation of the contaminant onto other particles. Note that in batch mode the attached contaminant is killed by sterilization operations carried out between successive runs while in continuous mode it can remain within the bioreactor and become a source for continuous contamination.

The situation would be more severe still in a well-mixed continuous fermenter, in which some particles have very long residence times. Of course, the severity of the problem would depend on the efficiency with which the contaminant was passed from particle to particle. However, it is possible that in this mode a contaminant may eventually conquer the whole bioreactor if it competes better than the process organism, even if the initial contamination level is very low.

Problems with contamination are often claimed to be less severe for SSF than for SLF. For example, it is often claimed that, in SSF processes using filamentous fungi, the water activity or pH of the substrate can be adjusted to values low enough to be unfavorable for most bacteria, although of course such conditions may not select against other fungi. With a fast-growing organism it may be sufficient to have a high density of vigorous inoculum and to provide optimum growth conditions early during the process in order to give the process organism a head start. In fact a large number of commercial SSF processes such as *koji* production and beet pulp protein enrichment are usually carried out under non-sterile conditions (Durand 2003). In those SSF processes in which the process organism has a selective advantage over any contaminants, contamination problems may in fact not be a serious barrier do continuous operation. However, for slow-growing microorganisms aseptic operation will be essential, and processes involving these organisms may be difficult to adapt to continuous operation due to contamination problems.

The acceptable degree of sterility depends also on the kind of product and legislative constraints. Pharmaceuticals should be produced under sterile conditions while *koji* can have a contamination of 10^9 bacteria per gram (Yokotsuka 1985).

So, from the point of view of contamination problems, continuous operation seems to be feasible for SSF processes in which fast-growing fungi are cultivated, provided that the product does not have to meet strict sterility standards. The degree of back mixing needs to be reduced to decrease the impact of any contamination. Moreover, internal surfaces should be highly polished. The temperature should be controlled near the optimum growth temperature of the process organism. If possible, water activity and pH should be kept as low as possible while not unacceptably retarding growth of the process organism.

11.4 Comparison by Simulation of the Three CSSFBs

Detailed experimental information on the performance of CSSFBs is not available. With this lack, simulation is a useful tool in understanding the potential of the various bioreactors. Note that almost no attention has been given to the modeling of the continuous operation of SSF bioreactors in the literature. The intention of the present section is to present simple models, while recognizing that many improvements in these models will be necessary in order for them to describe continuous performance reliably. For example, the models presented here for the mixed bioreactors do not take into account the fact that each particle is a batch micro-bioreactor and therefore will be most appropriate for very small particle sizes.

The different systems described above will be simulated using different flow models. The kinetic information has been taken from Ramos-Sánchez (2000), in which the logistic model is used to describe the growth kinetics of the yeast *Candida utilis* for the enrichment of sugarcane byproducts. The logistic model is frequently used to describe the growth kinetics in SSF (see Sect. 14.4) hence it is interesting to simulate the behavior of these systems using this kinetic model. The parameters of this model are the initial biomass content (X_o), the maximum possible biomass content (X_{max}), and the specific growth rate constant (μ). In these simulations X_o is set at 2.5 g kg-dry-matter^{-1}, X_{max} is set at 263 g kg-dry-matter^{-1}, and μ is set at 0.3 h^{-1}.

Constant temperature is assumed; heat and mass transfer phenomena are not modeled. The performance of each system is evaluated on the basis of the productivity of single-cell biomass (g-biomass kg-dry-matter^{-1} h^{-1}).

11.4.1 Continuous Tubular Flow Bioreactors (CTFBs) with Recycling

The operation of tubular flow SSF bioreactors with recycling, such as those shown in Figs. 11.3 and 11.5, can be simulated using a plug-flow bioreactor with a recycle stream (Fig. 11.6). The operating variables are the dilution rate (kg-solids kg-solids^{-1} h^{-1}), defined as in Sect. 11.2.1.1, and the ratio of recycled solid-flow to entrance mass-flow ($\gamma = f_R/F$), the so-called "recycle ratio" (dimensionless).

The results of the simulation are presented in Fig. 11.7 from which it is possible to conclude that:

- There is an optimal dilution rate above which the productivity falls rapidly with increasing dilution rate, as is characteristic of continuous SLF processes.
- Since the feed is inoculated with biomass, there is always biomass in the exit stream, regardless of dilution rate. Note that the graph therefore appears different from the graphs for "classical" continuous SSF processes in which there is no biomass in the feed and therefore above a critical dilution rate the steady state biomass concentration is zero. Of course, if a continuous SLF process were to have a certain level of inoculum in the inlet stream, then at dilution rates greater than the critical rate, the outlet stream would have a concentration of biomass equal to the inlet concentration, in the same manner as occurs for the situation in the CSSFB shown in Fig. 11.7 at high dilution rates.

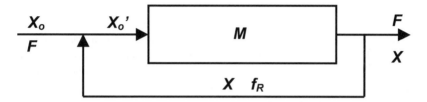

Fig. 11.6. Flow-model for screw conveyor and belt conveyor bioreactors with recycling, these being examples of Continuous Tubular Flow Bioreactors (CTFBs). M represents the mass of solids in the bioreactor, F represents the inlet flow rate, f_R the recycle flow rate, and X represents the biomass concentration

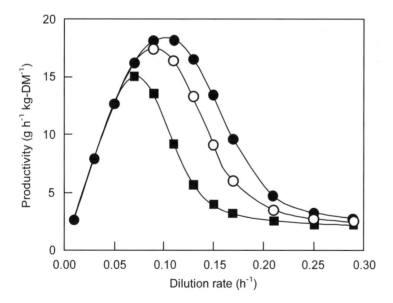

Fig. 11.7. Simulation of a Continuous Tubular Flow Bioreactor (CTFB) with recycling at different dilution rates and recycling ratios. Key: (■) $\gamma = 0.1$; (○) $\gamma = 0.3$; (●) $\gamma = 0.5$

- For dilution rates less than the optimal one, the fraction of mass-flow recycled back to the entrance has no influence on the productivity of the bioreactor.
- If high levels of the product are the main objective, the system can operate at a low dilution rate and with a low recycle ratio. In these cases the task is to find the optimal dilution rate, this being constrained by the minimum acceptable product concentration. The recycle ratio will not be of great importance.
- When high productivities are necessary at high dilution rates, the system will demand greater recycling ratios. The design problem in this case is more complicated and would include finding a combined optimum for both variables, namely the dilution rate and recycle ratio.

11.4.2 Continuous Rotating Drum Bioreactor (CRDB)

One of the many possible flow models for describing the micro-mixing inside a CRDB has been presented by Ramos-Sánchez et al. (2003). The pattern that describes the micro-mixing and, consequently, the behavior of this bioreactor, is a combination of a plug-flow reactor, a perfectly mixed reactor, and a recycle stream (Fig. 11.8). There are three main operating variables for such a system: The fraction of the flow that passes through the plug-flow bioreactor ($\alpha = f_p/f_m$), the fraction of the "in-bioreactor" mass that is contained by the plug-flow bioreactor ($\beta = M_p/M_m$), and the fraction of the flow that it is recycled back to the entrance of the CRDB ($\gamma = f_R/F$).

Figure 11.9 shows the simulations for a given set of α and β at different values of the dilution rate and recycled fraction γ. The behavior is similar to that shown in Fig. 11.7, but some important differences should be pointed out:

- Above the productivity maximum, the decrease in productivity with dilution rate is less pronounced than it was in the case of the former bioreactor (compare the profiles in Figs. 11.7 and 11.9). This means that the operation in this region is more stable, which is more desirable for practical purposes. In fact, for high dilution rates, for example, greater than 0.15 h^{-1}, the CRDB will have higher productivities than the CTFB.

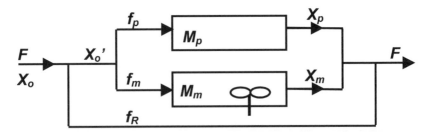

Fig. 11.8. Combined flow-model of a CRDB. M_p is the mass of solids in the plug-flow region while M_m is the mass of solids in the well-mixed region

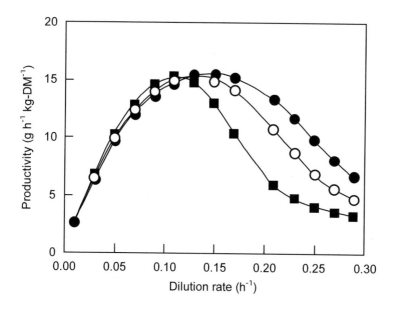

Fig. 11.9. Simulation of a Continuous Rotating Drum Bioreactor (CRDB) where $\alpha = 0.30$ and $\beta = 0.40$. Key: (■) $\gamma = 0.1$; (O) $\gamma = 0.3$; (●) $\gamma = 0.5$

- The maximum productivity is not as sensitive to increments in the recycle ratio (γ) as it was in the previous case, remaining between 15 and 16 g h^{-1} kg-dry-matter^{-1} as the recycle ratio is varied from 0.1 to 0.5.
- In the bioreactor simulated in Fig. 11.9, the maximum of productivity at a value of γ of 0.5 is 16% less than that in Fig. 11.7.

11.4.3 Continuous Stirred Tank Bioreactor (CSTB)

A real CSTB has a complex flow-pattern due to the solid nature of the system and the limitations on stirring imposed by the sensitivity of the microorganisms to shear damage. However, given that the flow patterns in such bioreactors have not been studied, a model assuming perfectly mixed flow is used for the simulations. Note that it is assumed that the particles are inoculated as they enter the bioreactor.

In the case in which there is no recycle stream, the dilution rate is the only operating variable. Figure 11.10 shows the results of the simulations for this bioreactor as a function of dilution rate, together with simulations of the two former bioreactors at a recycle ratio of 10%. The behavior of the CSTB is similar to that of the previous bioreactors but some important differences should be noted:

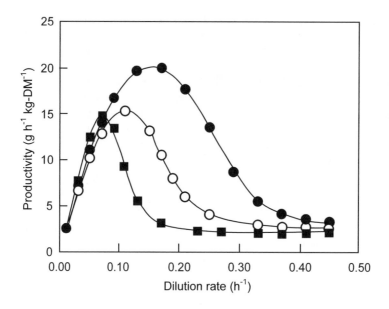

Fig. 11.10. Comparison of the performances of (●) a CSTB, (■) a CTFB ($\gamma = 0.1$), and (○) a CRDB ($\gamma = 0.1$)

- Surprisingly, the results of the perfectly-mixed CSTB are better than those the other two bioreactor types. Normally plug-flow bioreactors are better for simple reactions (Fogler 1999) but in the case in which the reaction rate increases with conversion, which could be the case for fermentation processes, which are autocatalytic, the perfectly mixed bioreactor can perform better.
- The CRDB has an intermediate behavior due to the fact that it combines perfectly-mixed flow with plug-flow. The greater the perfectly mixed component is, then the closer the performance of the CRDB will be to that of the CSTB.
- The CTFB tends to behave more and more like a CSTB as γ rises, which can be seen by comparing Figs. 11.7 and 11.10.

11.4.4 Evaluation of the Various CSSFB Configurations

Figure 11.11 shows the relationship between two important performance criteria, namely the bioreactor productivity and the biomass concentration in the product stream. The relationship is plotted for each of the three configurations of CSSFBs presented in Sects. 11.4.1 to 11.4.3, based on the results of the various simulations performed in those sections. Various points of interest are:

- The CSTB will have maximum productivity when the outlet biomass concentration is a half of X_{max}, while the maximum productivity of the CTFB occurs at greater biomass concentrations.

- The maximum productivity of the CSTB with perfectly mixed flow is 30% greater than that of the CTFB. However, the advantage of the CSTB over the CTFB becomes smaller as the recycle ratio of the CTFB is increased.
- For biomass concentrations up to 200 g kg-DM^{-1}, which is very close to the biomass concentration of 220 g kg-DM^{-1} that gives maximum productivity of the CTFB, the productivity of the CSTB is greater than that of the CTFB.
- The behavior of the CRDB is between these two ideal bioreactors. This is not surprising, because it represents a mixture of the two flow regimes. Indeed, the model of this bioreactor can represent the deviations of flow regimes from the ideal regimes assumed for the CTFB and the CSTB.

In the case of plug-flow through a tubular bioreactor, the reaction rate will be low at the entrance of the bioreactor because of the low concentration of biomass. As the solids flow through the bioreactor, the rate of the reaction will rise to a maximum level at a biomass concentration equal to $0.5X_{max}$, due to logistic growth kinetics, which cause the growth rate to decelerate as the biomass concentration rises from $0.5X_{max}$ towards X_{max}. Therefore at the exit of the plug-flow bioreactor, if the biomass concentration is close to X_{max}, the rate of the reaction will tend to be low. This means that the average reaction rate within the plug-flow bioreactor will always be lower than the maximum possible level; hence as a consequence, the overall productivity will never be as high as it would be for a CSTB in which the biomass concentration were maintained at $0.5X_{max}$.

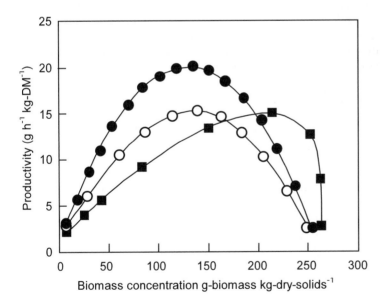

Fig. 11.11. Relation between productivity and biomass concentration in the simulation of a (●) CSTB, (■) a CTFB ($\gamma = 0.1$), and (○) a CRDB ($\gamma = 0.1$)

Finally, we should note a difference between the operation of CSTBs in SLF and SSF. In SLF it is practical to have a recycle stream for a CSTB, since it is possible to centrifuge or filter the stream exiting the bioreactor, such that the recycle stream has a higher biomass concentration than the stream exiting the bioreactor while the product stream leaving the process has a lower biomass concentration than the stream exiting the bioreactor. The important point is that in SLF it is possible to separate, at least partially, the biomass and the liquid. In contrast, in SSF the biomass in the stream exiting the bioreactor cannot be separated from the solids. If any recycling is done, then the composition of the recycle stream and the product stream leaving the process will have compositions identical to that of the stream leaving the bioreactor. Therefore, there are no advantages, in terms of productivity, in recycling solids in a CSTB. In fact, solids recycling is only useful for inoculation of the incoming fresh solids, and this will only be effective if there is efficient inter-particle transfer of biomass.

11.5 Scientific and Technical Challenges for CSSFBs

Continuous solid-state fermentation shares many of the challenges that are faced by SSF processes operated in the batch mode but it also has its own features. It has not received much attention in the literature. In order to understand the possible advantages and limitations of this mode of operation, it will be necessary to

- develop flow models that more realistically describe the flow patterns within the various designs;
- incorporate heat and mass balances into the models of continuous operation;
- recognize the fact that each particle acts much like a "batch micro-bioreactor", this being quite different from the situation in SLF where perfect mixing is assumed down to the molecular level;
- understand the dynamic behavior of these systems, in order to develop appropriate start-up strategies for continuous operation and also to control the process, minimizing oscillations in the product quality.

Further Reading

Contains a good treatment of the general principles of operation of continuous bioreactors for submerged fermentations
Nielsen J, Villadsen J, Liden G (2003) Bioreaction engineering principles, 2nd edn. Kluwer Academic/Plenum Publishers, New York

A recent example of use of a continuous process in SSF
van de Lagemaat J, Pyle DL (2004) Solid-state fermentation: A continuous process for fungal tannase production. Biotechnol Bioeng 87:924–929

12 Approaches to Modeling SSF Bioreactors

David A. Mitchell, Luiz F.L. Luz Jr, Marin Berovič, and Nadia Krieger

12.1 What Are Models and Why Model SSF Bioreactors?

The key message of this book is that mathematical modeling is a powerful tool that can help in the design of SSF bioreactors and in the optimization of their performance. It is not necessary for all workers in the area of SSF to know how to construct and solve models, because modeling can be done in collaboration with colleagues with the appropriate expertise. However, even if you have no intention of undertaking the modeling work yourself, it is useful to know what models are and what they can do, because this facilitates interactions with these colleagues. The aim of Chaps. 12 to 20 is to give you an understanding of how models of SSF bioreactors are developed. These chapters do not attempt to provide the necessary background in all the mathematical and computing skills required. Rather they attempt to convey the "modeling way of thinking". This will provide the basis for understanding the uses and limitations of the various models presented in the modeling case study chapters (Chaps. 22 to 25).

What is a mathematical model? The type of mathematical model that we are talking about in this book is a set of differential and algebraic equations that summarizes our knowledge of how a process operates. In other words, a model is a set of equations that describes how the various phenomena that occur within the system combine to control its overall performance, which, in the case of SSF bioreactors, will be evaluated in terms of growth and product formation. A model is a simplification of reality, and the equations therefore only describe the phenomena that are thought to be the most important in influencing the performance of the system. It is the modeler who, on the basis of experience with the system being modeled, decides which phenomena will be included and which will not be. As a simple example of this, amongst other factors, growth within an SSF bioreactor depends on both the O_2 concentration and the temperature experienced by the microorganism. However, in many models of SSF bioreactors the problem of controlling temperature is considered to be more difficult than the problem of supplying O_2, and therefore frequently equations describing energy generation and water transfer are written in order to predict temperatures, but equations to describe O_2 supply and consumption are not included within the model.

Of course, it is possible to make wrong decisions about which of the phenomena are most important, or to simply neglect to consider some phenomena that are important. If a model fails to describe the bioreactor performance well, it is essential to find out why it fails, and to then work to improve it.

The models of SSF processes that will be introduced in Chaps. 22 to 25 consist of differential equations that describe how key variables, such as biomass concentration or temperature, vary with over time and across space within a bioreactor during an SSF process. For example, a simple model of the operation of an SSF bioreactor might include equations to describe the rate of growth and heat production and the heat removal processes occurring. These equations would predict how the temperature of the substrate bed changes during the process, and the temperature would be taken into account in the calculation of the growth rate.

Models are a powerful way of summarizing our knowledge about how a system operates. When a system is as complex as an SSF process, we have a better chance of summarizing the complexity of the interactions with a model than if we simply looked at a large number of graphs of experimental results. However, models are more than simply a means of summarizing experimental data that describe system behavior. Models can be used to predict performance, and therefore can be used to identify optimal design parameters and operating conditions (Fig. 12.1). Consider the situation in which you are doing laboratory-scale work on an SSF process that is showing such promise that you intend to go to production scale. Models developed on the basis of this laboratory-scale work, combined with heat and mass transfer principles, can be used to forecast the performance of a large-scale bioreactor, before it is built. Even if the predictions are not fully accurate, this initial

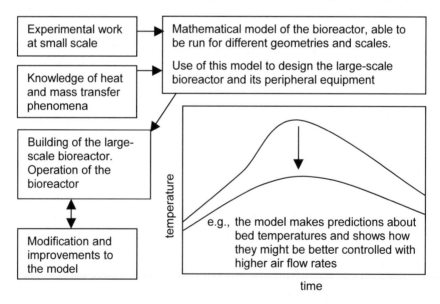

Fig. 12.1. An overview of how models can be used in the development of large-scale SSF bioreactors

modeling work has a better chance of leading to a large-scale bioreactor that operates successfully than do "best-guess" or "trial-and-error" approaches. Once the large-scale bioreactor is built and tested, the model can be modified with the new data generated at large scale, and the modified model can be used as a tool in optimizing bioreactor operation.

If powerful "off-the-shelf" bioreactor models were available, then you might never have to think about the "modeling process". However, the current SSF bioreactor models are simply not sufficiently sophisticated. Each research and development group will need to do its own modeling work, although of course this can be done by building on previous work. The point is that you will need to become involved in the modeling process, even if you do not undertake the mathematical and computing work yourself. The remainder of this chapter covers the very basic information that you need in order to understand what models are and how the modeling process operates.

12.2 Using Models to Design and Optimize an SSF Bioreactor

Figure 12.2 gives a more detailed view than Fig. 12.1 of how the design process should be carried out for production-scale SSF bioreactors, starting with the necessary laboratory-scale studies and ending with final optimization at large scale. It highlights the fact that it is ideally a process in which experimental and modeling work is undertaken simultaneously, with the mathematical model being refined constantly in the light of experimental evidence. The current section gives a broad overview of this bioreactor design process. It assumes that, after optimizing product formation by a particular organism on a particular solid substrate at laboratory scale, you have decided to develop a large-scale process.

12.2.1 Initial Studies in the Laboratory

Early studies will be needed in the laboratory to understand how the organism grows and how this depends on the environmental conditions that it experiences. On the basis of these studies, a growth kinetic model will be proposed (See Boxes 1 and 2 in Fig. 12.2). However, several questions must be asked before the experimental studies are planned. For example, what type of model will be used to model the growth kinetics? With what depth will it model the growth process? Will it simply describe biomass growth as a global value, such as g-biomass g-dry-solids^{-1}, or g-biomass m^{-3}? Or will it describe the spatial distribution of biomass at the particle level, for example, describing the biomass concentration as a function of height and depth above and below the particle surface? In answering these questions, it is important to consider that any decision that increases the complexity of the model may bring subsequent difficulties not only in solving it, but also in measuring all the necessary model parameters. These difficulties must be balanced against an evaluation of the potential advantages of improved predictive power that can be gained by describing the phenomena in greater detail. The appropriate level of detail for modeling growth kinetics within SSF bioreactors is discussed in greater depth in Chap. 13. It is only after these decisions have been made that the experimental program is planned. The experiments are planned

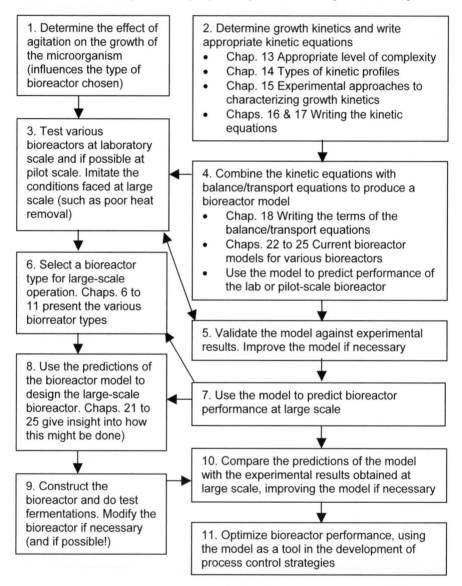

Fig. 12.2. Details of the strategy for using models as tools in the design and optimization of operation of SSF bioreactors

in such a way as to enable the development of mathematical expressions relating the growth rate to the various environmental variables. The way in which these experiments might be done and the types of mathematical expressions that might be used are described in Chaps. 14 to 17.

Ideally, various bioreactor types should be tested experimentally at laboratory scale, and, in fact, preferably at pilot scale, although few laboratories have sufficient resources to build laboratory-scale prototypes of all the possible bioreactor types, let alone pilot-scale prototypes (Fig. 12.2, Box 3). At the very least, experiments should be done in which some cultures are left static and others are submitted to various agitation regimes of different frequency, duration, and intensity. The results will be very useful in guiding bioreactor selection and determining the agitation regime to be used in the fermentation.

12.2.2 Current Bioreactor Models as Tools in Scale-up

Mathematical models have already been proposed for the various bioreactor types that are used in SSF. It makes sense to take advantage of these models, imperfect as they are (Fig. 12.2, Box 4). At this stage, it is quite likely that many of the parameters, such as transfer coefficients and substrate bed properties, will simply be based on literature values for similar systems. It may be appropriate to improve one or more of the models (Fig. 12.2, Box 5). Ideally laboratory-scale bioreactors should be operated in such a way as to mimic any limitations that will prevail at large scale, and the model predictions carefully validated against performance of these bioreactors. Disagreements between predicted and real performance should stimulate an investigation into the cause, which might be the mathematical form of the equations, but could also be the values used for some of the model parameters.

Simulations with the models will point to which bioreactor has the best potential to provide appropriate control of bed temperature and water content at large scale (Fig. 12.2, Box 6). Once a bioreactor has been selected, the appropriate model then represents a very useful tool for making decisions about design (e.g., geometric aspect) and operating conditions (e.g., air flow rate) (Fig. 12.2, Boxes 7 and 8). Careful attention must be given to the question as to whether the operating conditions necessary for good performance in the simulations are practical to achieve at large scale.

It is advisable to proceed to a scale that is intermediate between the laboratory scale and the final production scale, although this has not always been done. In any case, once a larger scale version of the selected bioreactor has been built, it is essential to validate the model again, since it is quite possible for the relative importance of the various heat and mass transfer phenomena to change with increase in scale (Fig. 12.2, Boxes 9 and 10). Phenomena that were not important at small scale and which were therefore not included in the model might suddenly become quite important at large scale. In this case the model will probably fail to describe large-scale performance with reasonable accuracy. If necessary, the model must be improved. Parameter values also must be determined with care. For example, it may be necessary to determine the bed-to-air mass transfer coefficient that is actually achieved within the production-scale bioreactor rather than to rely on estimates based on correlations given in the literature.

12.2.3 Use of the Model in Control Schemes

Once the bioreactor has been built with the help of the model, the model, improved in the light of data obtained at large scale, is still useful. It is highly likely that bioreactor performance will be significantly improved by implementing control strategies and the model can also play a useful role in the development of the control scheme (Fig. 12.2, Box 11). For example, the proposed control scheme can initially be tested and tuned with the model, which is obviously much cheaper than doing this initial testing and tuning with the bioreactor itself. The model may be embedded into the control system that is used to control the bioreactor.

12.3 The Anatomy of a Model

So models can and should play a central role in the development of large-scale SSF bioreactors. The remainder of this chapter gives an overview of the structure of mathematical models and the manner in which they are developed. The aim is not to teach those readers who do not have a background in modeling how to construct and solve models, but rather to increase their ability to interact with a modeling expert in the modeling process.

The structure of a model is presented in terms of a case study of a simple model of a well-mixed SSF bioreactor. Figure 12.3 shows the bioreactor, highlighting the various phenomena described by the model. Figure 12.4 shows the equations of this model, highlighting the fact that mathematical models of bioreactors contain two parts: the kinetic sub-model describes microbial growth kinetics, while the balance/transport sub-model describes transport phenomena and overall mass and energy balances. Work must be undertaken to generate data for both parts of the model.

Various symbols appear in these equations, representing different quantities. These quantities are of fundamentally different types, or, in other words, the various symbols represent a range of state variables, independent variables, operating variables and parameters. These are defined below.

State variables. These represent variable properties of the bioreactor, or the various phases within the bioreactor. For example, the state variables within the well-mixed SSF bioreactor model are the temperature of the substrate bed (T) and the amount of biomass in the bioreactor (X). They are called state variables because, together, the values for all these variables at a particular instant describe the state of the system at that instant. They are variables because their values vary as the independent variables change.

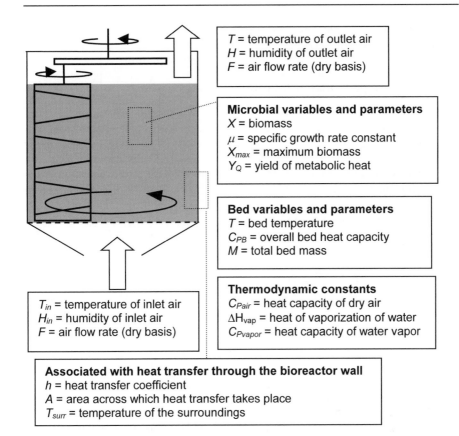

Fig. 12.3. A simple mathematical model for predicting the temperature within a well-mixed SSF bioreactor: The system modeled and the various variables and parameters involved in the model. Note that due to assumption of perfect mixing, the conditions within the bioreactor are equal to the outlet conditions

Independent variables. These represent variables that do not depend on the system and how it is operated. Rather the system depends on these variables. The independent variables that appear in models for SSF bioreactors are either time or space or both. In the current example it is assumed that the bioreactor is well mixed and therefore time is the only independent variable. In some cases the variations across space are significant while the variations in time occur only slowly. In this case, it might be appropriate to write the equations with space as the only independent variable, and the equation is referred to as a "pseudo-steadystate" equation. There are also bioreactors in which both the temporal and spatial variations are significant: the temperature at a specific position changes over time, and if the temperature is measured simultaneously at different locations within the substrate bed, the measured temperature varies with position. In this case both time and position appear as independent variables.

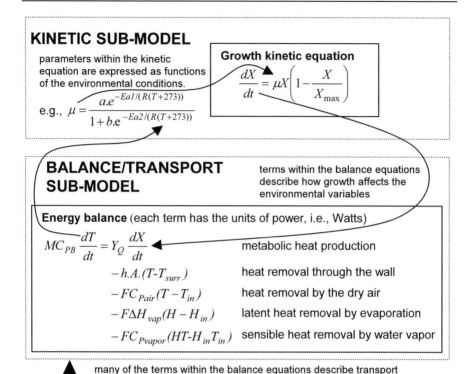

KINETIC SUB-MODEL

parameters within the kinetic equation are expressed as functions of the environmental conditions.

e.g., $\mu = \dfrac{a.e^{-Ea1/(R(T+273))}}{1+b.e^{-Ea2/(R(T+273))}}$

Growth kinetic equation

$$\frac{dX}{dt} = \mu X\left(1-\frac{X}{X_{max}}\right)$$

BALANCE/TRANSPORT SUB-MODEL

terms within the balance equations describe how growth affects the environmental variables

Energy balance (each term has the units of power, i.e., Watts)

$$MC_{PB}\frac{dT}{dt} = Y_Q\frac{dX}{dt}$$ metabolic heat production

$-h.A.(T-T_{surr})$ heat removal through the wall

$-FC_{Pair}(T-T_{in})$ heat removal by the dry air

$-F\Delta H_{vap}(H-H_{in})$ latent heat removal by evaporation

$-FC_{Pvapor}(HT-H_{in}T_{in})$ sensible heat removal by water vapor

many of the terms within the balance equations describe transport phenomena and these equations include the various operating variables

Operating variables
Variables, related with the aeration, agitation, and cooling systems, that can be manipulated by the operator. In this case the conditions of the inlet air (F, H_{in}, and T_{in}) and the temperature of the surroundings T_{surr} (which could be water in a cooling jacket)

Fig. 12.4. A simple mathematical model for predicting the temperature within a well-mixed SSF bioreactor: The model equations, showing the kinetic and balance/transport sub-models and their interrelations

Operating variables. These are variables that we can control the value of and which affect the performance of the bioreactor. We can use these in an attempt to control the state variables at their optimum values for the fermentation. In the current example, the operating variables are the conditions of the inlet air (F, H_{in}, and T_{in}) and the temperature of the surroundings T_{surr}.

Parameters. These represent various physical and biological properties of the system. They may be constants or their value at a certain time and position might depend on the state of the system (e.g., its temperature). In SSF systems there are various different types of parameters:

- design parameters, related to how the bioreactor was built. For example, in the current example, the area for heat transfer (A) is a design parameter.
- transport parameters, related to the transport of material and energy within and between phases. For example, in the current example, the coefficient for heat transfer between the bioreactor wall and the cooling water (h, J m^{-2} s^{-1} °C^{-1}) is a transport parameter.
- thermodynamic parameters, related to quantities of energy and the equilibrium state of materials. The enthalpy of vaporization of water (ΔH_{vap}) is one of the thermodynamic parameters in the current example.
- biological parameters, related to the behavior of the microorganism. In the current example, the maximum possible biomass content (X_{max}) and the yield of waste metabolic heat from growth (Y_Q) are biological parameters.

Figure 12.5 shows various variables and parameters that might be included within bioreactor models that are more complex than the simple model shown in Fig. 12.4. The biological parameters are addressed in detail in Chaps. 14 to 17 while the transport and thermodynamic parameters are addressed in Chaps. 19 and 20.

12.4 The Seven Steps of Developing a Bioreactor Model

In order to develop a mathematical model for your bioreactor from scratch, you would need to undertake 7 steps (Fig. 12.6). These steps were followed in the development of the various mathematical models presented in Chaps. 22 to 25. Of course, with the availability of these models, it is currently possible to start in the middle of the process. For example, you could use model equations from the literature for the same type of bioreactor and start at Step 4, with the determination of the parameter values for your particular system. However, even it this is done, it is necessary to check the original development of steps 1 to 3 in the literature model, to make sure that you agree with the decisions made by the authors during these steps.

You should also note that even though the steps are presented as a linear sequence here, the modeling process does not necessarily occur in a simple linear fashion. Frequently it is necessary to return and revise earlier decisions as the model is refined.

This section covers the 7 steps of modeling an SSF bioreactor, highlighting the tasks and questions that arise at each of the steps. It does not offer answers to these questions. Chapter 13 discusses how several of the key questions have been answered in the past, for example, in the development of the various bioreactor models that are presented in Chaps. 22 to 25.

Thermodynamic parameters
- the saturation humidity of the air
- the heat capacity of the moist substrate particles and the air
- the enthalpy of vaporization of water
- the equilibrium concentration of O_2 in the substrate

(each of these is a function of temperature)

State variables
- biomass concentration
- nutrient concentration?
- substrate bed, bioreactor wall, and headspace gas temperatures
- $[O_2]$ within the substrate particle?
- headspace O_2 concentration
- substrate bed bulk density
- overall dry matter

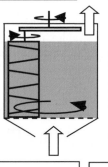

Design variables
bioreactor height, width, and depth

Operating variables
(not all will appear in all models)
- cooling water flow rate and temperature
- inlet air flow rate, temperature, and humidity
- frequency, duration, and intensity of agitation
- setpoints that are used to activate control schemes

Biological parameters
might include
- maximum specific growth rate
- maximum [biomass]
- fitting parameters in the equation used to describe how growth is affected by temperature
- yield coefficients (biomass/substrate, heat/biomass)

Transport parameters
- effective diffusivity of species such as O_2, nutrients, enzymes and hydrolysis products within the substrate particle?
- effective diffusivity of O_2 in the gas phase
- coefficient for heat transfer through the bioreactor wall
- thermal conductivity of the substrate bed

Independent variables
- time only – if the bed is well mixed
- time and space – if the bed is not well mixed
- space only – if we can make an assumption of a pseudo-steady state process, but there are spatial gradients

Fig. 12.5. Various parameters and variables that might be included in SSF bioreactor models. Not all these parameters and variables will appear within a particular model. Items marked with a question mark are typically not included within bioreactor models due to the complexity they would bring

Step 1 – Know what you want to achieve and the effort you are willing to put in to achieve it: Why develop the model? What level is appropriate to describe the microscale processes? Will intraparticle diffusion be described, or will simple empirical equations be used to describe the growth kinetics?

Step 2 – Draw the system at the appropriate level of detail and explicitly state assumptions: Which are the phenomena/processes that will be included in the model? Indicate them and their relationships in a diagram. What assumptions and simplifications will be made?

Step 3 – Write the equations: Balance equations will need to be written for which variables? How can the various phenomena that will be included in these equations be described? Which initial and boundary conditions must be specified? What equations are appropriate for the boundary conditions?

Step 4 – Estimate the parameters and decide on appropriate values for the operating variables and initial values: How can the parameter values (or equations that give their values as a function of the state of the system) be estimated? Are literature values acceptable? Must they be determined on the basis of experimental data?

Step 5 – Solve the model: What types of differential equations are present in the model, and what computer software will be used to solve them? What computing facilities are required?

Step 6 – Validate the model: Do the model predictions agree well with the experimental results? Is the model sufficiently accurate to be used as a design tool, or does it need to be revised? If the predictions do not agree well and the model needs to be revised, what specifically needs to be changed? What is the cause of the disagreement? Is it necessary to go back and redo or rethink an earlier step?

Step 7 – Use the model: What does the model say about the performance of the bioreactor? Does it allow the identification of better operating strategies? Are the predicted improvements obtained in practice? Does the model need further refinement?

Fig. 12.6. The seven steps of the modeling process

12.4.1 Step 1: Know What You Want to Achieve and the Effort You Are Willing to Put into Achieving It

You will typically want to construct a model that can be used as a tool in the bioreactor design process or in the optimization of operation of a bioreactor that has already been built. Models that have already been constructed with this motivation are described in Chaps. 22 to 25.

At this stage it is necessary to decide on the appropriate balance between the effort required (i.e., the work involved in writing the model equations, determining the values of the model parameters, and solving the model) and the "power" of the model, where the power of a model is defined by its ability to describe the performance of the system under a range of operating conditions, including conditions outside of the experimental range on which the model development was based. The greater the degree to which a model describes mechanistically the many phenomena presented in Chap. 2, the more likely it is to be more flexible. However, the description of fundamental phenomena can greatly increase the complexity of the model, and can require significant experimental effort to determine the parameters. If, in the particular bioreactor being modeled, there are significant temperature, water, and gas gradients across the bed, then clearly the model needs to describe the heat and mass transfer processes within the bed and to include position as an independent variable. A choice must then be made as to whether to describe the intra-particle gradients that arise. Doing so will lead to a highly complex model, because it will be simultaneously describing heterogeneity at the macroscale and heterogeneity at the microscale. Chapter 13 addresses this question in some detail.

The balance between model power and required effort may be decided from the outset, but it may also be decided later. Once the understanding of how the system functions is outlined in Step 2, the degree of complexity involved in a fully mechanistic approach becomes clearer, as do possible ways in which the mathematical description of the system can be simplified.

12.4.2 Step 2: Draw the System at the Appropriate Level of Detail and Explicitly State Assumptions

Once the aim of the modeling project is clear, the next step is to draw a diagram that summarizes the system and the important phenomena occurring within it. It is probably best to do this in two steps. Firstly, a detailed diagram should be drawn to include all the phenomena occurring within the system. Such a diagram might be similar to Fig. 2.6. Secondly, a simplified version should be drawn that includes only those phases and phenomena that have been selected as being sufficiently important to include in the model. For example, Fig. 12.3 shows a simplified diagram for a well-mixed bioreactor. An especially important question is as to whether the solid and air phases within the bed will be treated as separate subsystems, or whether the whole bed will be treated as a single pseudo-homogeneous subsystem that has the average properties of the solid and inter-particle air phases.

It will also need to be decided whether the bioreactor wall will be recognized as a separate subsystem. The diagram should clearly indicate the boundaries of the overall system and the various subsystems within it, the processes occurring within each subsystem, and the processes of exchange between different subsystems and between these subsystems and the surroundings of the bioreactor. It should be clearly annotated with the following information

- the state variables. In Fig. 12.3, these are the bed temperature and the biomass. Each of these should be given a symbol, which will be used in the equations;
- the interaction between the parameters and the state variables. For example, it should be noted that the growth rate of the organism will be modeled as depending on the bed temperature;

At the time of drawing these diagrams, the process of organizing the related information of assumptions, symbol definitions, and units should be started. All the symbols used to label the variables and parameters in the diagram should be listed and described, with their units. Also, all the assumptions and simplifications made should be carefully written down. As an example, for the well-mixed bioreactor in Fig. 12.3, it is assumed that:

- the substrate bed is well-mixed such that the whole bed can be represented by a single temperature, and the heat generation is uniform throughout the bed;
- the gas and solid phases are at temperature and moisture equilibrium, such that the air is saturated at the air outlet at the temperature of the bed;
- saturated air is used to aerate the bed;
- the loss of bed mass as CO_2 during the process is not significant, allowing the bed mass to be represented by a constant (M, kg);
- the water lost during the fermentation is replaced by a spray, such that the water content of the bed does not change during the fermentation;
- growth follows logistic growth kinetics;
- the specific growth rate constant depends only on the biomass concentration and the temperature and therefore growth is not limited by the supply of O_2 or nutrients;
- the thermal properties of the bed remain constant, even as the bed is modified by the growth process.

Of course many other assumptions are possible in order to reduce the complexity of models. Note that final decisions on the necessary variables and parameters and their appropriate units and the necessary assumptions might be made only at the stage of writing the equations.

12.4.3 Step 3: Write the Equations

This step builds on the foundation provided by the first two steps. The qualitative description of the system produced in Step 2 shows what equations need to be written and what terms should be included within these equations. The importance of the diagram drawn in Step 2 cannot be overstated. Lack of clarity in this diagram will lead to great difficulty in writing a coherent set of equations. The basic approach is to write:

- material and energy balance equations, usually in dynamic form (i.e., differential equations), with the state variables each expressed as:

$$\frac{d(variable)}{dt} = \text{system inputs - system outputs} +/- \text{ changes within the system}$$

- These balances must originally be written in terms of quantities that are conserved, although the equations can be rearranged later. For example, a balance on water would originally be written with each term having units of the mass of water per unit volume of bioreactor (i.e., kg-H_2O m^{-3}) and not the water content (kg-water kg-dry-substrate^{-1}). The differential term would therefore be d(WS)/dt and not dW/dt, where W is the water content and S is the kg of dry substrate per m^3 of bioreactor. If it were desired to predict the water content, then the differential terms in W and S would be separated using the product rule, such that in the final equation only dW/dt appeared on the left-hand-side;
- relevant thermodynamic relationships for important parameters of the equations (e.g., the saturation water content of the gas phase as a function of temperature, using the Antoine equation);
- relationships for other parameters that are functions of the state of the system (e.g., the specific growth rate may be expressed as depending on the temperature);
- other intrinsic relationships.

In writing the equations, it is necessary to know the mathematical forms appropriate for describing the various phenomena. These mathematical forms are presented in Chaps. 14 to 17 for empirical growth kinetic equations and in Chaps. 18 to 20 for the processes described in balance/transport equations. As an example from Fig. 12.4, convection of heat to the surroundings appears within the energy balance as "$h.a.(T-T_{surr})$", or, in other words: "the rate of heat removal through the bioreactor wall is equal to a heat transfer coefficient times the area for heat transfer times the difference in temperature between the bed and the surroundings".

Attention to detail is paramount in the writing of equations. All terms of an equation must have the same units. For example, each term in a material balance would have units of kg h^{-1} (or kg m^{-3} h^{-1}), while each term in an energy balance would have units of J h^{-1} (or J m^{-3} h^{-1}). In fact, the necessity for terms to have certain units can help to give insights into how a particular term is to be constructed. Careful attention must be given as to whether terms are to be added or subtracted within an equation.

The number of dependent state variables selected will depend on the decisions made in Steps 1 and 2. For example, in the simple model in Fig. 12.4, equations are not written to describe the change in either the total mass of dry solids in the bed or the total mass of water in the bed. If the aim were to describe product formation then an extra equation would be written for the product.

For systems that are both temporally and spatially heterogeneous, and which therefore involve partial differential equations, it is necessary to write equations to describe the "boundary conditions". For example, it may be necessary to write that the temperature at the inlet of the bed is maintained at a particular temperature, and it may be necessary to write an equation that says that the rate at which heat is removed from the side walls of the bioreactor by convection to the cooling water is equal to the rate at which heat reaches the wall by conduction from the bed.

12.4.4 Step 4: Estimate the Parameters and Decide on Values for the Operating Variables

In order to solve a set of differential equations, you must have values for all of the parameters of the model, and, in addition to this, initial values must be given for the dependent state variables for which the differential equations are written.

Parameters. The types of parameters that appear in the model depend on the particular bioreactor and the phenomena that the model is describing. The values of the parameters may be determined in separate experiments, although at times values from the literature may be used. Note that some parameters might be constants, in which case only a single value is required, or they may vary as the state of the system varies, in which case an equation is needed that relates the parameter value to the state of the system. In the model presented in Fig. 12.4, the parameters were determined as follows:

- The parameters in the equation describing the dependence of the specific growth rate constant on temperature were determined by Saucedo-Castaneda et al. (1990) on the basis of experimental results for the growth of *Aspergillus niger*, obtained by Raimbault and Alazard (1980), by non-linear regression of the equation against these experimental results.
- The heat transfer coefficient (h) was obtained from Perry's Chemical Engineer's Handbook (Perry et al. 1984), as a typical value for the transfer of heat across steel. However, it could also be determined experimentally for a particular bioreactor.
- The design parameter A (area for heat transfer to the water jacket) was calculated assuming that the water jacket is in contact with the curved outer surface of the cylindrical bioreactor.
- Thermodynamic parameters were obtained from reference books (e.g., heat capacities of water and water vapor, coefficients of the Antoine equation used in the calculation of humidities, the enthalpy of evaporation of water).
- The heat capacity of the bed (C_{PB}) was calculated on the basis of a starchy substrate of 50% moisture content.

Sometimes it is difficult to determine the value of a parameter in independent experiments. Although it is not particularly desirable, it is possible to allow this parameter to vary in the solving of the model, using an optimization routine to find the value of the parameter that allows the model to fit the data most closely.

Initial values of the state variables. The state variables that appear in the model depend on the combination of differential equations that make up the model. Their initial values will be determined by the way in which the bioreactor and inoculum were prepared. In the case of the well-mixed SSF bioreactor, it is necessary to give the initial mass of dry biomass and the initial temperature of the bed. Of course it is also possible to choose hypothetical initial values in order to explore the effect of the starting conditions on the predicted performance of the bioreactor.

Operating variables. The operating variables appearing in the model depend on the type of bioreactor and what manipulations it allows. The available operating variables for each SSF bioreactor type were presented in Chaps. 6 to 11. The values used for these variables in solving the model will either be experimental values, in the case of model validation, or hypothetical values, in the case where the model is being used to explore the effect of the operating conditions on the predicted performance of the bioreactor.

12.4.5 Step 5: Solve the Model

This book does not provide detailed information on how mathematical models are solved. Typically, numerical techniques will be used for solving differential equations. The amount of work that must be done to solve a model depends on the sophistication of the computer software available. In some cases it is necessary to write a program in a computer code such as FORTRAN or MatLab®, using pre-written subroutines as appropriate. With more sophisticated software packages, it may be sufficient simply to enter the equations and initial values in the appropriate fields and ask the computer to solve the equations.

Well-mixed systems will lead to a set of ordinary differential equations (ODEs), that is, equations in which the differential terms are only expressed as functions of time. Such a set of equations can be solved with well-known subroutines, such as the FORTRAN subroutine DRKGS, which is based on the Runge-Kutta algorithm. The solution of such models will be a graph, plotted against time, of the system variables that were described by the differential equations. In the case of the well-mixed SSF bioreactor model, the solution of the model is represented by temporal biomass and bed temperature profiles, such as the predictions presented in Fig. 12.7(a).

Systems with both spatial and temporal heterogeneity will lead to partial differential equations (PDEs), that is, equations that contain a mixture of differential terms that contain time in the denominator and differential terms that contain a spatial coordinate in the denominator. The solution methods involve transforming the PDEs into sets of ODEs, and then using numerical integration to solve these ODEs. Typically the transformation of the PDEs into sets of ODEs must be done by hand, and is not simple to do. The solution of such a model will be a graph against time of each of the state variables, with multiple curves, each curve representing a different position within the bed (Fig. 12.7(b)).

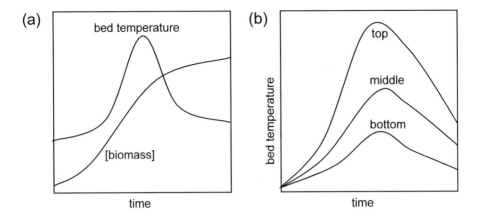

Fig. 12.7. What is the result obtained by solving a model? **(a)** For a model containing only ordinary differential equations, it is a predicted fermentation profile, or, in other words, a set of curves against time for each of the system variables. **(b)** For a model containing partial differential equations, fermentation profiles are predicted for various positions in the bed. For example, if the bioreactor shown in Fig. 12.3 were not mixed and the temperatures at various heights within the bed were predicted with an appropriate mathematical model, typical predictions would be as shown

12.4.6 Step 6: Validate the Model

If a model has been solved using independent estimates of all of the parameters, then it is of great interest as to whether the model manages to predict reasonably well the behavior of the system that is observed experimentally (Fig. 12.8). If it does, then this can be taken as supporting evidence, but not proof, that the mechanisms and phenomena included in the model are indeed those that are most important in determining the bioreactor behavior. Unfortunately, the validation of bioreactor models has only rarely been done well in the area of SSF to date.

As mentioned within Step 4, in some cases one or more of the parameters are determined during the solution step, by doing several simulations with different values for these parameters and seeing which solution agrees best with the experimental data (this being done most effectively by using an optimization routine to find the value of the parameter that gives the best statistical fit). The danger of this approach is that it might be possible to adjust the model to the data even if the mechanisms included in the model are inadequate. When this approach is used for parameter estimation, it is not possible to claim that the model has been validated, even if very close agreement is obtained.

A sensitivity analysis might be done at this stage (Fig. 12.9). This involves making changes one at a time to the various parameters in the model and seeing how large the effect is on the model predictions. The objective is to determine which parameters are most important in determining bioreactor performance:

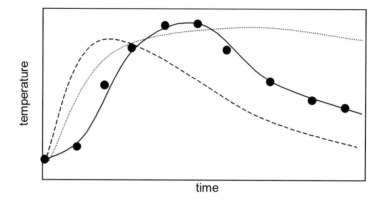

Fig. 12.8. Validation of the model. The graph illustrates three possible situations for a comparison between experimentally measured temperatures, represented by the solid circles, and the bed temperatures predicted by the model, represented by one of the curves. (—) Ideally there should be minimal deviation between the model predictions and experimental data; (- - -) At times general features of the experimental curve are described but are offset in magnitude and time. Possibly more accurate determination of one or more parameters is necessary; (·····) At times the predicted results are very different from the experimental results. A key phenomenon may have been omitted in the model

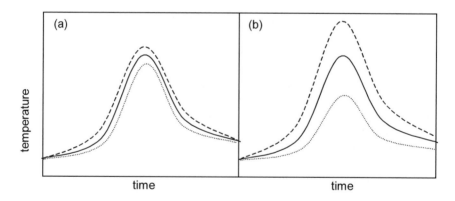

Fig. 12.9. Sensitivity analysis. In this example, the model presented in Fig. 12.4 is solved for various values of the heat transfer coefficient, h, associated with heat removal through the bioreactor wall. Key: (—) solution using h, the best estimate of the heat transfer coefficient; (·····) solution using $2.h$; (- - -) solution using $h/2$. Two possible situations are shown. **(a)** The variations in h have relatively little effect on the model predictions. Probably removal through the bioreactor wall makes only a relatively small contribution to overall heat removal. It might be appropriate to remove this term from the model. **(b)** The variations in h have a significant effect on the model predictions. The term describing heat removal through the bioreactor wall should be maintained in the model, since it is obviously an important contributor to overall heat removal, and it is important to have an accurate value for h if the model is to predict the experimental data well

- If relatively small changes in the value of a parameter significantly affect model predictions, then quite probably the phenomenon with which the parameter is associated is quite important in determining the system behavior and, furthermore, it is quite important to obtain accurate values for the parameter;
- If relatively large changes in the value of a parameter have a relatively small effect on model predictions, then possibly the phenomenon with which the parameter is associated is not very important in determining the system behavior, at least under the particular set of operating conditions used (the phenomenon might become more important under another set of operating conditions). The degree of accuracy needed for estimation of this parameter is not so great and possibly the term describing this parameter can be eliminated in order to simplify the model.

12.4.7 Step 7: Use the Model

The use to which the model is put will of course depend largely on the original motivation of the modeling work. For example, the model might be used to explore:

- how the same bioreactor will perform under operational conditions other than those for which experimental results were collected;
- how a different bioreactor geometry affects performance;
- how the size of the bioreactor affects performance.

Chapters 22 to 25 will show examples of such explorations for various different bioreactor types.

Of course there is no guarantee that the model will work well for a situation other than that for which it was validated. Predictions of the model about how performance can be improved must be checked experimentally. However, clearly an experimental program guided by use of a mathematical model has a good chance of optimizing performance more rapidly than a purely experimental program.

Deviations of the performance from predictions will lead to work to improve either or both of the model structure (the equations) and the parameter values. That is, it may be necessary to return to Steps 3 and 4. Such revisions lead to continual refinements of the model and to a greater understanding about how the various phenomena interact to control bioreactor performance.

Further Reading

The place of modeling in fermentation processes, argued in the context of submerged liquid fermentation
Anon (1997) Modelling is an indismissable tool to understand and control bioprocesses. J Biotechnol 52:173

Biwer A, Heinzle E (2004) Process modeling and simulation can guide process development: case study α-cyclodextrin. Enzyme Microb Technol 34:642–650

Schügerl K (2001) Progress in monitoring, modeling and control of bioprocesses during the last 20 years. J Biotechnol 85:149–173

13 Appropriate Levels of Complexity for Modeling SSF Bioreactors

David A. Mitchell, Luiz F.L. Luz Jr, Marin Berovič, and Nadia Krieger

13.1 What Level of Complexity Should We Aim for in an SSF Bioreactor Model?

This chapter addresses part of the first question that is raised in the modeling process (Sect. 12.4.1): "What can I hope to achieve by modeling my bioreactor and what effort am I willing to put in to achieve it"? Specifically, it explores the question of how decisions made about which phenomena should be described by the model will affect the complexity of the model and the difficulty of its solution.

The system within an SSF bioreactor is so complex that any attempt to describe it in full detail will lead to a highly complex model that will need much experimental work to determine the parameters and may require long solution times. We argue that there is much to gain from using "fast-solving" models that recognize the heterogeneity within the substrate bed at the macroscale, but beyond this, take a relatively simple view of the system. This is not to say that more advanced bioreactor models would not be useful tools, just that they need several years of development before they will be available as easy-to-use software packages.

It is convenient to consider the question in two parts. Firstly, what level of detail should be used to describe the growth kinetics in the kinetic sub-model? Secondly, what is the appropriate level for describing transport process in the balance/transport sub-model? The following sections explore these questions.

13.2 What Level of Detail Should Be Used to Describe the Growth Kinetics?

A crucial decision that must be made is whether to try to describe the spatial distribution of the system components at the microscale within the kinetic part of a model of an SSF bioreactor. The two key questions are:

1. Growth should be described as depending on which factors?
2. Should the spatial distribution of the biomass at the microscale be described?

13.2.1 Growth Should Be Treated as Depending on Which Factors?

Ideally, in a model describing the kinetics of microbial growth, the growth rate should be described as depending on those environmental factors that are important in influencing it. The problem in trying to meet this ideal in a model of an SSF bioreactor can be illustrated by comparing the implications, for both SSF and SLF, of a decision to include nutrient concentrations as one of the factors that determine the growth rate.

In SLF it is typically reasonable to assume that the fermentation broth is well mixed and therefore that the nutrient concentration is uniform throughout the broth. It is then a simple matter to use the Monod equation to describe the specific growth rate as a function of the nutrient concentration (Fig. 13.1(a)). In turn, it is also a simple matter to describe how the nutrient concentration changes during growth. In many cases such a simple model describes the growth curve quite well.

The situation is quite different in SSF. Mass transfer inside the substrate particle is limited to diffusion and, as a result of consumption of nutrients by the microorganism, concentration gradients will arise within the substrate (Fig. 2.8). As shown in Fig. 13.1(b), the Monod equation could be used to describe how growth depends on nutrient concentration, but it is not at all a simple matter to describe the nutrient concentration experienced by the microorganism. For example, even if a soluble nutrient is used, it is necessary to use equations that describe the diffusion of the nutrient within the substrate particle. If a polymeric nutrient is used, extra equations will be necessary in order to describe the processes of enzyme release, diffusion, and action. The problem is that the diffusion equations are written in terms of changes in both time and space, or, in other words, they are partial differential equations. Partial differential equations are significantly more difficult to solve than ordinary differential equations. As will be seen later, partial differential equations will typically arise in the modeling of the macroscale transport processes. A model with partial differential equations at both the microscale and macroscale would be highly complex and difficult to solve, requiring long solution times, as much as hours or days.

Actually, the situation is more complex still. SSF substrates typically involve a range of different carbon and energy sources and other nutrients, and these might be used sequentially or in parallel in a complex manner.

Besides the problems with the complexity introduced into the model, there is the extra problem of validating the model. In SLF it is a simple matter to withdraw a sample from the bioreactor and determine the nutrient and biomass concentrations. Plots of experimentally measured biomass and nutrient concentrations against time can then be compared with the model predictions. In SSF it is not a simple matter to determine experimentally the nutrient concentration experienced by the microorganism. Certainly it is not valid simply to homogenize a sample within a volume of water and then determine the nutrient concentration. This gives an average nutrient concentration that says nothing about the nutrient concentration gradients within the substrate particle. It would be necessary to use an analytical method that was capable of giving nutrient concentration as a function of position.

Given these modeling and experimental difficulties, if the aim is to develop a fast-solving model, the growth rate should not be expressed as a function of the intra-particle concentration of any component, neither of nutrients, nor of O_2, nor of protons (pH). Typically the growth kinetic equation would be empirical and the parameters of the equation would be described as functions of one or more of biomass concentration, temperature, and water activity.

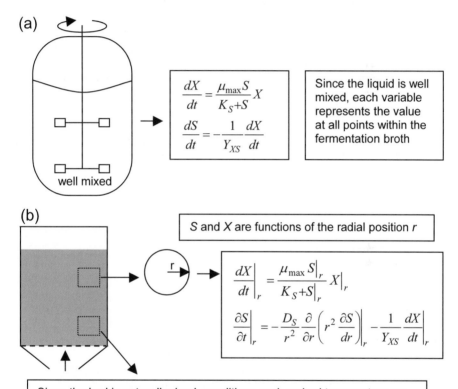

(a)

$$\frac{dX}{dt} = \frac{\mu_{max}S}{K_S+S}X$$

$$\frac{dS}{dt} = -\frac{1}{Y_{XS}}\frac{dX}{dt}$$

Since the liquid is well mixed, each variable represents the value at all points within the fermentation broth

well mixed

(b)

S and X are functions of the radial position r

$$\left.\frac{dX}{dt}\right|_r = \frac{\mu_{max}\left.S\right|_r}{K_S+\left.S\right|_r}\left.X\right|_r$$

$$\left.\frac{\partial S}{\partial t}\right|_r = -\frac{D_S}{r^2}\frac{\partial}{\partial r}\left(r^2\frac{\partial S}{\partial r}\right)\bigg|_r - \frac{1}{Y_{XS}}\left.\frac{dX}{dt}\right|_r$$

Since the bed is not well mixed, conditions such as bed temperature are different at different positions in the bed, affecting μ_{max}, such that:
- the same set of microscale equations must be applied to each different position within the substrate bed.
- the solution of the microscale equations will be different for each position within the bed

Fig. 13.1. The consequences, for model complexity, of a desire to model growth as a function of the concentration of a soluble nutrient. **(a)** A well-mixed submerged-liquid fermentation; **(b)** Solid-state fermentation in a static substrate bed

13.2.2 Is It Worthwhile to Describe the Spatial Distribution of the Biomass at the Microscale?

Modeling the spatial distribution of the biomass could potentially bring benefits. For example, a model that described how the microorganism grew into the inter-particle spaces could be used to predict how the pressure drop through the bed would change during the fermentation. However, any attempt to describe the spatial distribution of the biomass at the microscale will greatly increase the complexity of the model. Even for the relatively simple situation involving the growth of biofilms it would be necessary to describe the three dimensional arrangement of the particles. In fact, many SSF processes involve fungi, and it is not a simple matter to describe their growth in three dimensions (Fig. 13.2). One way to do this would be to describe the extension and branching of individual hyphae, however, it would be necessary to know the statistical distributions of branch frequencies and branch angles and a very large number of individual hyphae would need to be described. Furthermore, in an agitated bed, the effect that agitation would have on the spatial distribution of the biomass is not sufficiently understood to be able to describe it mathematically.

As a result of these complexities, and our current poor quantitative understanding of them, typically the biomass concentration is simply treated in bioreactor models as a global average over the particle, being expressed with units of "g dry biomass per g of dry solids".

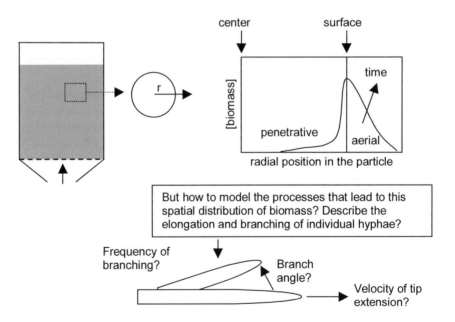

Fig. 13.2. The consequences, for model complexity, of a desire to model the spatial distribution of biomass at the microscale

13.2.3 Typical Features of the Kinetic Sub-models

As a result of the considerations above, the kinetic sub-models of SSF bioreactors often (although not always) are simple empirical equations and do not describe the spatial distribution of the biomass. Further, due to the difficulties of measuring biomass and metabolic states within SSF systems, many kinetic sub-models within SSF bioreactor models are "unstructured" and "non-segregated". Unstructured models do not describe any intracellular events, while unsegregated models do not divide the biomass into different subpopulations. The simple kinetic equations that result from these decisions will be explored in Chaps. 14 to 17.

13.3 What Level of Detail Should Be Used to Describe Transport Processes?

Several characteristics common to many SSF processes have implications for the balance/transport part of SSF bioreactor models:

- The vast majority of SSF processes are batch processes. Therefore almost all the models that have been developed to describe SSF bioreactors are dynamic models. This means that equations are written as differential equations and therefore need to be solved by numerical integration. Pseudo-steadystate models, in which the differential term is equated to zero and which can therefore be solved algebraically, are only rarely used.
- The presence of moisture and thermal gradients within the bed in many bioreactors means that in most cases it is necessary to write the model equations with differential terms that include both space and time, that is, as partial differential equations. These partial differential equations make the model more difficult to solve (see Chap. 13.5), but they are unavoidable if it is intended that the model describe bed heterogeneity.
- Even when bioreactors are mixed, the mixing will typically not be perfect. It is possible to model the mixing of a bed of solid particles using "discrete particle models", which describe the movement of individual particles as a result of the forces acting upon them and track the position of a population of thousands of particles (see Fig. 8.10(a)). However, it can take days for the model to solve even with a number of particles much smaller than a bioreactor really contains. Even if a general circulation pattern can be assumed (Fig. 13.3(a)), it is not a simple matter to characterize the flow pattern and to express the effectiveness of mixing as a function of operating conditions. Therefore, in those bioreactors in which the bed is mixed, it is usual to assume that the bed is well mixed.

As a result of these complicating factors, especially the appearance of partial differential equations in models of heterogeneous beds, the following decisions are often made in order to develop models that can be solved by a desktop computer within seconds to minutes, that is, fast-solving models:

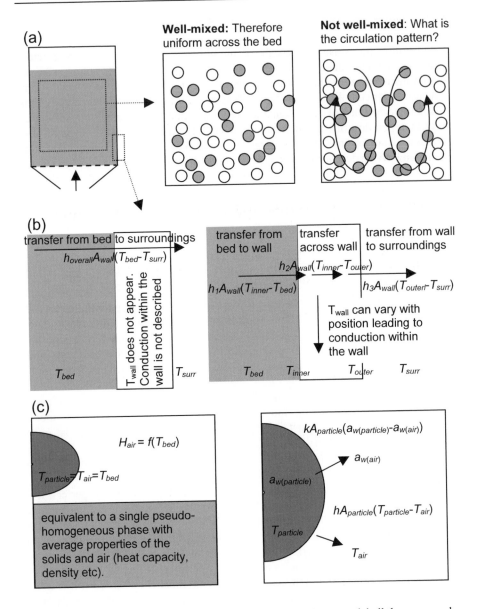

Fig. 13.3. The consequences, for model complexity, of a desire to model all the macroscale phenomena. **(a)** If the bed is mixed but is not well mixed, it is a complex matter to model the flow patterns. In the example on the right there is a region of circulating particles and a region of relatively-static particles. **(b)** Inclusion of the bioreactor wall as a separate subsystem complicates the model, as shown on the right. **(c)** If the substrate and air phases can be assumed to be in equilibrium, this simplifies the description of the system, as shown on the left

- **To neglect the bioreactor wall.** SSF bioreactors commonly contain several phases that could be treated as separate subsystems (Chap. 2). Equations must be written that describe the changes within each of these subsystems and the exchanges between subsystems. In order to reduce the overall number of equations, it is often decided not to treat the bioreactor wall as a separate subsystem (Fig. 13.3(b)). It may be simply ignored or it may be lumped together with the substrate bed. This removes the ability of the model to describe the changes in the temperature of the bioreactor wall, which might in fact have important influences on the process. Even when the bioreactor wall is recognized as a separate subsystem, typically it is assumed that the whole bioreactor body is at the same temperature, in order to not have to use partial differential equations to characterize the temperature gradients within it.

- **To treat the substrate bed as a single pseudo-homogeneous phase.** The substrate bed is often treated as though, at any particular point in the bed, the air and solid at that point were in equilibrium. The advantage is that in this case it is not necessary to describe the solids and inter-particle air as separate phases. Rather, the bed can be treated as though it were a single phase with the average properties of the air and solid (Fig. 13.3(c)). If this is done, the air and solid phases are assumed to have the same temperature and it is possible simply to write an equilibrium relationship to relate the air humidity with the temperature and the water activity of the solids. Of course, the suitability of this simplification depends on whether in actual practice this equilibrium is approached. The alternative is to treat the moist substrate particles and the inter-particle air as separate phases. This implies that equations must be written to describe heat and mass transfer between these two phases. Further, it implies that the solids-to-air heat and mass transfer coefficients must be determined.

- **To limit the number of key state variables.** It is possible to simplify the model by minimizing the number of macroscale state variables it describes. For example, the simplest models concentrate only on the substrate temperature, assuming that water levels are automatically controlled within the bioreactor. Some models include both energy and water balances. In some cases an O_2 balance might also be done. Of course models that contain balances for all three quantities (energy, O_2, and water) will be most flexible in describing what controls the rate of growth under a wide range of operating conditions.

13.4 At the Moment Fast-Solving Models Are Useful

The previous sections have pointed out that at present many simplifications are made, in both the kinetic and the balance/transport sub-models of a bioreactor model, in order to arrive at a fast-solving model. The question arises as to whether such fast-solving models are useful tools in bioreactor design, or whether we need to develop more sophisticated models before we can use them fruitfully.

Table 13.1 compares the characteristics of a fast-solving model with a model that attempts to describe as many phenomena as possible in a "fully-mechanistic" manner. Such a fully-mechanistic model has not yet been developed for an SSF bioreactor, although various of the models that have been proposed within the SSF literature have incorporated one or more of the characteristics listed for it. These two model types represent two extremes. Most of the presently available bioreactor models lie on a continuum between them, although, on the whole, they lie closer to the fast-solving model than they do to the fully-mechanistic model.

A fully-mechanistic model would be more likely than a fast-solving model to describe the performance of a bioreactor under a wide range of operating conditions. It would also give a much better insight into which phenomena were most responsible for limiting growth in different systems, under different operating conditions and at different times during the process. However, the amount of work necessary to establish a fully-mechanistic model seems prohibitive, at least to the present moment, as shown by the fact that such a model has not yet been developed! Further, even if such a model were developed, the description of both macroscale and microscale heterogeneity would lead to solution times of hours to days, even on a supercomputer. In contrast, the current fast-solving models can typically be solved on personal computers in less than a minute.

Will fast-solving models enable us to fulfill our objectives in modeling a bioreactor? Various objectives that we might have include:

1. To use models to contribute to decisions about which type of bioreactor to use;
2. To use the model that describes the selected bioreactor to contribute to decisions on the design parameters, such as how large the bioreactor should be and what geometric aspect it should have;
3. To use the model to help in the sizing of auxiliary equipment, such as the specifications of the blower, in terms of air flow rates and pressures;
4. To incorporate the model into process control strategies.

Certainly current fast-solving models are able to make worthwhile contributions to our attainment of all of these objectives, as will be demonstrated by the modeling case studies in Chaps. 22 to 25 and the process control case study in Chap. 28. Further, it is important to note that if a model is to be incorporated into a control system, it needs to be able to be solved reasonably rapidly, otherwise the control action will be unduly delayed.

Of course, in accepting the use of a fast-solving model, we must also accept its limitations. A fast-solving model will subsume many fundamental phenomena within simplified equations. As a result, the model will not be very flexible. As an example, a simple empirical kinetic equation such as the logistic equation might be used to describe growth. It may fit the growth data well, but it hides the interparticle phenomena that combine to cause the biomass profile to appear as a logistic curve. With a simple change from one substrate to another, there is no guarantee that the logistic equation will still describe the biomass profile adequately. Even if it does, it will be necessary to re-determine the parameters of the equation.

Table 13.1. Two extremes of approaches to modeling SSF bioreactors

Characteristic	A simple fast-solving model	A fully-mechanistic model
What is the general aim?	To concentrate on the processes in Fig. 2.6(a), simplifying as much as possible the processes in Fig. 2.6(b)	To describe as many of the processes in Fig. 2.6 as possible
How are the growth kinetics described?	• An empirical equation, not involving nutrient or O_2 concentrations, is used • Average biomass concentrations are used	• Intra-particle diffusion of nutrients and O_2 are described and the growth equation parameters depend on the local values of these variables • Spatial distribution of the biomass is described
How is the stoichiometry of growth described?	Experimentally determined overall growth yields are used. Elemental balances are not done.	The growth reaction is described by a stoichiometric equation that balances the various elements (C, N, S, P etc.)
How is the bed treated?	As a single pseudo-homogeneous phase with the average properties of the solid and gas phases in equilibrium	The solids and inter-particle gas phase are treated as different sub-systems and equations are written for heat and mass transfer between them
Is the bioreactor wall recognized as a separate system?	No. Transfer across the wall is subsumed in a global equation describing transfer from bed to surroundings	Yes. Conduction across the wall is described as well as conduction within the wall from hotter to colder regions
If the bed is mixed, how is mixing modelled?	Perfect mixing is assumed	Real mixing patterns are described
If the bed is static, are gradients recognized?	Macroscale temperature and moisture gradients are described	Macroscale temperature and moisture gradients are described
Are pressure drops described?	No	Yes. The effect of the growth of the biomass on the pressure drop through the bed is described

13.5 Having Decided on Fast-Solving Models, How to Solve Them?

Having argued that fast-solving models have an important role to play, it is worthwhile to make some comments about the solution of such models. Note that this book does not address or teach methods for solution of mathematical models based on differential equations. Suffice to say that readers without the appropriate training in mathematics and computing should seek help from engineers or mathematicians with the appropriate skills.

The models may contain either ordinary differential equations (in the case of well-mixed systems) or partial differential equations (in situations where the system cannot be treated as well-mixed):

- Typically for ordinary differential equations it is possible to find software packages that only require the user to input: (1) equations; (2) parameter values; and (3) the initial values of each of the state variables. The writing of computer programs is typically not necessary. Such software packages typically operate on the basis of numerical integration according to the method of Runge-Kutta, although other integrating algorithms are also available.
- Techniques are available for the numerical solution of partial differential equations, such as "orthogonal collocation" and "finite differences". Readers with mathematical abilities interested in these methods can find more information elsewhere (see the further reading section at the end of this chapter). Unlike the case with ordinary differential equations, unless a highly sophisticated software package is available, it is not a simple matter of inserting the equations into the appropriate place within a program. Typically various lines of code need to be written.

13.6 Conclusions

While attention should certainly be given to furthering the development of fully-mechanistic models, at the moment fast-solving models are sufficiently accurate to be useful tools in the design of bioreactors and the optimization of their operation. The rest of the book concentrates on fast-solving models. Chapters 14 to 17 describe approaches to establishing and modeling the growth kinetics in a manner appropriate for incorporation into fast-solving models. Chapters 18 to 20 show how the heat and mass transfer phenomena within bioreactors can be described at an appropriate level of detail for a fast-solving model. Chapters 22 to 25 then present several fast-solving models and show how they can be used to give insights into optimal design and operation. We are confident that readers, with relatively little effort, can adapt these models to their own systems, and obtain useful results from doing so.

Further Reading

Modeling studies of SSF that illustrate the complexity introduced by attempts to describe the intra-particle phenomena
Georgiou G, Shuler ML (1986) A computer model for the growth and differentiation of a fungal colony on solid substrate. Biotechnol Bioeng 28:405–416
Mitchell DA, Do DD, Greenfield PF, Doelle HW (1991) A semi-mechanistic mathematical model for growth of *Rhizopus oligosporus* in a model solid-state fermentation system. Biotechnol Bioeng 38:353–362
Rajagopalan S, Modak JM (1995) Evaluation of relative growth limitation due to depletion of glucose and oxygen during fungal growth on a spherical solid particle. Chem Eng Sci 50:803–811
Rajagopalan S, Rockstraw DA, Munson-McGee SH (1997) Modeling substrate particle degradation by *Bacillus coagulans* biofilm. Bioresource Technol 61:175–183
Rottenbacher L, Schossler M, Bauer W (1987) Modelling a solid-state fluidized bed fermenter for ethanol production with *Saccharomyces cerevisiae*. Bioprocess Eng 2:25–31

A model that segregates the biomass into several different physiological states
Georgiou G, Shuler ML (1986) A computer model for the growth and differentiation of a fungal colony on solid substrate. Biotechnol Bioeng 28:405–416

A review of modeling of intra-particle phenomena in SSF, demonstrating the complexity of the equations involved
Mitchell DA, von Meien OF, Krieger N, Dalsenter FDH (2004) A review of recent developments in modeling of microbial growth kinetics and intraparticle phenomena in solid-state fermentation. Biochem Eng J 17:15–26

Spatial distribution of hyphae in artificial SSF systems and the complexity of trying to describe the extension and branching of individual hyphae
Nopharatana M, Mitchell DA, Howes T (2003) Use of confocal microscopy to follow the development of penetrative hyphae during growth of *Rhizopus oligosporus* in an artificial solid-state fermentation system. Biotechnol Bioeng 81:438–447
Nopharatana M, Mitchell DA, Howes T (2003) Use of confocal scanning laser microscopy to measure the concentrations of aerial and penetrative hyphae during growth of *Rhizopus oligosporus* on a solid surface. Biotechnol Bioeng 84:71–77
Yang H, Reichl U, King R, Gilles ED (1992) Measurement and simulation of the morphological development of filamentous microorganisms. Biotechnol Bioeng 39:44–48

Use of the discrete particle modeling approach
Schutyser MA, Padding JT, Weber FJ, Briels WJ (2001) Discrete particle simulations predicting mixing behavior of solid substrate particles in a rotating drum fermenter. Biotechnol Bioeng 75:666–675
Schutyser MAI, Briels WJ, Rinzema A, Boom RM (2003) Numerical simulation and PEPT measurements of a 3D conical helical-blade mixer: A high potential solids mixer for solid-state fermentation. Biotechnol Bioeng 84:29-39

Use of numerical methods for solving partial differential equations

Petzold LR, Brenan KE, Campbell SL (1989) Numerical solution of initial-value problems in differential-algebraic equations. Elsevier, New York

Villadsen J, Michelsen ML (1978) Solution of differential equation models by polynomial approximation. Prentice Hall, Englewood Cliffs, New Jersey

14 The Kinetic Sub-model of SSF Bioreactor Models: General Considerations

David A. Mitchell and Nadia Krieger

14.1 What Is the Aim of the Kinetic Analysis?

As pointed out in Chap. 12, a mathematical model of an SSF bioreactor requires two sub-models, a sub-model that describes the growth kinetics of the microorganism and a sub-model that describes the energy and mass balances and transport phenomena. Each of these sub-models is written at an appropriate level of detail, depending on what simplifications and assumptions have been made. Chapter 13 argued for the use of simple empirical equations within the kinetic sub-model, in order not to make it too difficult to solve the bioreactor model. Chapters 14 to 17 address various questions related to the establishment of kinetic sub-models of this type (Fig. 14.1).

The aim is to write a kinetic equation in which the change in the amount of biomass, or a variable associated with it, is described by a differential equation, with the parameters of this differential equation taking into account the effect on growth of the key state variables that will be included in the bioreactor model, such as the temperature and water activity of the substrate bed. This is achieved as shown in Fig. 14.2. Note that the experiments done for the purpose of selecting the kinetic equation should be done after some efforts have been made to find a medium on which the organism grows well and to identify the optimal environmental conditions. This book does not address the optimization of the medium and environmental conditions (see the further reading section at the end of this chapter). A kinetic profile is constructed by measuring the biomass, or some indirect indicator of the biomass, in samples removed over the time course of the fermentation (Fig. 14.2(a)). Various kinetic equations are fitted to the data by regression and the one that fits best to the data is selected. Later, experiments are done in which different environmental conditions are imposed, such that, after analysis of the growth profile in each condition, plots can be made that relate the parameters of the kinetic equation to the environmental variable (Fig. 14.2(b)). Each kinetic parameter will then be expressed as an empirical function of the environmental parameter.

The current chapter covers some of the issues that must be addressed before beginning the process of kinetic modeling and then goes on to explain how the basic kinetic equation is selected.

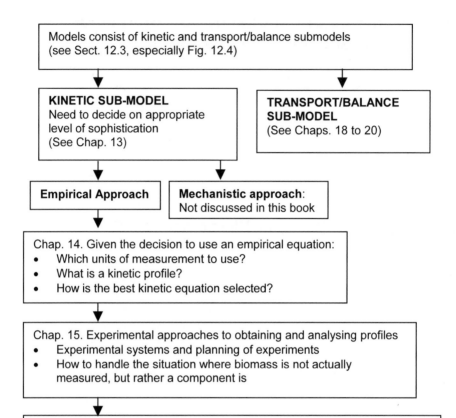

Fig. 14.1. An overview of how to go about establishing the kinetic sub-model of a mathematical model of an SSF bioreactor, showing how these issues are covered within various chapters of this book

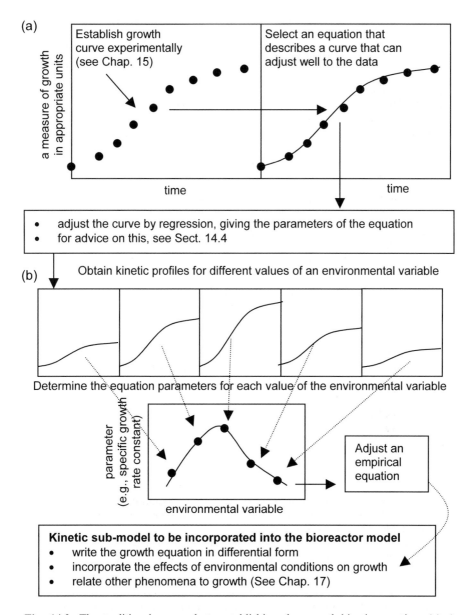

Fig. 14.2. The traditional approach to establishing the growth-kinetic equation. **(a)** A growth curve is established under optimal conditions and an empirical kinetic equation is selected that describes the curve well. **(b)** The parameters of the equation are expressed in terms of key environmental conditions (e.g., temperature) by repeating the growth curve experiment in different conditions, determining the growth parameters for each curve, and expressing the parameters as empirical functions of the environmental variable

14.2 How Will Growth Be Measured Experimentally?

14.2.1. The Problem of Measuring Biomass in SSF

The first experimental step in developing a kinetic model is to undertake a fermentation and plot the biomass content of the fermenting solid substrate against time (Fig. 14.2(a)). However, this immediately raises an experimental difficulty that is not faced in typical SLF processes (Fig. 14.3). In order to measure the dry weight of biomass directly, it is necessary to separate the biomass from solids. Many SSF processes involve filamentous fungi and, due to the penetration of the mycelium into the solid substrate, it is often impossible to remove the biomass quantitatively from the substrate, meaning that indirect methods of biomass measurement have to be used. Even in fermentations that involve unicellular organisms, although it may be possible to suspend many of the cells that are adhered to the particle surface and let the solid material sediment, the measurements are likely to be inaccurate (see Sect. 14.2.2).

The difficulty in measuring biomass dry weight in SSF raises the question of whether it is really necessary to use the dry weight of biomass as an indicator of growth. In fact, in SLF it is usually so simple to measure the dry weight of biomass (Fig. 14.3) that thought is often not given to whether this is the best parameter. So why do we need to measure the dry weight of biomass? Our aim in writing the kinetic sub-model of the bioreactor model is to write an equation that describes changes in a key variable to which we can relate other key processes that have important effects on bioreactor performance, such as metabolic heat production and O_2 consumption. However, does this variable have to be the dry weight of biomass? Are heat production and O_2 consumption actually related to the amount of dry biomass in the system? Or are they related to the amount of actively metabolizing biomass in the system? Given that we are typically limited to indirect measurements of growth in SSF, is it really necessary to convert the indirect measurement into dry weight? The answer is that no, it is not essential to write the kinetic sub-model in terms of the dry weight of biomass; we can use any growth-related parameter to which the important growth-related processes can be linked. For example, it may be possible to couple all the important growth-related activities to experimentally determined respiration kinetics.

Having said this, it is important to note that many of the current bioreactor models do in fact base their kinetic sub-models on changes in the dry weight of biomass. Therefore this book recognizes that indirect measures of growth will typically be converted into estimates of the dry biomass. The point is that the approach presented in this book is not the only possibility; other approaches to modeling the kinetics are possible. These other approaches will follow the general principles that we develop here in terms of dry biomass measurements.

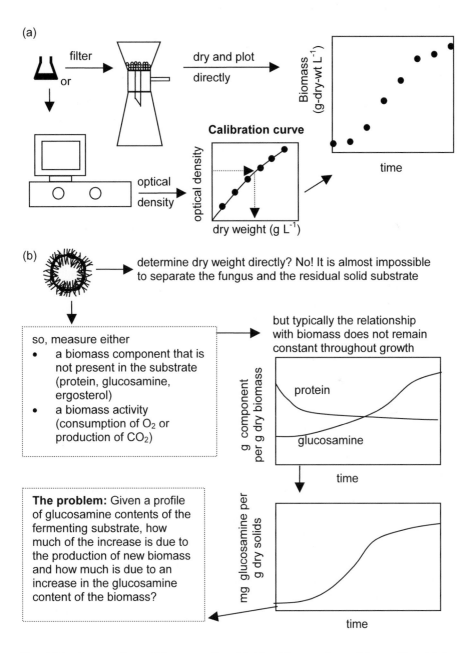

(a)

filter → dry and plot directly

or

optical density

Calibration curve

optical density

dry weight (g L^{-1})

Biomass (g-dry-wt L^{-1})

time

(b) determine dry weight directly? No! It is almost impossible to separate the fungus and the residual solid substrate

so, measure either
- a biomass component that is not present in the substrate (protein, glucosamine, ergosterol)
- a biomass activity (consumption of O_2 or production of CO_2)

but typically the relationship with biomass does not remain constant throughout growth

g component per g dry biomass

protein

glucosamine

time

The problem: Given a profile of glucosamine contents of the fermenting substrate, how much of the increase is due to the production of new biomass and how much is due to an increase in the glucosamine content of the biomass?

mg glucosamine per g dry solids

time

Fig. 14.3. A comparison of the ease of establishing biomass dry weight profiles. **(a)** In submerged liquid fermentation. **(b)** In solid-state fermentation

14.2.2 Indirect Approaches to Monitoring Growth

This section briefly mentions some of the direct approaches and various indirect approaches that can be used for monitoring growth in SSF systems. It is not intended to be an exhaustive review and it does not give protocols for the various methods. These should be searched for in original references. Some useful sources are given in the further reading section at the end of this chapter.

In some cases direct separation of the biomass is possible. With unicellular organisms it may be possible to dislodge the cells from the solid particles during a homogenization step and then to separate the solids from the suspended cells by sedimentation. However, some cells will remain adhered to the sedimented solids while some fine solid particles ("fines") liberated from the solids will not sediment. These fines will cause problems for determination of dry weight by filtration of the supernatant, since they will be erroneously counted as dry biomass. If viable count measurements are done on the supernatant, it is quite probable that the fines will have various cells adsorbed onto them, and these will give rise to only one colony per particle instead of one colony per cell.

In fungal fermentations, it is sometimes possible to digest the solid substrate within an aqueous enzyme solution, thereby allowing the mycelial biomass to be recovered by filtration. For example, this may be possible if the solid substrate is based on starch and contains little fiber, in which case the substrate can be hydrolyzed with amylases. However, some of the dry weight of biomass may be lost in this procedure and some solid residues may remain in the filtered biomass fraction. The efficiency of the recovery could be checked by submitting known masses of fungal mycelium, for example, from membrane filter culture (Chap. 15.3.1), to the hydrolysis and recovery procedure.

Various indirect methods rely on measurement of biomass components such as:

- **Ergosterol.** This is the predominant sterol in the cell membrane of many fungi, and is typically not found in plant material. It can be quantitatively measured by gas chromatography, HPLC, or UV spectrometry.
- **Glucosamine.** This is produced by the hydrolysis of chitin, which many fungi contain in their cell wall. It is typically not found in materials of plant origin. The hydrolysis of the biomass and subsequent determination of glucosamine by the chemical method can be quite tedious. It may be preferable to determine the glucosamine in the hydrolysate by HPLC.
- **Protein.** Protein is a major cell component. However, it is present in many plant materials and, if present, it will be impossible to know the proportion of protein in the sample that comes from the substrate, and the proportion that comes from the biomass, since the microorganism will typically hydrolyze the protein during growth. Therefore use of protein determination as an indicator of growth is restricted to cases in which the substrate contains negligible protein.

Unfortunately, the content of all of these components within the biomass can vary with culture conditions and with the age of the fungal mycelium. This greatly complicates the conversion of indirect measurements into estimates of the dry weight of biomass.

Other indirect methods rely on detecting activities of the biomass. Of these, the consumption of O_2 and production of CO_2 are most important. Gas metabolism is potentially a very important growth activity, especially since the rate of heat evolution will typically be directly proportional to the O_2 consumption rate, at least for an aerobic process. Further, the overall O_2 consumption within a bioreactor can be used for on-line monitoring of the growth process, even though it is not necessarily a simple matter to convert the O_2 consumption profile into a trustworthy biomass growth profile. Due to the importance of O_2 uptake measurements, the experimental use of this method in growth kinetic studies is discussed in Chap. 15.

The above discussion shows that several questions must be answered when selecting an appropriate indirect method for estimating growth:

* Is the component that is to be measured also present in the substrate?
* What time and resources are required for processing of the samples?
* To what degree does the relationship between the activity or component and the amount of biomass present change during the fermentation?

It may or may not be desired to convert an indirect measurement into an estimation of the biomass itself. If it is desired to do so, then the measurement method must be calibrated. In other words, the organism must be grown in a system that allows the dry weight of biomass to be measured in addition to the component or activity. These issues are discussed in Chap. 15.3.

14.3 What Units Should Be Used for the Biomass?

Once a direct or indirect measurement method has been selected, it will be used to give an estimate of the amount of biomass in samples removed over the course of the fermentation, allowing the construction of a kinetic profile. However, there remains a question: "What units will be used to express the biomass concentration in the kinetic profile?" The importance of this question becomes apparent when it is realized that various different units have been used to construct kinetic profiles in the past. These various methods are compared in Fig. 14.4, which also indicates the meaning of the various symbols used below:

* **grams of biomass or component per gram fresh sample.** In this case the sample is removed and weighed directly, the amount of biomass or component then being measured and divided by the fresh weight of the sample (i.e., X/M);
* **grams of biomass or component per gram dry sample.** In this case the sample is removed, dried in an oven at around 50-70°C and then weighed, after which the biomass or component is measured and divided by the dry weight of the sample (i.e., X/D). Note that if the analytical method used would be adversely affected by a drying step, the sample can be divided, with the water content being determined by drying of one fresh sub-sample and the other fresh sub-sample being used for biomass determination.

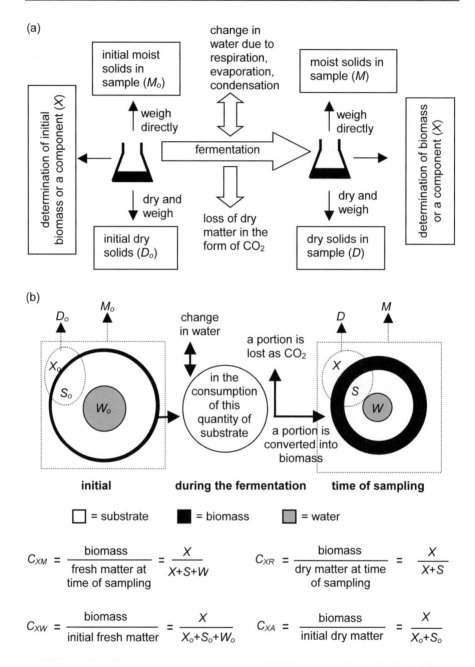

Fig. 14.4. Various manners in which the biomass content can be expressed. **(a)** The various measurements that can be made. **(b)** The biomass content will be calculated as a different number depending on what is included in the denominator

- **grams of biomass or component per gram initial fresh sample.** In this case the sample is removed and the amount of biomass or component is determined. To calculate the biomass content, the amount of biomass is divided not by the mass of fresh solids in the sample, but by the mass of fresh solids present at the time of inoculation (i.e., X/M_o);
- **grams of biomass or component per gram initial dry substrate.** In this case the sample is removed and the amount of biomass or component is determined. To calculate the biomass content, the amount of biomass is divided not by the mass of dry solids in the sample, but by the mass of dry solids present at the time of inoculation (i.e., X/D_o).

Of course, if sufficient data is available about how the water content and total dry solids vary during the fermentation, it is possible to calculate the biomass concentration in any of the above units. It is easy to obtain sufficient data to do this in laboratory experiments, but not so easy within a bioreactor.

So which is the most appropriate set of units to use in analyzing kinetics? This question will be addressed in Sect. 14.3.5 after considering the consequences of using each set of units.

14.3.1 Grams of Biomass per Gram of Fresh Sample

Expressing the biomass concentration per gram of fresh sample (C_{XM}) means that the denominator depends on changes in three factors, the mass of biomass (X), the mass of residual dry substrate (S), and the mass of water (W):

$$C_{XM} = \frac{X}{X+S+W} = \frac{X}{D+W} = \frac{X}{M}.$$ (14.1)

The sum of X and S is the total mass of dry solids (D). The sum of the dry solids and the water gives the total mass of the moist solids (M).

A biomass content expressed in these terms will not only be influenced by the consumption of dry matter, but will also be influenced by changes in the water content of the substrate, these changes arising from metabolic water production and evaporation. At the extreme, even if the organism is neither growing nor consuming substrate, C_{XM} can increase due to evaporation of water from the substrate.

14.3.2 Grams of Biomass per Gram of Dry Sample

Expressing the biomass in terms of the amount of dry sample removes the effect of changes in the water content on the apparent biomass concentration. However, due to the conversion of solid organic matter into CO_2 during the fermentation, the amount of solid material in the bioreactor can change significantly during the fermentation. In this case, the change in the biomass content expressed on the basis of "g of biomass per g of dry sample" arises from two sources: increase in the mass of biomass and decrease in the mass of solids. It is possible to have a situation where the microorganism is not growing, but is metabolizing to maintain itself. In such a situation the biomass concentration expressed per mass of dry sample will increase due to the loss of dry matter as CO_2, despite the fact that the biomass is not increasing.

The symbol C_{XR} (g-dry-biomass g-dry-solids^{-1}) can be used to represent a biomass content of this kind. It is given by:

$$C_{XR} = \frac{X}{X+S} = \frac{X}{D}. \tag{14.2}$$

14.3.3 Grams of Biomass per Gram of Initial Fresh or Dry Sample

The effect of water and dry matter loss on the apparent biomass concentration can be removed by expressing the biomass on the basis of an initial quantity of solids. One possibility would be to define the biomass concentration in terms of the initial mass of moist solids:

$$C_{XW} = \frac{X}{X_o + S_o + W_o} = \frac{X}{D_o + W_o} = \frac{X}{M_o}, \tag{14.3}$$

where the subscript "o" indicates initial masses of the various components. However, this has been used only rarely, since it is more common to work in terms of dry solids:

$$C_{XA} = \frac{X}{X_o + S_o} = \frac{X}{D_o}. \tag{14.4}$$

C_{XA} has the units of g-biomass g-initial-dry-solids^{-1}, these units typically being written as g-biomass g-IDS^{-1}.

Unlike the other methods of expressing biomass concentration, these measures will only change in response to changes in the amount of biomass. In the absence of growth, they will not change as a result of changes in either moisture content or dry solids content. Therefore they will be referred to as "absolute biomass concentrations". Concentrations expressed in the manner shown in Eqs. (14.1) and (14.2) will be referred to as "relative biomass concentrations".

There are other absolute measures of biomass:

- the amount of biomass per gram of inert support material, which can be used in some cases where an inert support matrix is impregnated with nutrients;
- the absolute amount of biomass within the bioreactor;
- the amount of biomass per unit volume of the substrate bed. Note that this is only an absolute biomass concentration in those cases in which the bed volume does not change significantly during the process. Biomass per unit volume is typically not used to express biomass concentration in laboratory-scale experiments, but biomass concentrations may be expressed in this manner within mathematical models of bioreactors.

In this book we will use the symbol X to represent either the absolute mass of biomass in the bioreactor or the mass of biomass per m^3 of substrate bed. Other symbols will be used to represent a concentration based on a denominator that does not change, such as grams initial dry solid, or grams of inert support material.

14.3.4 Which Set of Units Is Best to Use for Expressing the Biomass?

It is probably best, in the kinetic studies undertaken in the laboratory, to express the biomass concentration on an absolute basis. This is because key phenomena that will be included in the bioreactor model (such as the production of waste metabolic heat, the consumption of O_2, and the production of CO_2) depend directly on the absolute amount of biomass.

However, as will become obvious in the following section, absolute biomass contents have not always been used in growth profiles reported in the literature (Viccini et al. 2001). This must be kept in mind when analyzing kinetic profiles taken from the literature. In any case, it is not difficult to convert between absolute and relative concentrations if the yield and maintenance coefficients are known. A method of doing this is presented in Chap. 16.2.

14.4 Kinetic Profiles and Appropriate Equations

This section summarizes the various shapes of kinetic profiles that have observed in the literature, the empirical equations that have been used to describe them, and the manner in which the parameters of the equations are estimated.

Four differently shaped kinetic profiles have been reported in various SSF systems: "linear", "exponential", "logistic", and "deceleration". The general shapes of these kinetic profiles are shown in Fig. 14.5 (Viccini et al. 2001).

The equations that describe these curves are shown in Table 14.1. The task is to select the curve that is best able to fit the particular experimental results for biomass, or some indicator of biomass. Note that other shapes of growth curves are possible, in which case it is necessary to propose a new equation that describes the shape of the new curve. Curve selection and fitting will typically be done by regression. In regression analysis the model parameters are adjusted until the sum of squares of deviations between the experimental results and the corresponding values on the fitted curve are at a minimum (Fig. 14.6). There are many software packages that can be used to do regression. After doing the regression for each of the different equations, the curve chosen will typically be the one for which the sum of squares of deviations is the smallest. However, there may also be reasons for preferring a specific equation, even if it does not give the best fit to the data. For example, the logistic equation is usually preferred because often it is possible to use it to describe the whole growth curve adequately, whereas with the other kinetic equations the growth cycle needs to be broken up into intervals, each with a different equation. The regression analysis also gives the values for the

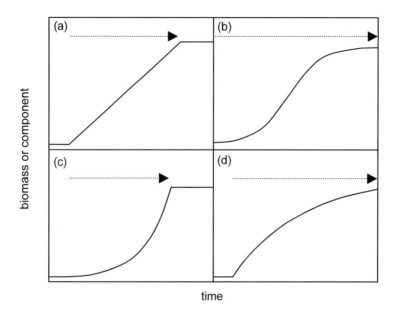

Fig. 14.5. Various types of kinetic profiles that have been found in SSF. The arrows indicate the parts of the profile that correspond to the kinetic type. **(a)** linear; **(b)** logistic;
(c) exponential; **(d)** deceleration

parameters of the kinetic equation, at least for the conditions under which the experiment was done. Note that, as will be discussed in Chap. 16, the parameters
will typically not appear in the final kinetic equation as constants, but rather as
functions of key environmental variables.

The logistic equation fits reasonably well to around 75% of the literature profiles obtained in SSF systems (Viccini et al. 2001). The other 25% of profiles are
described acceptably by one of the other three equations. Note that many of the
experimental growth profiles obtained in the past were not done with kinetic
analysis in mind. As a consequence, often there are relatively few data points during the period of rapid growth. This can lead to a situation in which several of the
equations can adjust reasonably to the data, it not being possible to determine
which gives the best fit. Chapter 15 gives some advice about how to plan experiments to avoid such problems.

Other important issues related to the kinetic analysis that you would need to do
for your own system are presented in the following paragraphs.

Use absolute concentrations. As noted in Sect. 14.3.4, it is advisable to undertake the experiments in such a manner as to be able to plot the data in terms of absolute concentrations and to fit an equation to this absolute concentration data.

Table 14.1. Equations that have been used to describe growth profiles or parts of growth profiles in SSF systems[a]

Name	Equation	Equation number	Parameters to be found by regression
Linear	$C = C_o + kt$	(14.4)	C_o, k
Exponential	$C = C_o e^{\mu t}$	(14.5)	C_o, μ
Logistic	$C = \dfrac{C_m}{1 + \left(\dfrac{C_m}{C_o} - 1 \right) e^{-\mu t}}$	(14.6)	C_o, C_m, μ
Deceleration	$C = C_o \exp\left(A(1 - e^{-kt}) \right)$	(14.7)	C_o, A, k

[a] In the past these equations have been used for biomass concentrations expressed on both absolute and relative bases.

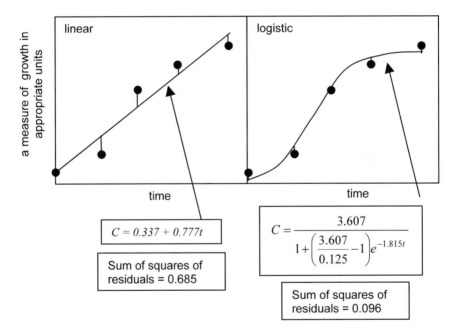

$$C = 0.337 + 0.777t$$

Sum of squares of residuals = 0.685

$$C = \frac{3.607}{1 + \left(\dfrac{3.607}{0.125} - 1 \right) e^{-1.815t}}$$

Sum of squares of residuals = 0.096

Fig. 14.6. How regression analysis is used to determine the most appropriate kinetic equation and the values of the parameter of this equation that give the best fit to the experimental data. In this case the logistic equation would be selected since it gives a better fit, as indicated by the smaller sum of squares of residuals. The residuals are the vertical lines that represent, for a particular time, the difference between the experimental value and the value predicted by the equation

Take care to select the appropriate interval for the regression analysis. The kinetic equations in Table 14.1 may apply to only part of the overall kinetic profile. There might be lag and stationary phases not described by these equations, in which case it is necessary to select carefully the region of the growth profile to which the kinetic equation will be fitted. For example:

- None of these equations explicitly describe a lag phase. However, the exponential and logistic equations may give apparent lag phases on a linear-linear plot if the initial biomass concentration is very low.
- The exponential and linear equations do not describe any limitation on growth. Of course if the growth curve is followed for long enough, the biomass profile must eventually show a maximum concentration (C_m). For these equations it may be appropriate to define a separate stationary phase. The logistic and deceleration equations can describe a stationary phase, which occurs at C_m for the logistic equation and at $C_o . e^A$ for the deceleration equation. These equations make no assumptions about the mechanism of limitation. In different systems limitations on the maximum amount of growth might be related to the exhaustion of essential nutrients, to the accumulation of inhibitory end products of metabolism, or to steric considerations (i.e., through the biomass "filling" the physical space available, noting that, even at their maximum packing density, fungal hyphae occupy only about 34% of the available volume (Auria et al. 1995)). Therefore the significance of C_m may vary from system to system. Typically it will be treated as a simple empirical parameter.
- There may even be a decline or death phase, which is not described by any of these equations. The modeling of death kinetics is discussed in Chap. 16.4.

Keep the environmental conditions constant. The parameters of the equation will change for cultures grown in different conditions, for example, at different temperatures, on different substrates, or with different O_2 concentrations in the gas phase. Therefore, during the fermentation the conditions should be held as constant as possible. This may not be simple, even at small scale. Difficulties in maintaining constant conditions and experimental strategies to minimize deviations are discussed in Chap. 15. Note that in more sophisticated studies, in which the effects of varying conditions on growth are investigated, it may actually be desirable to vary the conditions in a deliberate manner during the fermentation.

14.5 Conclusions

So far we have addressed the graphical and mathematical issues associated with constructing and analyzing the kinetic profile. The next chapter gives advice about the experimental techniques that may need to be used.

Further Reading

A survey of kinetic profiles in SSF systems
Viccini G, Mitchell DA, Boit SD, Gern JC, da Rosa AS, Costa RM, Dalsenter FDH, von Meien OF, Krieger N (2001) Analysis of growth kinetic profiles in solid-state fermentation. Food Technol Biotechnol 39:271–294

Use of response surface analysis for determining the optimum medium and conditions for growth and product formation
Kalil SJ, Maugeri F, Rodrigues MI (2000) Response surface analysis and simulation as a tool for bioprocess design and optimization. Process Biochem 35:539–550
Lekha PK, Chand N, Lonsane BK (1994) Computerized study of interactions among factors and their optimization through response surface methodology for the production of tannin acyl hydrolase by *Aspergillus niger* PKL 104 under solid state fermentation. Bioprocess Eng 31:7–15

General reviews of methods for determination of biomass in SSF
Desgranges C, Vergoignan C, Georges M, Durand A (1991) Biomass estimation in solid state fermentation I. Manual biochemical methods. Appl Microbiol Biotechnol 35:200–205
Desgranges C, Georges M, Vergoignan C, Durand A (1991) Biomass estimation in solid state fermentation II. On-line measurements. Appl Microbiol Biotechnol 35:206–209
Matcham SE, Wood DA, Jordan BR (1984) The measurement of fungal growth in solid substrates. Appl Biochem Biotechnol 9:387–388
Mitchell DA (1992) Biomass determination in solid-state cultivation In: Doelle HW, Mitchell DA, Rolz CE (eds) Solid Substrate Cultivation. Elsevier Applied Science, London, pp 53–63

Specific methods for determination of biomass in SSF
Acuna G, Giral R, Thibault J (1998) A neural network estimator for total biomass of filamentous fungi growing on two dimensional solid substrate. Biotechnol Techniques 12:515–519
Cordova-Lopez J, Gutierrez-Rojas M, Huerta S, Saucedo-Castaneda G, Favela-Torres E (1996) Biomass estimation of *Aspergillus niger* growing on real and model supports in solid state fermentation. Biotechnol Techniques 10:1–6
Davey CL, Penaloza W, Kell DB, Hedger JN (1991) Real-time monitoring of the accretion of *Rhizopus oligosporus* biomass during the solid-substrate tempe fermentation. World J Microbiol Biotechnol 7:248–259
Dubey AK, Suresh C, Umesh Kumar S, Karanth NG (1998) An enzyme-linked immunosorbent assay for the estimation of fungal biomass during solid-state fermentation. Appl Microbiol Biotechnol 50:299–302
Ebner A, Solar I, Acuna G, Perez-Correa R, Agosin E (1997) Fungal biomass estimation in batch solid substrate cultivation using asymptotic observation. In: Wise DL (ed), Global Environmental Biotechnology, Kluwer Academic Publishers, Dordrecht, pp 211–219
Matcham SE, Jordan BR, Wood DA (1985) Estimation of fungal biomass in a solid substrate by three independent methods. Appl Microbiol Biotechnol 21:108–112

Ooijkaas LP, Tramper J, Buitelaar RM (1998) Biomass estimation of *Coniothyrium minitans* in solid-state fermentation. Enzyme Microbial Technol 22:480–486

Penaloza W, Davey CL, Hedger JN, Kell DB (1992) Physiological studies on the solid-state quinoa tempe fermentation, using on-line measurements of fungal biomass production. J Sci Food Agr 59:227–235

Ramana Murthy MV, Thakur MS, Karanth NG (1993) Monitoring of biomass in solid state fermentation using light reflectance. Biosensor Bioelectronics 8:59–63

Roche N, Venague A, Desgranges C, Durand A (1993) Use of chitin measurement to estimate fungal biomass in solid state fermentation. Biotechnol Adv 11:677–683

Rodriguez Leon JA, Sastre L, Echevarria J, Delgado G, Bechstedt W (1988) A mathematical approach for the estimation of biomass production rate in solid state fermentation. Acta Biotechnol 8:307–310

Terebiznik MR, Pilosof AMR (1999) Biomass estimation in solid state fermentation by modeling dry matter weight loss. Biotechnol Techniques 13:215–219

Weber FJ, Tramper J, Rinzema A (1999) Quantitative recovery of fungal biomass grown on solid kappa-carrageenan media. Biotechnol Techniques 13:55–58

Wiegant WM (1991) A simple method to estimate the biomass of thermophilic fungi in composts. Biotechnol Techniques 5:421–426

Wissler MD, Tengerdy RP, Murphy VG (1983) Biomass measurement in solid-state fermentations using ^{15}N mass spectrometry. Dev Ind Microbiol 24:527–538

Wood DA (1979) A method for estimating biomass of *Agaricus bisporus* in a solid substrate, composted wheat straw. Biotechnol Lett 1:255–260

15 Growth Kinetics in SSF Systems: Experimental Approaches

David A. Mitchell and Nadia Krieger

15.1 Experimental Systems for Studying Kinetics

In order to establish the kinetic profile, a small-scale experimental system should be used so that heat transfer and inter-particle mass transfer will not be limiting. The idea is that the conditions within the substrate bed are those that you wish the organism to experience; therefore heat and mass transfer limitations should not cause significant deviations from these conditions. In other words, the aim is to characterize the growth kinetics of the organism without interference from bulk transport phenomena, to the extent that is possible. Of course, when empirical equations are used to describe the kinetics, the intra-particle transport phenomena are subsumed in the overall kinetic equation. This is impossible to avoid, since intra-particle transport limitations are an intrinsic characteristic of SSF (Chap. 2).

As mentioned in Chap. 14, you will undertake these kinetic studies once you have identified a substrate composition and environmental conditions that allow reasonably good growth of the organism. The most important conditions to control are the gas phase composition and the temperature and the water activity of the substrate bed. The two basic experimental strategies available are: (1) The use of multiple erlenmeyer flasks (or similar vessels) within an incubator and (2) the use of multiple columns within a waterbath.

Kinetic studies are typically done in these systems, rather than in laboratory-scale bioreactors, because such bioreactors are commonly not well-mixed, and therefore it is difficult to remove representative samples from them (Fig. 15.1). The problem is most evident in the case where it is desirable to leave the bed totally static, in which case it is impossible to avoid heterogeneity in beds containing even as little as a few hundred grams of substrate. There will be differences between inner and outer regions of the bed, and samples cannot be removed from anywhere other than the exposed surface without disrupting the bed. This disruption will affect growth of the microorganism in the part of the bed left behind after the sample is removed. In systems that involve multiple flasks or columns, individual units can be sacrificed at each sampling time. Even though within individual flasks or columns with less than 100 g of substrate there might still be some heterogeneity in the substrate bed (for example, from top to bottom of a column or from the inner to the outer regions within a flask), each flask or column should be identically heterogeneous, and therefore representative of all the other flasks.

Even though each flask or column should be identical with the others, there will always be some variation. Therefore it is important to establish, before the fermentation, the order in which the flasks or columns will be removed. If the decision were made at the time of sampling, then it would be possible to be influenced by the relative appearance of the different flasks or columns. Also, given the possibility that the conditions in a waterbath or incubator might vary with position due to imperfect circulation patterns, the pattern of removal should be random (Fig. 15.1(b)).

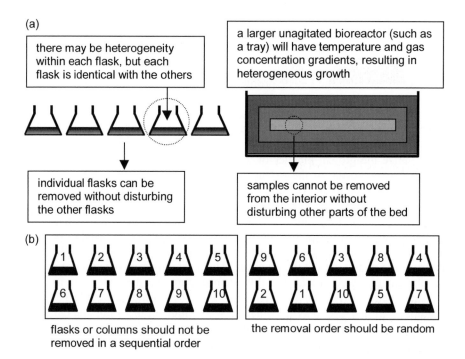

Fig. 15.1. Basic considerations about kinetic studies. **(a)** It is better to use multiple small containers in which individual containers are sacrificed at each sampling time rather than to remove subsequent samples from a larger mass; **(b)** The individual containers should be removed in random order

15.1.1. Flasks in an Incubator

This system is very commonly used. Its basic features are shown in Fig. 15.2. Ideally the substrate layer should not be thicker than 1 to 2 cm, although even with this thickness growth at the bottom of the layer may be limited by poor O_2 supply.

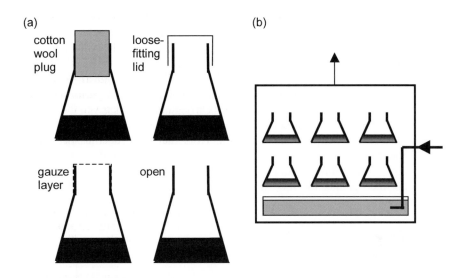

Fig. 15.2. Considerations in the use of flasks for kinetic studies. **(a)** The manner in which the flask is "closed" must be considered carefully; **(b)** It may be appropriate to bubble air through a tray in the bottom of the incubator in order to maintain a high humidity and reduce evaporative losses from the substrate

It is important to consider the control of water content of the substrate and the O_2 concentration in the gas phase. In order to provide a well-oxygenated gas phase, it would be preferable to leave the flasks open, allowing the headspace within the flask to communicate directly with the airspace in the incubator. However, if the relative humidity of the incubator atmosphere is not controlled, then it is likely that this will promote evaporation and drying out of the substrate. Also, open flasks provide no barrier to the entry of contaminants.

If necessary, water can be added to flasks at various intervals during the fermentation. In this case it would be desirable to mix the substrate bed in order to distribute the water evenly. If mixing is undesirable (due to adverse effects on the microorganism) then it is more challenging to add the water in a uniform manner. This might be achieved by adding the water as a fine spray over a thin layer of substrate.

In order to minimize water losses and therefore minimize the need to add water, it may be desirable to close the flask. However, in this case the O_2 concentration in the headspace will fall quickly. Note that cotton wool plugs can provide a significant barrier to O_2 transfer, so it would be wrong to assume that the headspace gas in plugged flasks has the composition of air. It would be necessary to take gas samples and analyze them.

15.1.2. Columns in a Waterbath

This system, involving small columns submerged in a waterbath (Fig. 15.3), has come to be known as "Raimbault columns", after their use by Maurice Raimbault in the 1980s. This system allows for forced aeration of the substrate bed, and therefore provides better control of the composition of the gas phase within the bed than can be obtained with fermentations carried out in flasks.

The columns need to be relatively thin, possibly only of 1-2 cm width, in order to minimize radial temperature gradients. Note that Saucedo-Castaneda et al. (1990) found significant radial temperature gradients in a 6-cm-diameter column. The height of the column could vary between 10-20 cm or even more, although the higher the column, the more likely that a special waterbath will have to be constructed.

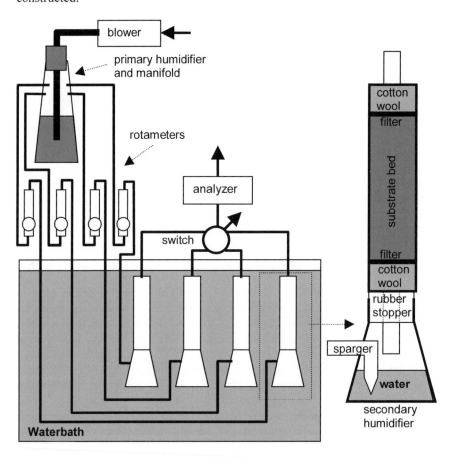

Fig. 15.3. Basic features of the Raimbault column system. Only four columns are shown, but the number is in fact only limited by the size of the waterbath. The diagram at the right shows detail of an individual column

It is important to saturate the incoming air to minimize the drying out of the bed in the column, because it is not practical to add water to the bed during the fermentation. Typically this will require at least two humidification steps. Note that temperature control is also important in minimizing evaporation, since if the bed temperature were allowed to rise above the inlet air temperature, the air would heat up and evaporate water from the bed, even if it were saturated at the air inlet.

It is also important to regulate the airflow to each column independently, with a separate rotameter on each line. If a single manifold were used then, in the absence of individual controls, any differences in the resistance to flow through the various columns would lead to different columns receiving different flow rates.

Whole columns are removed from the bed and sacrificed as samples for analysis. Of course, when the column is removed its air line must be closed, and it may be necessary to regulate the airflow through the remaining columns. In the analysis of the column that is removed as a sample, it may be interesting to check how homogeneous the growth is with height, dividing the sample into various bands that are analyzed separately, instead of mixing the whole bed contents together.

The outlet gas from each column can be analyzed separately, allowing the growth process to be monitored on the basis of O_2 consumption or CO_2 evolution. This will be most simple if an automatic switching system is available to cycle each of the outlet air flows through the analyzer in turn.

15.1.3. Comparison of the Two Systems

Most laboratories will already have an incubator, so the flask/incubator system is typically the cheaper to apply. However, the control of the air phase in the bed is obviously better in the column/waterbath system.

If sufficient incubator space is available, it is easier to increase the number of flasks than the number of columns, which may require the construction or purchase of more waterbaths and, even if waterbaths are available, will require the construction of more columns. Due to these considerations, the flask system typically allows more replicates in the same fermentation than does the column system.

The on-line monitoring of growth through gas metabolism that the column system allows is advantageous, since gas metabolism, especially O_2 consumption, is intimately linked with heat production, and data on the heat production rate will be needed within the energy balance in the bioreactor model.

15.2 Experimental Planning

Once the experimental system has been selected, it is necessary to decide how to carry out the experiment, namely, how many samples will be removed, at what times, and what will be analyzed. The flexibility to decide how many samples will be removed during the fermentation might be limited by the resources available, especially in the case of the Raimbault columns. In terms of the later mathematical analysis, it would be desirable for the data points to be evenly spaced over the whole of the active growth phase, including the acceleration and deceleration phases. It may be necessary to undertake the first fermentation simply to see when things happen, before undertaking a second experiment with better planning of the times at which samples are to be removed (Fig. 15.4).

It is advisable to record sufficient information for the biomass profile to be plotted in terms of both grams of biomass per gram of initial dry substrate (g-biomass g-IDS^{-1}) (i.e., on an absolute basis) and grams of biomass per gram of dry solids (i.e., on a relative basis). See Sect. 14.3 for an explanation of these terms. The following procedure, shown schematically in Fig. 15.5, explains how this can be done. Figure 15.5 also shows the meaning of the various symbols used in the equations presented below. It talks in terms of "flasks", but the same procedure applies if Raimbault columns are used. It assumes that:

- two samples are removed from the same flask (after mixing of the contents): one for moisture content determination and the other for analysis of the amount of biomass or a component of the biomass. Another strategy would be to sacrifice two flasks at each sampling time, using the entire contents of one for biomass determination and the entire contents of the other for determination of the moisture content.
- the biomass determination is done on a moist sample and not a dried sample.

The mass of moist substrate added initially to each flask should be measured, and a sample of this substrate should also be dried in order to determine the initial water content (*IWC*, expressed as a percentage and on a wet basis). The initial water content should be determined after inoculation, especially if the inoculum brings a significant amount of water into the system. It can be calculated from the moist and dry weights of the removed sample (Fig. 15.5):

$$IWC = 100\frac{(1-d_o)}{m_o}. \tag{15.1}$$

Since the original amount of moist substrate added to each flask is known (M_{oi} g added to the "ith" flask), it is simple to calculate the amount of "initial dry substrate" (*IDS*) that the ith flask holds:

$$IDS_i = \left(1 - \frac{IWC}{100}\right)M_{oi}. \tag{15.2}$$

At the time of sampling of the ith flask, after mixing of the substrate in the flask, a sample should be removed, weighed, dried, and reweighed, allowing the moisture content at the time of sampling (*WC*) to be calculated:

$$WC_i = 100\frac{(1-d_i)}{m_i}. \tag{15.3}$$

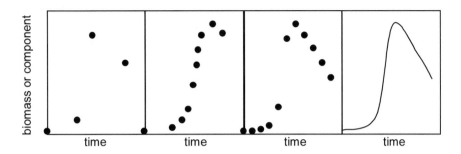

Fig. 15.4. Planning of sampling times. **(a)** A preliminary experiment should be done to identify at approximately what time the rapid growth period occurs; **(b)** This information can be used to plan for several samples to be removed during the period of most rapid growth, in order to characterize the growth curve well; **(c)** The profile that would be obtained if evenly-spaced samples were planned without first identifying the period of rapid growth. Insufficient samples would be removed during the rapid growth period; **(d)** The underlying fermentation profile assumed in this example. Of course, this profile is an unknown when the initial experimental planning is being undertaken

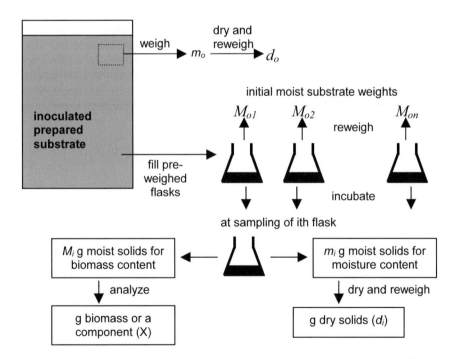

Fig. 15.5. A scheme that allows biomass concentrations to be expressed on either an absolute basis or a relative basis

The dry solids in the moist sample used for biomass determination, this sample being of fresh mass M_i, can then be calculated as:

$$D_i = \left(1 - \frac{WC_i}{100}\right)M_i .$$
(15.4)

The amount of biomass (or biomass component) within the sample will then be determined. The biomass content on a relative basis can then be calculated as:

$$C_{XR} = \frac{X}{D_i} .$$
(15.5)

The biomass content on an absolute basis is given by:

$$C_{XA} = \frac{X}{IDS_i} .$$
(15.6)

Regarding the number of samples to be removed, a reasonably good characterization of the kinetics will typically be achieved with around 10 sampling times, chosen to include the whole growth curve, with many of these taken during the rapid growth phase (Fig. 15.4). The best interval between samples will then depend on how fast the organism grows, and is not necessarily uniform. For a fast-growing organism, it may be appropriate to remove samples every two hours for twenty hours after the lag phase. For a slower grower, a sampling interval of 6 hours to 1 day, or even longer, might be appropriate. It would be desirable to remove triplicate samples at each sampling time; however, the number of replicates will depend on the resources available. With Raimbault columns, it is not uncommon for workers to remove only a single column at each sampling time.

15.3 Estimation of Biomass from Measurements of Biomass Components

It is highly likely that the estimation of biomass will be based on the measurement of a component of the biomass. It may be desired to convert the measured value for the component into an estimation of the dry weight of biomass. The current section shows how this can be done. Note that this conversion is not necessarily essential. As mentioned in Chap. 14.2.1, it might be decided to relate all other growth-related processes to the component and not directly to the biomass.

15.3.1 Suitable Systems for Undertaking Calibration Studies

The indirect method of biomass estimation must be calibrated in a system that allows biomass measurement. Due to the difficulty in direct biomass measurements (Chap. 14.2.1), normally this cannot be done in the SSF system itself, especially for processes that involve fungi. If it could, then there would be no need to use the indirect method to monitor the fermentation process in the first place!

Note that, given the possibility that cell composition can change throughout the growth cycle, it is not appropriate simply to determine the relationship between the biomass and the component for a single sample. Rather, several samples must be removed throughout the growth cycle and the relationship between biomass and the component determined for each sample.

Given that the growth conditions can affect cell composition, it is desirable for the system in which the calibration is done to mimic as closely as possible the conditions in SSF. For this reason, it is not a good idea to calibrate the indirect method using results from SLF. It is highly likely that the biomass composition in SLF will be different from that in SSF. Two systems that are better for the calibration are "impregnated ion exchange resin" and "membrane filter culture". Both mimic, to some extent, the conditions in SSF.

In the membrane filter culture method, a medium is prepared that mimics the solid substrate as closely as possible, but in such a manner as to give a slab of substrate with a flat surface (Fig. 15.6). For example, a gel can be used to solidify a

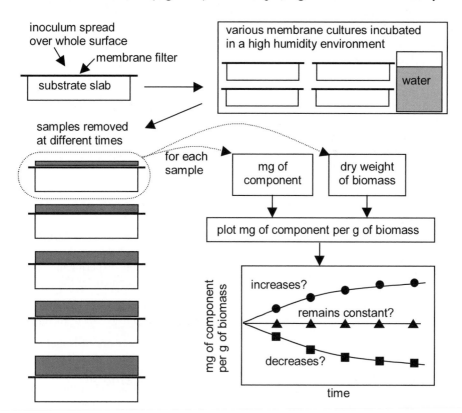

Fig. 15.6. The use of membrane filter culture for calibrating methods of biomass determination that involve the measurement of a component of the biomass

medium containing the key nutrients of the solid substrate, or it may even be possible simply to press the moist solid substrate into a compact slab. Various slabs within appropriately-sized petri dishes are then overlaid with sterilized or pasteurized membrane filters, inoculated evenly across their surfaces with a spore suspension, and placed in an incubator. Typically it will be necessary to provide a high humidity within the incubator to prevent the cultures from drying out. In more sophisticated systems, each individual culture has its own chamber. At various sampling times, one or more of these replicate plates can be removed and processed.

The membrane filter is chosen with a pore size (e.g., 0.2 μm) that is sufficiently small to prevent the fungus from penetrating through the membrane into the substrate slab. It is then a simple matter to peel the mat of biomass off the filter. It can then be processed as desired, for example, it can be dried for dry weight determination. Alternatively, it might be processed for the determination of a biomass component such as glucosamine, ergosterol, or protein. As long as the membrane allows extracellular enzymes to diffuse into the substrate, this method can be used with polymeric carbon sources.

Beads of an ion exchange resin, impregnated with nutrient medium, have also been used as a system that allows direct biomass measurement, since the biomass can be easily dislodged from the beads (Auria et al. 1990). However, it has only been used with soluble nutrients, presumably because it is difficult to impregnate the resin with macromolecules such as starch.

15.3.2 Conversion of Measurements of Components of the Biomass

If the biomass composition remains constant for samples removed at different times during the growth curve, then it is a simple manner to convert an indirect measurement into an estimate of the biomass.

$$C_{XA} = \frac{C_{CA}}{C_F},$$
(15.7)

where C_{CA} is the concentration of the component in a sample removed during the fermentation (mg-component g-IDS^{-1}), C_F is the relationship between the component and the biomass, as determined in the calibration experiments (mg-component g-biomass^{-1}), and C_{XA} is the calculated biomass content of the sample (g-biomass g-IDS^{-1}).

If the level of the component in the biomass varies during the fermentation, it will be necessary to use different conversion factors for samples removed at different times. One example where this has been done is in the work of Nagel et al. (2001b). They measured the glucosamine content as a function of time in a system where their fungus was grown on a membrane overlaid on a slab of pressed, ground wheat. They used non-linear regression of a curve plotted in the manner described in Fig. 15.6 to obtain the following equation for the glucosamine content of the biomass (G_x, mg-glucosamine mg-dry-biomass^{-1}) as a function of time:

$$G_x = 44.61 + \cfrac{43.65}{\left(1 + \exp\left(\cfrac{(t - \lambda) - 61.70}{12.34}\right)\right)}, \tag{15.8}$$

where λ is the lag time (h). Obviously this kind of approach can be adapted to other systems.

15.3.3 Limitations of these Calibration Methods

A problem with these calibration methods is that it is not possible to be sure that the conversion factor (or temporal relationship) that holds in the calibration system will be followed in the real SSF system. The more the substrate used in the calibration system can be made to mimic the substrate used in the SSF system then the more likely it is that the conversion factor will be reliable, but some doubt will always remain. Also, the calibration must be redone, even with the same microorganism, if the substrate is changed, or if environmental or nutritional conditions are varied significantly from those under which the relationship was determined.

15.4 Conclusion

This chapter has shown how it is possible to plan experiments in order to determine the kinetic profile and, in the case that the biomass is estimated on the basis of measurements of a biomass component, how these measurements can be processed in order to obtain an estimated biomass profile. The next chapter shows how the kinetic equation determined from this profile can be written in differential form within a kinetic model of growth.

Further Reading

An experimental study that demonstrates that even a relatively thin column can have significant temperature gradients
Saucedo-Casteneda G, Gutierrez-Rojas M, Bacquet G, Raimbault M, Viniegra-Gonzalez G (1990) Heat transfer simulation in solid substrate fermentation. Biotechnol Bioeng 35:802–808

Membrane filter culture method
Gqaleni N, Smith JE, Lacey J (1996) A novel membrane overlay cultivation technique for the measurement of growth and production of mycotoxins by *Penicillium commune* and *Aspergillus flavus*. Biotechnol Techniques 10:783–788
Huang SY, Hsu SW (1993) Growth estimation of solid-state koji by covering a cellophane membrane on the mash. Biotechnol Adv 11:685–699

Mitchell DA, Doelle HW, Greenfield PF (1989) Suppression of penetrative hyphae of *Rhizopus oligosporus* by membrane filters in a model solid-state fermentation system. Biotechnol Techniques 3:45–50

Rahardjo YSP, Korona D, Haemers S, Weber FJ, Tramper J, Rinzema A (2004) Limitations of membrane cultures as a model solid-state fermentation system. Lett Appl Microbiol 39:504–508

16 Basic Features of the Kinetic Sub-model

David A. Mitchell, Graciele Viccini, Lilik Ikasari, and Nadia Krieger

16.1 The Kinetic Sub-model Is Based on a Differential Growth Equation

Chapters 14 and 15 have shown how experiments should be done in order to select an appropriate empirical equation to describe the growth kinetics. This involves working with experimental growth curves and fitting the integrated form of the appropriate kinetic equation, which could be, for example, one of the equations from Table 14.1. However, the integrated form is not appropriate for direct incorporation into the bioreactor model. This can be understood by considering a simple model such as that shown in Fig. 12.4. In the logistic equation used to describe the growth kinetics in this model, the parameter μ is expressed as a function of temperature. If the integrated form of the equation were to be used (Eq. (14.6) in Table 14.1), then μ would have to be maintained constant, which is not consistent with the fact that the temperature and therefore μ vary during the fermentation. On the other hand, this does not present any problem for the numerical integration of the differential equation (Eq. (16.3) in Table 16.1), since μ can take on a new value for each step in the integration process.

The current chapter concentrates on how the differential form of the kinetic equation is incorporated into the bioreactor model. The kinetic sub-model expresses the various parameters in the growth equation as functions of the local conditions: This is the link that allows the bioreactor model to describe how growth is restricted by poor macroscale transport, since such transport limitation will lead to unfavorable local conditions for growth. The manner in which this is done is covered in the current chapter. The question of how to describe the manner in which growth in turn affects the local conditions is considered in Chap. 17.

Note that this chapter and the next consider growth in terms of the dry biomass itself. However, if the kinetic equation is determined in terms of a biomass component, the same considerations can be applied. Of course, the units must be changed appropriately. For example, biomass has the units of g-dry-biomass g-substrate^{-1}. If glucosamine were used, it would be necessary to write a term for it with units of mg-glucosamine g-substrate^{-1}, and this will affect the significance and units of other parameters, such as yield coefficients.

16.2 The Basic Kinetic Expression

The various types of growth profiles that have been found in SSF systems were presented in Table 14.1. Section 14.3 pointed out that biomass profiles in SSF can be plotted on two different bases, referred to as relative biomass concentrations (kg-biomass kg-dry-solids^{-1}) and absolute biomass concentrations (kg-dry-biomass kg-initial-dry-solids^{-1}). It also argued that the basic kinetic profile should be plotted in terms of "absolute concentration", since various of the effects of growth on the environment will depend on the absolute and not the relative concentration. Assuming that this has in fact been done, the integrated form of the equation selected from Table 14.1 by regression analysis will be expressed in terms of absolute biomass concentration. The corresponding differential form of the equation will then be selected from Table 16.1, for incorporation into the kinetic sub-model of the bioreactor model. Note that, in order to describe the whole profile, it may be necessary to use several equations. Further, an integrated equation other than the four presented in Table 14.1 may have been used, in which case it will be necessary to differentiate the equation. Each of these equations has one or more parameters. It may be interesting to express some of these parameters as functions of key environmental variables such as the temperature and the water activity of the substrate. Experimental approaches to doing this are described later (Sect. 16.4).

However, even though it is desirable to determine the kinetic profile based on absolute biomass concentrations, the bioreactor model should be able to predict the relative biomass concentration, in order to allow comparison between the model predictions and experimental results obtained in the bioreactor, which are typically obtained in terms of relative biomass concentrations. In order to convert

Table 16.1. Differential forms of the equations that have been used to describe growth profiles or parts of growth profiles in SSF systems

Name	Equation[a]	Equation number	Parameters[b]
Linear	$\dfrac{dC_{XA}}{dt} = k$	(16.1)	k
Exponential	$\dfrac{dC_{XA}}{dt} = \mu C_{XA}$	(16.2)	μ
Logistic	$\dfrac{dC_{XA}}{dt} = \mu C_{XA}\left(1 - \dfrac{C_{XA}}{C_{XAM}}\right)$	(16.3)	C_{XAM}, μ
Deceleration	$\dfrac{dC_{XA}}{dt} = kAC_{XA}e^{-kt}$	(16.4)	k, A

[a] The integrated form of these equations are given in Table 14.1. These equations are expressed in terms of absolute biomass concentration (e.g., g-dry-biomass g-IDS^{-1}).
[b] These parameters may later be expressed as functions of the environmental conditions.

a relative concentration to an absolute basis, it would be necessary to know to what initial dry weight of substrate the removed sample corresponded. To do this it would be necessary to weigh the whole bioreactor contents and determine the moisture content of the bed just before each sampling time. It is not a simple matter to weigh the whole bioreactor, especially at large scale. It is easier to use the kinetic sub-model to predict the relative biomass concentration.

Such a conversion can be done in the following manner. If the total dry weight of solids in the bioreactor (D, kg) is given as:

$$D = X + S, \tag{16.5}$$

where X is the total dry weight of biomass (kg) and S the total dry weight of residual substrate (kg), then for the absolute amount of biomass in the bioreactor (X, kg) we have:

$$\frac{dX}{dt} = \frac{d(C_{XA}D_o)}{dt} = D_o \frac{dC_{XA}}{dt}, \tag{16.6}$$

while for the "relative concentration" we have:

$$\frac{dX}{dt} = \frac{d(C_{XR}D)}{dt} = D \frac{dC_{XR}}{dt} + C_{XR} \frac{dD}{dt}. \tag{16.7}$$

Equation (16.6) can be substituted into Eq. (16.7) in order to eliminate the term dX/dt. The resulting equation can be rearranged to be explicit in dC_{XR}/dt:

$$\frac{dC_{XR}}{dt} = \frac{D_o}{D} \frac{dC_{XA}}{dt} - \frac{C_{XR}}{D} \frac{dD}{dt}. \tag{16.8}$$

Equation (16.8) says that the change in relative concentration (kg-dry-biomass kg-dry-solids^{-1}) during growth occurs due to growth itself in absolute terms, as described by the first term on the right-hand side, and due to the decrease in dry solids that occurs during growth, as described by the second term on the right-hand side. Growth leads to an overall loss of dry solids, and therefore dD/dt will be negative; given that this term is subtracted, its effect is to increase the relative concentration.

The rate of change in the total dry weight of solids is the sum of the rates of change in dry biomass and residual dry substrate:

$$\frac{dD}{dt} = \frac{dX}{dt} + \frac{dS}{dt}. \tag{16.9}$$

The rate of consumption of the residual dry substrate is related to the rate of growth by the following equation:

$$\frac{dS}{dt} = -\frac{1}{Y_{XS}} \frac{dX}{dt} - m_S X, \tag{16.10}$$

where Y_{XS} is the true growth yield (kg-dry-biomass kg-dry-substrate^{-1}) and m_S is the maintenance coefficient (kg-dry-substrate kg-dry-biomass^{-1} h^{-1}).

Substituting Eq. (16.10) into Eq. (16.9) and using the distributive law to separate out dX/dt on the right hand side gives:

$$\frac{dD}{dt} = \left(1 - \frac{1}{Y_{XS}}\right)\frac{dX}{dt} - m_S X .$$
(16.11)

Equation (16.11) can be rewritten in terms of the absolute biomass concentration by replacing X with $C_{XA}D_o$

$$\frac{dD}{dt} = D_o\left(\left(1 - \frac{1}{Y_{XS}}\right)\frac{dC_{XA}}{dt} - m_S C_{XA}\right).$$
(16.12)

Given a kinetic equation written in terms of the absolute biomass concentration, such as one of the equations from Table 16.1, it is possible to use Eqs. (16.8) and (16.12) to predict the growth profile that would be obtained for measurements made on a relative basis (C_{XR}). Figure 16.1 shows how this is done.

In order to undertake this conversion, it is necessary to have values for Y_{XS} and m_S. One method of estimating these parameters is to obtain experimental data in the initial kinetic studies in terms of both the absolute and the relative biomass concentrations. Figure 16.2 shows how this data can be used to obtain estimates for these two parameters.

This conversion is not limited to biomass. It is possible to use the model to convert measurements of biomass components between absolute and relative measurement bases. In this case X will represent the component, Y_{XS} will have the units of kg-component kg-dry-substrate^{-1}, and m_S will have the units of kg-dry-substrate kg-component^{-1} h^{-1}.

16.3 Incorporating the Effect of the Environment on Growth

The kinetic sub-model needs to describe how growth depends on the key environmental variables, since these variables typically cannot be simply controlled at their optimum values in an SSF bioreactor. The bioreactor model will be most useful if it can be used to explore how the operating variables affect the values of the key environmental variables, and how changes in the environmental variables in turn affect the overall performance of the bioreactor.

So which environmental variables are the "key environmental variables"? This question was raised in Sect. 13.2.1, where it was recommended that, at the very least, the effects of temperature and water activity on growth should be described. During a fermentation, these variables can change quite significantly. For example, the temperature might start at the optimum temperature for growth, but it can increase quite substantially during the mid parts of the fermentation, falling again

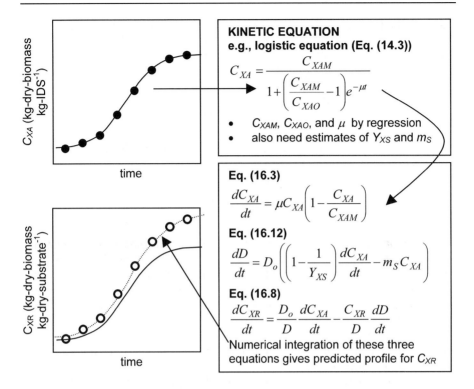

Fig. 16.1. How kinetics determined on an absolute basis can be converted to a relative basis, in order to allow comparison between the model predictions and experimental results. This is necessary since samples removed from a bioreactor are processed to give biomass contents on a relative basis. Note that, even though growth has finished by the end of the fermentation in absolute terms, the relative biomass concentration continues to rise through the conversion of substrate into CO_2 due to maintenance metabolism

as the growth decelerates at the end of the process. In addition, the water activity of the substrate bed may start at the optimum but may then decrease during the fermentation due to the evaporation of water from the bed. Further, these two variables can be influenced significantly by the manner in which the bioreactor is operated, and bioreactor models that describe the effects of these two variables on growth can be used to explore strategies of bioreactor operation that attempt to minimize the deviation of these variables from the optimum values for growth and product formation.

In kinetic models, the effect of these varying environmental variables on growth is taken into account by expressing the parameters in the kinetic equation as functions of the local conditions. Table 16.1 indicates, for each of the kinetic equations, which of the parameters might be expressed as functions of the environmental variables.

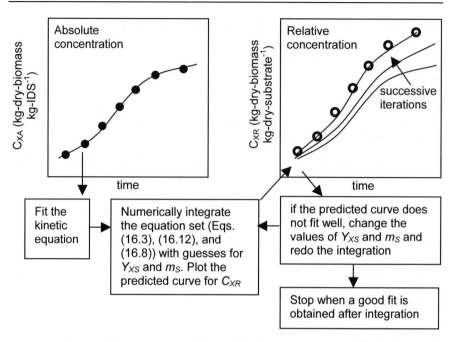

Fig. 16.2. How estimates of Y_{XS} and m_S can be obtained if, during the initial laboratory studies (See Chaps. 14 and 15), growth profile data is obtained in both the absolute and relative concentrations. Note that optimization programs can be used to undertake the iterative fitting of the relative biomass curve

The sections below present experimental approaches that can be used to gather experimental data, and approaches to developing appropriate equations, for the case of temperature and the case of water activity. Note that the recommendations are for "isothermal" and "isohydric" studies, in which conditions are maintained constant throughout the growth cycle, whereas in real SSF processes the temperature and the water activity change during the process. It is possible that expressions for the effects of temperature and water activity that are obtained on the basis of the isothermal and isohydric approaches will not describe the true effect on growth of the time-varying conditions that are encountered by the organism in SSF processes at large scale (Ikasari et al. 1999). The advantage of the isothermal and isohydric approaches is that they are easy to carry out. Possible approaches to determining the effects of temporal variations in the environmental variable are also discussed.

16.3.1 Incorporating the Effect of Temperature on Growth

16.3.1.1 The "Isothermal Approach"

This experimental approach is as follows (see Fig. 14.2):

1. A small-scale experimental system is used so that heat transfer will not be limiting (see Sect. 15.1) and therefore the substrate will be at the temperature of the incubator or waterbath used;
2. Cultures are incubated at various different temperatures, with the temperature experienced by each culture being held constant during the entire growth cycle;
3. The growth profile for each culture is then plotted and the appropriate kinetic equation is fitted to each profile, allowing determination of the values of the parameters of the kinetic equation for each temperature. For example, if the growth curve is logistic, the integrated form of the logistic equation is fitted by non-linear regression to the growth profile. This will yield a specific growth rate constant and a maximum biomass concentration for each temperature;
4. The parameters that are sensitive to temperature are then plotted against temperature and an empirical equation is used to describe this curve, being fitted to the curve by non-linear regression.

16.3.1.2 Equations that Have Been Developed Using this Approach

Equations that have been used to describe the effect of temperature on growth are presented below. All are simply empirical fits to the data.

Saucedo-Castaneda et al. (1990) used a "double Arrhenius" equation to describe the effect of temperature on the specific growth rate constant:

$$\mu_T = A \exp\left(\frac{-E_{a1}}{R(T+273)}\right) \bigg/ \left(1 + B\exp\left(\frac{-E_{a2}}{R(T+273)}\right)\right), \qquad (16.13)$$

where A (h^{-1}), B (dimensionless), and E_{a1} and E_{a2} (J mol^{-1}) are simply fitting parameters, R is the universal gas constant (J mol^{-1} °C^{-1}), μ_T is the specific growth rate parameter (h^{-1}), and T is the temperature (°C). The symbol μ_T is used to denote that the equation describes specifically the effect of temperature on the specific growth rate parameter. Note that this equation does not describe a maximum temperature for growth, since the value of μ_T is always positive and greater than zero. The shape of this curve is shown in Fig. 16.3(a).

The maximum biomass concentration (C_m, g-biomass 100-g-dry-matter^{-1}), which is a parameter in the logistic growth equation, was modeled with a polynomial equation:

$$C_m = a_o + a_1 T + a_2 T^2 + a_3 T^3 + a_4 T^4, \qquad (16.14)$$

for temperature T in °C. The parameters a_o to a_4 are simply fitting parameters.

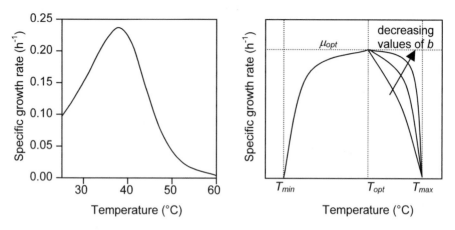

Fig. 16.3. The dependence of the specific growth rate parameter (μ_T) on temperature, as described by two different equations. **(a)** The "double-Arrhenius" equation of Saucedo-Castaneda et al. (1990). Their values for the parameters of the equation were used to plot the curve, being $A = 2.694 \times 10^{11}$ h^{-1}, $B = 1.3 \times 10^{47}$, $E_{a1} = 70225$ J mol^{-1}, $E_{a2} = 283356$ J mol^{-1}. Adapted from Saucedo-Castaneda et al. (1990) with kind permission from John Wiley & Sons, Inc. **(b)** The general shape of the profile described by the equation set of Sangsurasak and Mitchell (1998). The parameter b allows the model to describe greater or lesser sensitivities of μ_T to increases in temperature above the optimum

The advantage of modeling the effect of temperature is not as obvious for C_m as it is for μ_T. In Eq. (16.14) the maximum biomass concentration depends only on the actual temperature. Therefore C_m varies throughout the fermentation and, if the temperature falls back to the value that gives the maximum value for C_m, then the biomass is predicted to reach this value, regardless of the previous high temperatures that the culture may have suffered. In this manner, the effect of Eq. (16.14) (in combination with the kinetic equation) is simply to modify the instantaneous growth rate, not the maximum biomass concentration obtained.

It is highly likely that the temperature history affects the value of C_m. However, there is simply not sufficient data available in the literature to enable an equation to be proposed to describe this effect. One possibility might be to use Eq. (16.14), but only to allow decreases in C_m as the temperature varies above the optimum temperature. That is, once the temperature begins to fall from the maximum temperature reached during the fermentation, the value of C_m then remains fixed at the value it had at the time when the maximum temperature was reached. Experimental validation will be necessary to confirm whether this approach is appropriate.

Sangsurasak and Mitchell (1998) developed a set of empirical equations, which, although being more cumbersome than the equation used by Saucedo-Castaneda et al. (1990), does describe minimum and maximum temperatures for growth. Below the minimum temperature for growth (T_{min}, °C) and above the maximum temperature for growth (T_{max}, °C) the specific growth rate parameter was set to zero. Between the minimum temperature and the optimum temperature (T_{opt}, °C) the following equation was used:

$$\mu_T = \mu_{opt} \left(F_1 + F_2 (T+273) + F_3 (T+273)^2 \right), \tag{16.15}$$

where F_1, F_2, and F_3 are simply fitting constants, determined by non-linear regression of the appropriate part of the curve. Between the optimum and the maximum temperature the following equation was used:

$$\mu_T = \mu_{opt} \left(\frac{b + (T_{\max} - T_{opt})}{(T_{\max} - T_{opt})} \right) \left(\frac{T_{\max} - T}{b + (T_{\max} - T)} \right), \tag{16.16}$$

where μ_{opt}, T_{max}, and T_{opt} were determined by visual inspection of the plot of μ_T against temperature, and the fitting parameter b determines the degree of curvature (Fig. 16.3(b)).

16.3.1.3 Is the "Isothermal Approach" Valid?

The dependence of the growth rate on temperature that is predicted by an equation developed using data obtained by the isothermal approach might not actually be the behavior demonstrated during an actual SSF process (Ikasari et al. 1999). There is a significant difference between the "isothermal approach" and a large-scale SSF process: the temperature in the SSF process does not remain constant; rather, it varies as a function of time. It typically begins at the optimal temperature for growth, and during the early periods the temperature is near the optimal temperature. An organism experiencing a temperature rise from the optimum to say 5°C above the optimum would very likely be healthier than an organism reaching the same temperature during the later stages of the fermentation (Fig. 16.4(a)). In the latter case the organism has recently been exposed to temperatures of as much as 10°C above the optimum, which very likely have had deleterious effects on cell structure and metabolism. The isothermal approach does not predict this, rather it assumes that the specific growth rate constant at any given instant is simply a function of the temperature at that instant (Fig. 16.4(b)).

It is highly likely that the recent history of temperatures experienced by the microorganism influences its current growth rate. For example, intracellular enzymes may denature at high temperatures, and it may take some time to replace them, meaning that high growth rates cannot immediately be re-established, even if the organism is returned to the optimum temperature. Another possibility is that senescence or sporulation may be triggered and, once triggered, may be irreversible, even if in the meantime the organism is returned to the optimal temperature. On the other hand, microorganisms do have mechanisms of adaptation to higher temperatures. Various heat shock proteins are produced and processes are induced that lead to a change in the lipid composition of the membrane. These might take several hours after an elevation of temperature to come into effect, but then growth might accelerate. Unfortunately, there is very little information available in the literature about the effect on growth kinetics of what might be called "sub-lethal temperature excursions". In the absence of more information, the best current strategy is to use the isothermal approach.

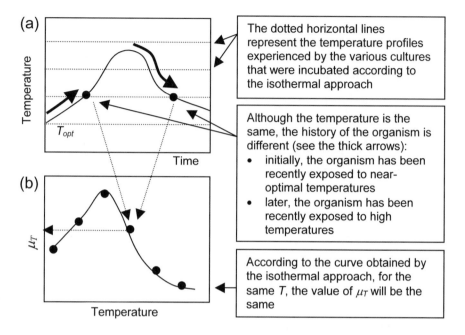

Fig. 16.4. Is the isothermal approach valid? **(a)** A typical temperature profile that might occur in a large-scale bioreactor, demonstrating how the same supra-optimal temperature will be reached twice, once before the temperature peak and once after the temperature peak; **(b)** The isothermal approach gives the same value for μ_T, regardless of the recent temperature history of the microorganism

Recently, a model has been proposed that is capable of describing delayed temperature effects (Dalsenter et al. 2005). The model describes the effect of temperature on the relative rates of synthesis and denaturation of a pool of key metabolic enzymes (Fig. 16.5). In turn, the growth rate of the microorganism depends on the state of this enzyme pool. At the moment this model has not been sufficiently validated to have confidence that it will accurately predict growth rates under a wide range of conditions, however, it does suggest a general strategy by which future models might be developed.

16.3.2 Incorporating the Effect of Water Activity on Growth

16.3.2.1 The Experimental Approach to Collecting Data

A similar concept to the isothermal approach for determining temperature effects has been used to determine the effect of water activity on growth. Various cultures are incubated in various atmospheres of controlled relative humidity (in which the substrate is pre-equilibrated, such that its water activity is equal to the percentage

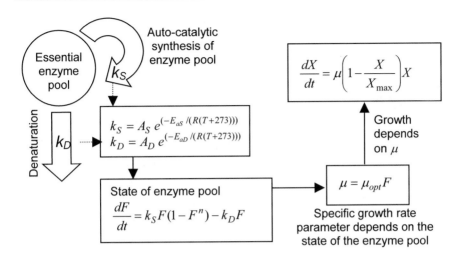

Fig. 16.5. Schematic representation of a model that can describe the effects of the recent temperature history on the growth rate (Dalsenter et al. 2005). F is a nondimensional variable representing the state of the intracellular "essential enzyme pool" and its value varies between 0 and 1. The coefficient of the autocatalytic synthesis reaction (k_S) depends on temperature (T, °C) according to the Arrhenius equation (with frequency factor A_S and activation energy E_{aS}). The coefficient of the denaturation reaction (k_D) depends on temperature according to the Arrhenius equation (with frequency factor A_D and activation energy E_{aD})

relative humidity divided by 100). The growth profile for each culture is analyzed to determine the parameters of the kinetic equation. These parameters are plotted against water activity (see Fig. 14.2) and an empirical equation is fitted to this plot. This approach is referred to here as the "isohydric approach".

In fact, the effect of water activity on growth rates in real SSF systems has been relatively little studied. Instead of this, many studies that involve fungi characterize the effect of water activity on the radial expansion rate of colonies. Furthermore, no effort has been made to look at the effect on growth of variations in the water activity during the growth cycle.

16.3.2.2 Equations that Have Been Developed Using this Approach

A simple empirical equation was used by von Meien and Mitchell (2002):

$$\mu_W = \mu_{opt}\ \exp\left(D_1 a_{ws}^3 + D_2 a_{ws}^2 + D_3 a_{ws} + D_4\right), \tag{16.17}$$

where D_1 to D_4 are fitting parameters and a_{ws} is the water activity of the solid substrate phase. The symbol μ_W is used to denote that the equation describes specifically the effect of water activity on the specific growth rate parameter. von Meien and Mitchell (2002) fitted this equation to data for two different fungi, presented by Glenn and Rogers (1998) (Fig. 16.6).

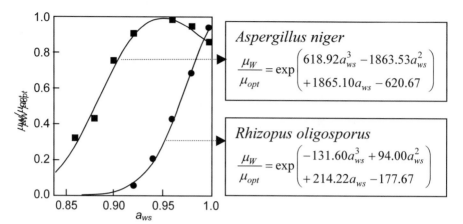

Fig. 16.6. The dependence of the specific growth rate parameter (μ_W) on water activity, as described by the equation of von Meien and Mitchell (2002) for two different organisms. The experimental data is from Glenn and Rogers (1988) and is reproduced with kind permission from the authors

16.3.3 Combining the Effects of Several Variables

If the kinetic model attempts to take into account the effect of both temperature and water activity on growth, the question arises as to how best to combine the effects of simultaneous variations in both variables. The best approach might be to determine the specific growth rate parameter at a large number of different combinations of water activity and temperature and simply use regression against two independent variables to determine an empirical equation (Fig. 16.7). However, to date most studies in which both water activity and temperature have been varied have not explored a sufficiently large number of combinations to allow such equations to be proposed. In the absence of this data, simple rules have been proposed for combining the effects determined in studies in which the variables are varied one-by-one, typically one variable being varied at the optimum value of the other. The maximum value of the specific growth rate constant, determined at the optimum values of water activity and temperature, is denoted μ_{opt}. During the experiments to determine the effect of each environmental variable on growth, the specific growth rate can be expressed as a fraction of this optimum:

$$f = \frac{\mu_{measured}}{\mu_{opt}}. \tag{16.18}$$

For example, in the case of temperature effects, using Eq. (16.13) gives:

$$f_T = \frac{\mu_T}{\mu_{opt}} = \frac{1}{\mu_{opt}} A \exp\left(\frac{-E_{a1}}{R(T+273)}\right) \Big/ \left(1 + B \exp\left(\frac{-E_{a2}}{R(T+273)}\right)\right), \tag{16.19}$$

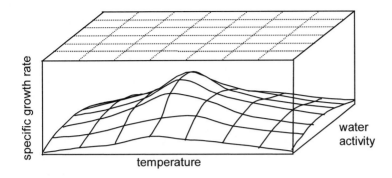

Fig. 16.7. One strategy for determining the combined effect of temperature and water activity on the specific growth rate parameter would be to determine the "response surface", that is, to determine the specific growth rate parameter at various different combinations of temperature and water activity. An equation, involving two independent variables, can then be fitted to this surface. Such a strategy was recently used by Hamidi-Esfahani et al. (2004). The disadvantage is the number of experiments required. This example involves all possible combinations of 8 temperatures and 7 water activities, that is, a total of 56 different experiments.

where the subscript "T" in f_T denotes that this is the fractional specific growth rate based on variations in temperature. Similarly, in the case of water activity effects, using Eq. (7.18) gives:

$$f_W = \frac{\mu_W}{\mu_{opt}} = \frac{1}{\mu_{opt}} \exp\left(D_1 a_{ws}^3 + D_2 a_{ws}^2 + D_3 a_{ws} + D_4\right), \qquad (16.20)$$

If equations are written for all of the environmental variables that are taken into account in the model, then the overall fractional specific growth rate can be calculated on the basis of the geometric mean of the individual fractional specific growth rates (Sargantanis et al. 1993). In the case in which only temperature and water activity are taken into account, the equation for the combined effect on the specific growth rate would be:

$$\mu = \mu_{opt} \sqrt{f_T f_W} \ . \qquad (16.21)$$

16.4 Modeling Death Kinetics

16.4.1 General Considerations in Modeling of Death Kinetics

Given the difficulty in controlling the fermentation conditions, especially the temperature, in large-scale SSF bioreactors, it is quite possible that conditions will occur that cause cells to die. Therefore it might be of interest to describe death kinetics within the kinetic sub-model of the bioreactor model. Note that this has often not been done. In various bioreactor models the kinetics are written in terms of viable biomass only, with the growth rate reflecting the net increase in viable biomass, that is, the true growth rate minus the death rate. In other words, the equation only describes the overall outcome of growth and death, and does not segregate the biomass into live and dead biomass. Note that such an approach can lead to inaccuracies, since if there is significant death then the increase in viable biomass does not represent the overall growth activity. In this case growth-related activities such as metabolic heat generation would be underestimated.

In cases were death is taken into account explicitly, the growth equation is written in terms of the underlying true growth rate and a separate equation expresses the death rate. Note that many SSF processes involve fungi, and it is not necessarily a simple matter to measure fungal death experimentally. The difficulty can be seen by comparing the situation with that of studies of the death of unicellular organisms. In this case, the total cell number can be determined from total counts done in a Neubauer chamber, while the number of viable cells can be determined by viable counts, that is, agitating the culture well to separate the cells, then plating the culture out and counting the number of colonies that arise. In the case of fungi, it is not possible to separate out individual cells in this manner, since they are linked together in the mycelium. Death is often inferred by indirect means, such as a decrease in the specific O_2 uptake rate. As a result, only relatively few attempts have been made to model fungal death kinetics in SSF. Further, no attempts have been made to validate the model predictions about the relative populations of live and dead biomass, rather the growth equations have simply been empirically adjusted to agree with observed growth curves.

Another factor needs to be considered. If the model describes the dry weight of the biomass, death will only cause this dry weight to decrease if the model describes a process of autolysis. In a model in which biomass dies and is converted into dead biomass, which then remains stable, it is not possible for the model to describe decreases in the overall biomass.

16.4.2 Approaches to Modeling Death Kinetics that Have Been Used

The simplest assumption is that death is a first order process, giving the equation:

$$\frac{dC_{XAD}}{dt} = r_d = k_d C_{XAV} , \qquad (16.22)$$

where C_{XAV} and C_{XAD} are the absolute concentrations of viable and dead biomass, respectively, and k_d is the specific death rate coefficient (h^{-1}).

This term might simply be subtracted from the equation for the production of viable biomass. In the case in which growth follows logistic kinetics then the equation for total biomass production might be:

$$\frac{dC_{XAT}}{dt} = \mu C_{XAV}\left(1 - \frac{C_{XAT}}{C_{XAM}}\right) , \qquad (16.23)$$

where C_{XAT} is the absolute concentration of total biomass (i.e., both viable and dead). C_{XAT} appears in the numerator of the term within the parentheses since it is assumed that the biomass-associated limitation of growth is due to the total biomass concentration and not simply the viable biomass concentration. This could be true for the case in which growth is limited by the availability of nutrients.

Subtracting the rate of death (Eq. (16.22)) from the overall rate of biomass production (Eq. (16.23)) gives the rate of increase of viable biomass:

$$\frac{dC_{XAV}}{dt} = \mu C_{XAV}\left(1 - \frac{C_{XAT}}{C_{XAM}}\right) - k_d C_{XAV} . \tag{16.24}$$

Arrhenius equations can be used to express the effect of environmental conditions such as temperature on growth:

$$\mu = A_g \exp\left(\frac{-E_{ag}}{R(T+273)}\right), \tag{16.25}$$

and on death:

$$k_d = A_d \exp\left(\frac{-E_{ad}}{R(T+273)}\right), \tag{16.26}$$

where T is the temperature (°C), A_g and A_d are the frequency factors for growth and death (h^{-1}) and E_{ag} and E_{ad} are the activation energies for growth and death (J mol^{-1}). Typical profiles that could be expected for these two rate constants against temperature are shown in Fig. 16.8.

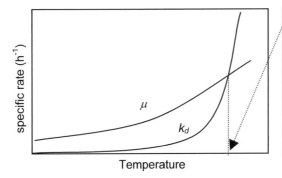

Critical temperature: the amount of viable biomass cannot increase at or above this temperature. The effect of the term $(1-C_{XAT}/C_{XAm})$ in Eq. (16.24) means that, as C_{XAT} increases, this critical temperature falls to lower and lower values

Fig. 16.8. Typical behavior that might be expected for the specific growth rate and specific death rate parameters as a function of temperature according to the Arrhenius equations (Eqs. (16.25) and (16.26))

16.5 Conclusion

This chapter has shown how the basic empirical kinetic equation is written, and how the parameters of the equation can be written as functions of the key environmental variables. The next chapter extends the discussion to how we can model the effects that growth has on the environment of the organism.

Further Reading

A detailed development of a system of equations for converting between biomass profiles expressed on relative and absolute bases
Viccini G, Mitchell DA, Krieger N (2003) A model for converting solid state fermentation growth profiles between absolute and relative measurement bases. Food Technol Biotechnol 41:191–201

Alternative approaches to modeling temperature effects
Dalsenter FDH, Viccini, G, Barga MC, Mitchell DA, Krieger N (2005) A mathematical model describing the effect of temperature variations on the kinetics of microbial growth in solid-state culture. Process Biochemistry 40:801–807

Combined temperature and moisture effects on growth kinetics
Hamidi-Esfahani Z, Shojaosadati SA, Rinzema A (2004) Modelling of simultaneous effect of moisture and temperature on *A. niger* growth in solid-state fermentation. Biochem Eng J 21:265–272

17 Modeling of the Effects of Growth on the Local Environment

David A. Mitchell and Nadia Krieger

17.1 Introduction

In a mathematical model of a bioreactor, it will be necessary to write terms in the mass and energy balance equations to describe the changes that the microorganism causes in its local environment since, in turn, the changes in the local environment affect the rate of growth of the microorganism (Fig. 17.1). The mass and energy balance equations and the terms in them that are related to transport phenomena will be presented in later chapters. The current chapter concerns itself with those terms within these balance equations that are related to growth and maintenance metabolism by the microorganism.

Various effects that we may wish to include in a bioreactor model are:

- the liberation of waste metabolic heat;
- the consumption of substrate (i.e., overall residual substrate) or particular nutrients;
- the consumption of O_2 and production of CO_2;
- the production of water;
- the formation of products.

Typically the aim will be to link each of these growth-related activities with the kinetic equation. Note that it is not necessary for the kinetic equation to be expressed in terms of dry biomass. If the appropriate parameters are used, the growth-related activities can be related to a kinetic equation expressed in terms of a biomass component or activity.

The equations in this chapter are written in absolute terms, that is, the symbol X represents the total mass of biomass within the bioreactor (kg). The equations can also be expressed in terms of the "absolute biomass concentration" by dividing the entire equation by the initial mass of dry solids within the bioreactor (D_o, kg). Once this is done, the ratio X/D_o can be replaced by the symbol C_{XA} (the meanings of these symbols are explained in Chap. 15). In this case, the corresponding rate term will have the units of kg or mol per kg-IDS (initial dry solids).

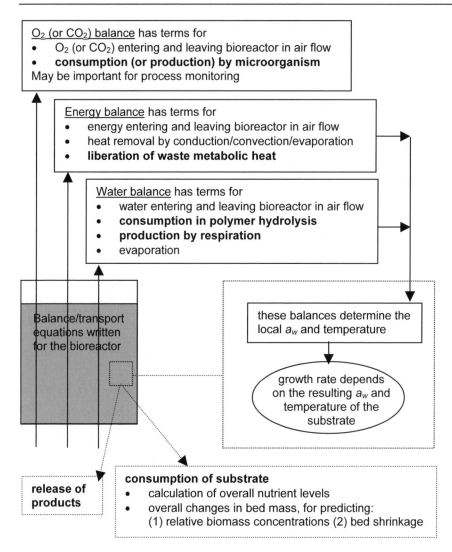

Fig. 17.1. The importance of describing changes in the environment caused by growth and maintenance activities. These changes in turn affect growth (see also Fig. 12.4)

In this chapter all rates (given the symbol "r" and an appropriate subscript, and with units of kg or mol per hour) are calculated as positive numbers, regardless of whether they represent components that are produced or consumed. When the time comes to incorporate these terms within balance equations, it will be necessary to subtract the term if it is for a component that is consumed and to add the term if it is for a component that is produced. Also note that stoichiometric coefficients within these equations can be written in two ways. For example, to express the stoichiometric relationship between A and B, we can use Y_{AB} (kg-A kg-B^{-1}) or Y_{BA} (kg-B kg-A^{-1}). These can be easily interchanged by remembering that one is the reciprocal of the other (i.e., $Y_{AB} = 1/Y_{BA}$).

17.2 Terms for Heat, Water, Nutrients, and Gases

The following sections cite some typical values for the relevant coefficients that have been used in the literature. However, it must be realized that the values of these parameters will be affected to a large extent by the organism and substrate used and the conditions under which the fermentation is carried out.

17.2.1 Metabolic Heat Production

Waste heat production will be associated with both growth-related and maintenance metabolism:

$$r_Q = Y_{QX} \frac{dX}{dt} + m_Q X , \qquad (17.1)$$

where r_Q is the overall rate of metabolic waste heat production (J h^{-1}). As will be seen in the bioreactor modeling section, this waste heat production term appears within the overall energy balance equation, the equation that is used to describe changes in the temperature of the bed. Y_{QX} is the yield of heat from the growth reaction (J kg-dry-biomass^{-1}), and m_Q is the coefficient for heat production associated with maintenance metabolism (J kg-dry-biomass^{-1} h^{-1}).

Direct determination of Y_Q and m_Q requires careful calorimetric studies, although reasonable estimates can probably be obtained by relating these values to O_2 consumption, given that, for aerobic growth of a number of different microorganisms, the heat yield based on O_2 consumption is about 519 kJ mol-O_2^{-1} (Cooney et al. 1968). Various values that have been used in the SSF literature are shown in Table 17.1.

Table 17.1. A selection of reported metabolic heat yield and maintenance coefficients

Symbol	Organism	Value[a,b]	Reference
Y_{QX}	*Rhizopus oligosporus*	8.366×10^6 J kg-X^{-1}	Sargantanis et al. (1993)
m_Q	*Rhizopus oligosporus*	329.3 J kg-X^{-1} h^{-1}	Sargantanis et al. (1993)
Y_{QX}	*Gibberella fujikuroi*	1.54×10^7 J kg-X^{-1}	Pajan et al. (1997)
m_Q	*Gibberella fujikuroi*	3.3×10^5 J kg-X^{-1} h^{-1}	Pajan et al. (1997)
Y_{QC}	*Gibberella fujikuroi*	7.0233×10^6 J kg-CO_2^{-1}	Lekanda and Pérez-Correa (2004)

[a] Where appropriate, values have been converted from the units used by the authors.
[b] "X" in the units stands for "dry-biomass" (this also applies to other tables in this chapter).

17.2.2 Water Production

The equation for water production is:

$$r_W = Y_{WX} \frac{dX}{dt} + m_W X, \qquad (17.2)$$

where r_W is the overall rate of metabolic water production (kg h^{-1}). This term may be included in a water balance equation within the bioreactor model. Y_{WX} is the yield of water from the growth reaction (kg-H$_2$O kg-dry-biomass^{-1}) and m_W is the coefficient for water production associated with maintenance metabolism (kg-H$_2$O kg-dry-biomass^{-1} h^{-1}).

The parameters Y_{WX} and m_W need to be determined in careful studies, because the production of metabolic water may be small compared to the total water in the system, and there is the possibility of condensation or evaporation occurring. Further, it is not possible to distinguish experimentally between water that was produced by respiration and water that already was in the medium. Estimated values may include the contribution of the consumption of water by extracellular hydrolysis reactions, although this is generally a minor contributor to the overall water balance and is often neglected.

Estimates of the yield coefficient can be made on the basis of assumptions about the stoichiometry of the growth reaction, which will be significantly affected by the types of nutrients being consumed and whether growth is aerobic or anaerobic. In the same way, the maintenance coefficient can be simply correlated to m_o, on the basis of an assumed stoichiometry of the maintenance reaction, rather than being measured experimentally. Values that have been used in the literature are shown in Table 17.2.

Table 17.2. A selection of reported metabolic water yield and maintenance coefficients

Symbol	Organism	Value	Reference
Y_{WX}	*Rhizopus oligosporus*	0.304 kg-H$_2$O kg-X^{-1}	Sargantanis et al. (1993)
m_W	*Rhizopus oligosporus*	0.0106 kg-H$_2$O kg-X^{-1} h^{-1}	Sargantanis et al. (1993)
Y_{WO}	*Aspergillus oryzae*	1.22 mol-H$_2$O mol-O$_2$$^{-1}$	Nagel et al. (2001b)
Y_{WO}	*Gibberella fujikuroi*	0.79 mol-H$_2$O mol-O$_2$$^{-1}$	Lekanda and Pérez-Correa (2004)

17.2.3 Substrate and Nutrient Consumption

Section 16.2 presented an equation for consumption of the overall residual dry substrate (Eq. 16.10), it being necessary to calculate this in order to convert absolute biomass concentrations into relative values.

Even though it is not appropriate to try to model intra-particle diffusion of nutrients in bioreactor models (Sect. 13.2.1), it may be of interest to calculate the residual quantity of a specific nutrient. In this case the basic equation is:

$$r_N = \frac{1}{Y_{XN}}\frac{dX}{dt} + m_N X , \qquad (17.3)$$

where r_N is the overall rate of nutrient consumption (kg-nutrient h^{-1}), which will be subtracted in the appropriate place within a balance equation. Y_{XN} is the yield of biomass from that nutrient (kg-dry-biomass kg-nutrient^{-1}) and m_N is the maintenance coefficient for that nutrient (kg-nutrient kg-dry-biomass^{-1} h^{-1}).

Given that solid substrates are typically complex mixtures of various carbon-containing nutrients, it is not necessarily a simple matter to determine the yield and maintenance coefficients experimentally. In the case of overall residual dry substrate it may be possible to obtain estimates of Y_{XS} and m_S by the method shown in Fig. 16.2. Values reported in the literature for parameters related to the overall residual dry substrate (Y_{XS} and m_S) and specific nutrients (Y_{XN} and m_N) are shown in Table 17.3.

Table 17.3. A selection of reported metabolic nutrient yield and maintenance coefficients

Symbol	Organism	Value	Reference
m_S	Coniothyrium minitans	1.82×10^{-4} kg-S kg-X^{-1} day^{-1} 2.62×10^{-6} kg-S kg-X^{-1} day^{-1} 0.020 kg-S kg-X^{-1} day^{-1}	Ooijkaas et al. (2000) (values obtained in different conditions)
Y_{XS}	Rhizopus oligosporus	0.625 kg-S kg-X^{-1}	Sargantanis et al. (1993)
m_S	Rhizopus oligosporus	0.01932 kg-S kg-X^{-1} h^{-1}	Sargantanis et al. (1993)
Y_{XS}	Gibberella fujikuroi	0.55 kg-X kg-S^{-1}	Thibault et al. (2000b)
Y_{XS}	Gibberella fujikuroi	0.35 kg-X kg-glucose^{-1}	Pérez-Correa and Agosin (1999)
m_S	Gibberella fujikuroi	0.0110 kg-glucose kg-X h^{-1}	Pajan et al. (1997)
m_S	Gibberella fujikuroi	0.0125 kg-S kg-X^{-1} h^{-1}	Thibault et al. (2000b)
m_S	Gibberella fujikuroi	0.028 kg-glucose kg-X^{-1} h^{-1}	Pérez-Correa and Agosin (1999)
Y_{XN}	Gibberella fujikuroi	14.397 kg-X kg-nitrogen^{-1}	Pajan et al. (1997)
Y_{XN}	Gibberella fujikuroi	13.5 kg-X kg-nitrogen^{-1}	Thibault et al. (2000b)
Y_{NO}	Gibberella fujikuroi	0.56 kg-nitrogen kg-O$_2$$^{-1}$	Lekanda and Pérez-Correa (2004)

17.2.4 Oxygen Consumption and Carbon Dioxide Production

O_2 consumption and CO_2 evolution are of particular interest, since they represent the most convenient way of on-line monitoring of the growth in a bioreactor. Furthermore, in aerobic systems, as mentioned in Sect. 17.2.1, the heat generation rate is typically directly proportional to the O_2 consumption rate. It is therefore of great interest to model either or both of O_2 consumption and CO_2 production within the bioreactor model.

The expression for O_2 consumption within an O_2 balance equation will be:

$$r_O = OUR = \frac{1}{Y_{XO}} \frac{dX}{dt} + m_o X \ , \tag{17.4}$$

where r_O, also called the oxygen uptake rate (OUR), is the overall rate of O_2 consumption (mol-O_2 h^{-1}), Y_{XO} is the yield of biomass from O_2 (kg-dry-biomass mol-O_2^{-1}), and m_o is the maintenance coefficient for O_2 (mol-O_2 kg-dry-biomass^{-1} h^{-1}).

The expression for CO_2 evolution within a CO_2 balance equation will be:

$$r_C = CER = Y_{CX} \frac{dX}{dt} + m_c X \ , \tag{17.5}$$

where r_C, also called the carbon dioxide evolution rate (CER), is the overall rate of CO_2 production (mol-CO_2 h^{-1}), Y_{CX} is the yield of CO_2 from biomass (mol-CO_2 kg-dry-biomass^{-1}), and m_c is the maintenance coefficient for CO_2 (mol-CO_2 kg-dry-biomass^{-1} h^{-1}).

Typical values of the gas metabolism parameters that have been reported in the literature are shown in Table 17.4. The following subsections outline a procedure by which they may be determined experimentally during the initial studies of growth kinetics. It is necessary to do these experiments in a system in which both the O_2 consumption and the biomass can be measured, such as "membrane filter culture" (Sect. 15.3.1). The procedure is based on the case in which the O_2 uptake rate is measured, but exactly the same steps can be taken if measurements are done with the CO_2 evolution rate.

Table 17.4. A selection of reported gas metabolism yield and maintenance coefficients

Symbol	Organism	Value	Reference
Y_{CX}	*Rhizopus oligosporus*	0.76394 kg-CO_2 kg-X^{-1}	Sargantanis et al. (1993)
m_c	*Rhizopus oligosporus*	0.031 kg-CO_2 kg-X^{-1} h^{-1}	Sargantanis et al. (1993)
Y_{XC}	*Gibberella fujikuroi*	1.0496 kg-X kg-CO_2^{-1}	Thibault et al. (2000b)
m_c	*Gibberella fujikuroi*	0.0150 kg-CO_2 kg-X^{-1} h^{-1}	Thibault et al. (2000b)
m_o	*Gibberella fujikuroi*	0.0130 kg-O_2 kg-X^{-1} h^{-1}	Pajan et al. (1997)
m_c	*Gibberella fujikuroi*	0.0140 kg-CO_2 kg-X^{-1} h^{-1}	Pajan et al. (1997)
Y_{XO}	*Gibberella fujikuroi*	0.9510 kg-X kg-O_2^{-1}	Pajan et al. (1997)
Y_{XC}	*Gibberella fujikuroi*	0.625 kg-X kg-CO_2^{-1}	Pajan et al. (1997)
m_c	*Gibberella fujikuroi*	0.12 kg-CO_2 kg-X^{-1} h^{-1}	Perez-Correa and Agosin (1999)
Y_{XC}	*Gibberella fujikuroi*	0.19 kg-X kg-CO_2^{-1}	Perez-Correa and Agosin (1999)
Y_{XO}	*Aspergillus oryzae*	1.06-1.16 Cmol-X mol O_2^{-1}	Nagel et al. (2001b)[b]

[a] In this case biomass was measured indirectly on the basis of glucosamine.
[b] Cmol stands for "carbon mol".

17.2.4.1 Experimental Approach for Parameter Estimation

The first challenge is to measure the O_2 uptake rate (OUR, mol-O_2 h^{-1}) experimentally. There are various possibilities. The most reliable method is to grow a culture within an enclosed headspace that is continuously aerated with a known flow rate of air and to measure the inlet and outlet O_2 concentrations (C_{in} and C_{out}, respectively), as shown in Fig. 17.2. The OUR can then be calculated as:

$$OUR = F(C_{in} - C_{out}) . \tag{17.6}$$

The variables on the right hand side of Eq. (17.6) can have various different units, as long as the units used combine to give the correct units for the OUR. For example, F, the dry air flow rate, might originally be measured in L h^{-1}, which will need to be converted to mol-dry-air h^{-1}, taking into account the temperature and pressure of the air. C_{in} and C_{out} will typically be measured as volume percentages, in air that has been dried to prevent water from interfering with the measurement (which is important if a paramagnetic O_2 analyzer is used). At the low pressures used in SSF processes, the air will behave as an ideal gas, and therefore the %(v/v) is also equal to the mol% of O_2 in the gas. It is then simple to express C_{in} and C_{out} in terms of mol-O_2 mol-dry-air^{-1} (i.e., dividing the %(v/v) by 100).

Another possibility is to remove a sample of a culture from a fermentation and place it in an enclosed headspace with an O_2 electrode of the type that is capable of measuring O_2 concentrations in a gas phase. It is also possible to undertake the culture in a sealed chamber and remove and analyze samples by gas chromatography, although in this case the O_2 level in the chamber will decrease significantly during the growth cycle and this might influence growth.

17.2.4.2 Treatment of the Data for Parameter Estimation

If the experiments are done in a system that allows biomass measurement, it is possible to determine the values of Y_{XO} and m_o. Once the growth kinetic equation has been determined from the biomass profile, both the integrated and differential forms of the kinetic equation can be substituted into Eq. (17.4) (the differential forms of various kinetic equations were presented in Table 16.1, while the integral forms of these equations were presented in Table 14.1). The resulting equation can be integrated to give an equation that gives the cumulative O_2 uptake (the total mol of O_2 consumed since growth commenced, called "COU") as a function of time. For example, if the organism shows logistic growth kinetics, then substitution of the differential form (Eq. 16.3) and the integral form (Eq. 14.6) into Eq. (17.4) and subsequent integration will give Ooijkaas et al. (2000):

$$COU = X_m \left[\frac{\frac{1}{Y_{XO}}}{1 + \left(\frac{X_m}{X_o} - 1 \right) e^{-\mu t}} - \frac{\left(\frac{1}{Y_{XO}} \right)}{\left(\frac{X_m}{X_o} \right)} \right] + \frac{X_m m_o}{\mu} \ln \left[\frac{1 + \left(\frac{X_m}{X_o} - 1 \right) e^{-\mu t}}{\left(\frac{X_m}{X_o} - 1 \right) e^{-\mu t}} \right], \tag{17.7}$$

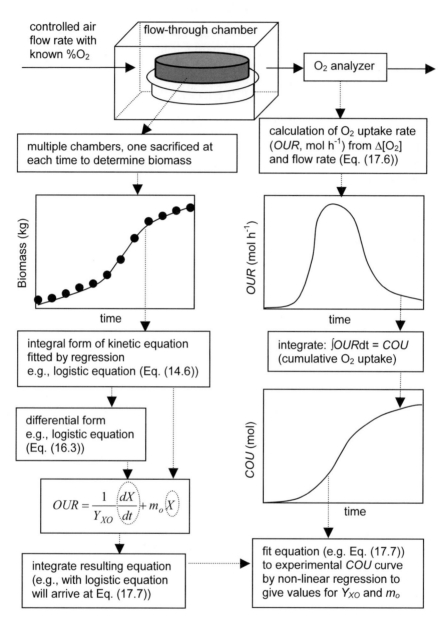

Fig. 17.2. Procedure by which Y_{XO} and m_o can be determined experimentally during the kinetic studies. The example shows direct biomass measurement by membrane filter culture. However, the same procedure can be undertaken in fermentations of the substrate that is used in the SSF process if a component of the biomass is measured and the kinetics are expressed in terms of this component

assuming a *COU* of 0 at zero time. This equation appears complex, but the values of X_m, X_o, and μ are already known from the regression analysis of the biomass profile. The only two remaining unknowns are Y_{OX} and m_o, and these can be determined by non-linear regression of Eq. (17.7) against the experimental profile for *COU* against time.

Similar equations could be derived for exponential and linear growth kinetics. However, this would be more complicated, because these simple equations typically only describe a part of the growth profile in a SSF process, so a "piecewise fitting" of the growth and profile might need to be made.

This analysis can also be done in a system in which biomass can only be measured indirectly, through determination of a component of the biomass. In this case X would represent the absolute amount of component and Y_{XO} and m_o would express the relationships between O_2 and the component.

17.2.5 General Considerations with Respect to Equations for the Effects of Growth on the Environment

Within the various bioreactor models that have been proposed to date, the maintenance term (the second term on the right-hand side of Eqs. (17.1) to (17.5)) has often been omitted. This has been done to simplify the equation and not because maintenance metabolism is negligible. In fact, due to the physiological stress that the microorganism experiences during SSF, maintenance metabolism is often significant.

The procedure outlined in Fig. 17.2 for the determination of Y_{XO} and m_o can potentially be adapted to determine the yield and maintenance parameters for heat, nutrients, and CO_2 (the case of water is more difficult due to the possibility of evaporation occurring). This procedure is somewhat more complicated than the determination of yield and maintenance coefficients in continuous culture in an SLF system. However, these parameters should not be determined in continuous SLF since the yield and maintenance coefficients in liquid culture will likely be quite different from those in SSF.

This procedure for determining yield and maintenance coefficients would typically be undertaken at the optimal values of temperature and water activity. However, in reality, these coefficients will be functions of temperature and water activity, and the temperature and water activity vary during the fermentation within a large-scale SSF bioreactor. A large amount of effort would be required to obtain sufficient experimental data to allow these coefficients to be expressed as functions of the environmental conditions. In the great majority of cases this has not been done, rather, these coefficients have been treated as constants, determined under controlled conditions at laboratory-scale. One exception is the work of Smits et al. (1999) where the maintenance coefficients were expressed as a function of the fermentation time. For example, their equation for O_2 consumption was written as:

$$r_O = OUR = \frac{1}{Y_{XO}} \frac{dX}{dt} + (m_o - D)X , \tag{17.8}$$

where the value of D changes over time, causing the apparent maintenance coefficient to change. Note that this could represent the situation in which the true maintenance coefficient was constant but biomass was dying, such that not all of the biomass was contributing to maintenance activity. Smits et al. (1999) investigated various forms of expressions for D, noting that realistic predictions for their system were given by the following:

$$0 \le t < t_d; D = 0 \tag{17.9a}$$
$$t_d \le t < t_r; D = m_d(t\text{-}t_d) \tag{17.9b}$$
$$t \ge t_r; D = m_d(t_r\text{-}t_d). \tag{17.9c}$$

This model says that before t_d there is no death, that is, all the biomass contributes to maintenance metabolism, then between t_d and t_r there is a linear decline in maintenance activity. After time t_r there is a new and constant level of specific maintenance activity, equal to $(m_o\text{-}m_d(t_r\text{-}t_d))$. They determined the values for m_d, t_d, and t_r by comparing experimental profiles for the biomass with the experimental O_2 uptake rate results.

17.3 Modeling Particle Size Changes

Particle size reduction may occur as a result of the growth process. For example, if the microorganism degrades a polymer that is responsible for the particle structure, then the particle size will decrease. It may be of interest to describe the reduction in particle size within the bioreactor model. For example, the reduction in particle size might be used to estimate the decrease in overall bed volume.

To date the modeling of particle size reduction has not been done in association with bioreactor models, although there is no reason why it could not be done.

17.3.1 An Empirical Equation for Particle Size Reduction

Nandakumar et al. (1994) derived an equation for the length of the residual substrate particle flakes. The equation was derived on the basis of the assumption that the consumption of the substrate particle at the substrate/biomass interface was limited by diffusion of O_2 through the biomass film, which was assumed to grow in such a manner as to keep the overall particle size constant. Their assumptions are unlikely to be true in practice and therefore the equation should be considered as empirical. Other empirical equations might also adjust well to experimental data. The equation was (Nandakumar et al. 1994):

$$\frac{t}{T} = 1 + \frac{l_c^2}{L^2} - 2\frac{l_c}{L} , \tag{17.10}$$

where L is the initial particle length, l_c is the residual particle length at time t, and T is the time for complete particle degradation. Based on their model, they showed how T could be expressed in terms of fundamental constants, however, in practice they determined T by regression of Eq. (17.10) against experimental data for residual particle size versus time. This regression can be done by treating t as the dependent variable and the fractional particle length $(\lambda=l_c/L)$ as the independent variable, in which case the equation is:

$$t = T(1+\lambda^2-2\lambda). \tag{17.11}$$

To measure residual particle size experimentally, it is necessary first to remove the biomass layer. Nandakumar et al. (1994) simply sieved their wet fermented substrate. The process organism was a bacterium, and they claimed that it was easily removed during the wet sieving. Such studies would be more difficult with a fungus, which would bind more tightly to the residual substrate particle.

17.3.2 How to Model Particle Size Changes in Bioreactor Models?

As noted above, it may be interesting to model particle size changes in bioreactor models in order to predict changes in the overall bed volume. This requires an understanding of various factors:

- degradation of the residual substrate particle;
- expansion of the biofilm, and how this is affected by reduction in the size of the residual substrate particle;
- interactions between particles caused by biomass and gravity and how this affects bed structure.

Particle degradation has not yet been taken into account in bioreactor models in SSF. More understanding of the phenomenon and how it affects the bed structure is needed before it can be incorporated into a model in a meaningful way. A useful model would not only predict changes in bed volume but also in the bed porosity. Bed porosity is important for bioreactors that are forcefully aerated, since it is one of the factors that determine the pressure drop through the bed.

Such models may or may not attempt to model the development of the biomass structure above the particle surface. The work of Rajagopalan et al. (1997) gives an idea of how this might be done, at least for a residual substrate particle of constant size. They modeled the expansion of a biofilm of constant density, for a single particle. Applying such a model to the situation in a bioreactor would be more complex since the spatial distribution of particles would prevent the biomass from expanding freely. Such a model could predict the filling in of void spaces, and therefore would be useful for predicting changes in bed porosity.

17.4 Product Formation – Empirical Approaches

Given the difficulties in modeling biomass growth in SSF systems even with simple empirical equations, it is not surprising that much attention has been paid to overcoming these difficulties, and that little attention has been paid to modeling the kinetics of product formation.

In fact, it may be very difficult to use other than simple empirical equations for product formation within a bioreactor model, especially for products like enzymes or secondary metabolites. The production of these products can depend on variables such as the rate of nutrient uptake. As argued in Sect. 13.2, nutrient uptake is often controlled by the rate at which the nutrient diffuses to the surface, and this can only be predicted by a model that describes intra-particle diffusion processes. Such models are too complex to include in fast-solving bioreactor models.

The empirical equation of Leudeking and Piret might be used (Ooijkaas et al. 2000):

$$r_P = \frac{dP}{dt} = Y_{PX}\frac{dX}{dt} + m_P X , \qquad (17.12)$$

where r_P is the overall rate of product formation (kg h^{-1}). Y_{PX} is the yield of product from the growth reaction (kg-product kg-dry-biomass^{-1}) and m_P is the coefficient for product formation related to maintenance metabolism (kg-product kg-dry-biomass^{-1} h^{-1}).

In order to determine the yield and maintenance coefficients of Eq. (17.12) it may be necessary to fit an integrated version of this equation to the product profile, in a manner similar to that shown in Fig. 17.2. As in that case, it is necessary to substitute the integral and differential versions of the kinetic equation into Eq. (17.12). For example, with logistic growth kinetics, it is possible to derive the following integrated equation (Ooijkaas et al. 2000):

$$P = P_o + X_m\left[\frac{Y_{PX}}{1+\left(\frac{X_m}{X_o}-1\right)e^{-\mu t}} - \frac{Y_{PX}}{\left(\frac{X_m}{X_o}\right)}\right] + \frac{X_m m_P}{\mu}\ln\left[\frac{1+\left(\frac{X_m}{X_o}-1\right)e^{-\mu t}}{\left(\frac{X_m}{X_o}-1\right)e^{-\mu t}}\right], \qquad (17.13)$$

where P_o is the product present at time zero (kg). As with Eq. (17.7), if X_o, X_m, and μ are determined from the biomass profile, then the only unknowns in this equation are m_P and Y_{PX}, and these can be determined by a least-squares fitting of Eq. 17.13) to the product profile.

Ooijkaas et al. (2000) used this approach to characterize spore production kinetics, with two minor differences. Firstly, the maintenance-associated spore production was assumed to be zero, meaning that the term involving m_p disappeared from the equation and, secondly, there was a lag in the appearance of the first spores, which was taken into account by subtracting the lag phase from the total fermentation time. Additionally, the initial spore number was taken as zero. For a product that follows these conditions, the equation is:

$$t \le \lambda \qquad P = 0; \tag{17.14a}$$

$$t > \lambda \qquad P = X_m \left[\frac{Y_{PX}}{1 + \left(\frac{X_m}{X_o} - 1 \right) e^{-\mu(t-\lambda)}} - \frac{Y_{PX}}{\left(\frac{X_m}{X_o} \right)} \right]. \tag{17.14b}$$

17.5 Conclusions

Chapters 14 to 17 have given an overview of the experimental and mathematical steps in the development of the kinetic sub-model of an SSF bioreactor model. Chapters 18 to 20 will address basic principles related to the balance/transport sub-model.

Further Reading

Oxygen uptake measurements and calculations
Sato K, Nagatani M, Nakamura KI, Sato S (1983) Growth estimation of *Candida lipolytica* from oxygen uptake in a solid state culture with forced aeration. J Ferment Technol 61:623–629

A classical paper demonstrating the close relationship between oxygen consumption and metabolic heat production
Cooney CL, Wang DIC, Mateles RI (1968) Measurement of heat evolution and correlation with oxygen consumption during microbial growth. Biotechnol Bioeng 11:269–281

Modeling particle size changes
Rajagopalan S, Rockstraw DA, Munson-McGee SH (1997) Modeling substrate particle degradation by *Bacillus coagulans* biofilm. Bioresource Technol 61:175–183

General reviews of modeling of growth kinetics that address of the issue of how the effects of growth on the environment can be described
Mitchell DA, Berovic M, Krieger N (2000) Biochemical engineering aspects of solid state bioprocessing. Adv Biochem Eng/Biotech 68:61–138
Mitchell DA, von Meien OF, Krieger N, Dalsenter FDH (2004) A review of recent developments in modeling of microbial growth kinetics and intraparticle phenomena in solid-state fermentation. Biochem Eng J 17:15–26

A model that describes water metabolism quite carefully, demonstrating how the effects of growth on the water content of the substrate could be described
Nagel FJI, Tramper J, Bakker MSN, Rinzema A (2001b) Model for on-line moisture-content control during solid-state fermentation. Biotechnol Bioeng 72:231–243

18 Modeling of Heat and Mass Transfer in SSF Bioreactors

David A. Mitchell, Oscar F. von Meien, Luiz F.L. Luz Jr, and Marin Berovič

18.1 Introduction

Chapters 22 to 25 present case studies in which mathematical models are used to explore the design and operation of various SSF bioreactors. Chapters 18 to 20 address the basic principles of the balance/transport sub-models of these bioreactor models.

The various phenomena that need to be described by the balance/transfer sub-model, such as conductive and convective heat transfer, were covered in a qualitative manner in Chap. 4. The current chapter shows the mathematical expressions that are used to describe these phenomena. The aim is not to teach heat and mass transfer principles to a depth that will allow readers to construct the appropriate mathematical expressions themselves. Rather, it is to enable readers to inspect a mathematical model of an SSF bioreactor and recognize which transport phenomena are described by the model, on the basis of the various terms that appear within the model equations. These terms include various system, thermodynamic, and transport parameters. Chapters 19 and 20 quote some typical values that have been used for these parameters and give some general advice as to how they might be determined experimentally. However, please note that detailed experimental instructions are not provided.

18.2 General Forms of Balance Equations

The transport/balance part of a mathematical model of a bioreactor consists of mass and energy balance equations. Such an equation expresses how a key system variable changes over time and includes terms that describe various phenomena that affect that variable.

Regardless of what the units of the variable of interest are, the balance equation should initially be written in such a way that all of its terms have units of either kg h^{-1}, in the case of a mass balance, or J h^{-1}, in the case on an energy balance. After this the equation can be rearranged if necessary to isolate the variable of interest.

As an example, an energy balance will appear in the form:

$$m_{bed} C_{Pbed} \frac{dT_{bed}}{dt} = \pm Q_A \pm Q_B \pm Q_C + ... + r_Q, \qquad (18.1)$$

where m_{bed} is the mass of the bed (kg), C_{Pbed} is the overall heat capacity of the bed (J kg^{-1} °C^{-1}), T_{bed} is the bed temperature (°C), r_Q is the rate of metabolic heat production (J h^{-1}) (see Eq. (17.1)), and Q_A, Q_B, and Q_C represent expressions that describe the rates at which different heat transport phenomena occur (all in J h^{-1}). Whether they are added or subtracted will depend on whether they tend to increase or decrease the energy of the bed. The current chapter addresses the question of how these various "Q-terms" can be written mathematically. Equation (18.1) says that the rate of change in the amount of energy stored within the bed (in J h^{-1}), which is represented by the left hand side of the equation, depends on the rates of the various processes that either add energy to the bed or remove energy from it. Equation (18.1) is written in terms of energy, because this is a conserved quantity, whereas temperature is not. Later on, this equation will be rearranged to leave only dT_{bed}/dt on the left hand side, since this is actually the system variable of interest.

The construction of the left hand side of Eq. (18.1) can be understood by assuming that initially a substrate bed is at a temperature $T_{initial}$, and during the fermentation a part of the metabolic heat released by growth remains in the bed, increasing its temperature. The amount of "extra energy" held within the substrate bed due to this increase in temperature is given by the product of the mass of the bed, the heat capacity of the bed and the temperature difference:

$$\text{"Extra Energy"} = m_{bed} C_{Pbed} (T_{bed} - T_{initial}), \qquad (18.2)$$

which can be shown by determining the units of the result of the calculation (i.e., kg × J kg^{-1} °C^{-1} × °C simplifies to give J).

On the other hand, a mass balance, for example, a balance on the water in the bed, will appear in the form:

$$\frac{dM_{water}}{dt} = \pm R_A \pm R_B \pm R_C + ... + r_W, \qquad (18.3)$$

where M_{water} is the overall mass of water in the bed (kg), r_W is the rate of metabolic water production (kg h^{-1}) (See Eq. (17.2)), and R_A, R_B, and R_C represent the rates of various mass transfer phenomena that involve water (all in kg h^{-1}). Whether they are added or subtracted will depend on whether they tend to increase or decrease the amount of water in the bed. The current chapter addresses the question of how these various "R-terms" can be written mathematically. Equation (18.3) says that the rate of change in the mass of water in the bed (in kg h^{-1}), which is represented by the left hand side of the equation, depends on the rates of the various processes that either add water to the bed or remove water from it.

Note that it may be desirable to have an equation that expresses directly the rate of change of the water content of the bed (W, kg-water kg-dry-solids^{-1}), and not the total mass of water in the bed. Even in this case, the equation should initially be written in the form shown in Eq. (18.3). The term W can then be separated out by realizing that the total amount of water in the bed is the product of the water content W and the total mass of dry solids in the bed (D, kg-dry-solids). In other words, "kg-water kg-dry-solids^{-1} × kg-dry-solids" simplifies to give units of "kg-water". Of course, since both the water content and the total mass of dry solids in the bed are changing over time, W must be isolated using the product rule of differentiation:

$$\frac{dM_{water}}{dt} = \frac{d(WD)}{dt} = D\frac{dW}{dt} + W\frac{dD}{dt}. \tag{18.4}$$

Substituting the right hand side of Eq. (18.4) into the left hand side of Eq. (18.3) and rearranging gives:

$$\frac{dW}{dt} = \frac{1}{D}\left(\pm R_A \pm R_B \pm R_C + \dots + r_W - W\frac{dD}{dt} \right). \tag{18.5}$$

Note that the solution of the problem, which will be done by numerical integration, is not unduly complicated by the appearance of the variable W and the differential term dD/dt on the right hand side of the equation.

The aim of this chapter is therefore to give an insight into how the rates of the various different heat and mass transfer phenomena that appear within these balance equations can be expressed mathematically. Once you are able to recognize the mathematical forms, it is possible to inspect a bioreactor model and deduce which heat and mass transfer processes it describes. This section will show that the same phenomenon can appear in slightly different mathematical forms, depending on where in the bioreactor it is occurring. For example, the expressions describing heat conduction in a static bed and heat conduction between the bed and the wall have different forms. Note that equations will not be given for O_2 balances, since they will not appear in the modeling case studies presented later. In any case, the mathematical forms of the terms of an O_2 balance are similar to those that will be presented for water balances.

These mathematical expressions include various parameters. The values of these parameters will need to be known in order to be able to use the mathematical model of the bioreactor to make predictions about how the bioreactor will perform. Chapters 19 and 20 will give advice about how the values of these parameters can be estimated.

The sections below will talk in terms of the typical directions of transfer during the rapid growth phase, namely when both heat and water are being removed from the bed. However, the processes are freely reversible: The direction in which they occur simply depends on the direction of the driving force. This is taken into account automatically in the form of the equations, since the driving force calculated will be either positive or negative, and the sign will determine the direction of transfer.

18.3 Conduction

Conduction occurs in several places within subsystems of SSF bioreactors:

- within the solid bed (both within the solid and gas phases of the bed);
- within the headspace gas;
- across the bioreactor wall, usually treated as occurring only directly from the inside surface to the outside surface of the wall and not along the wall.

The mathematical forms for describing these processes are presented below.

18.3.1 Conduction Across the Bioreactor Wall

The rate of heat transfer across the bioreactor wall (Q_{cond}, J h^{-1}) depends on:

- the difference in temperature between the bed in contact with the wall and the phase on the other side of the wall (°C);
- the area of the wall across which heat transfer is taking place (A, m^2);
- the heat transfer coefficient for conduction through the wall, representing the Joules of energy that will be transferred per unit of time per area of wall per degree of temperature difference (i.e., J h^{-1} m^{-2} °C^{-1});
- the heat transfer coefficients for transfer from the bed to the inner surface of the wall and for transfer from the outer surface of the wall to the surroundings (i.e., J h^{-1} m^{-2} °C^{-1}).

It is common to treat the three steps in heat removal (that is from the bed to the wall, through the wall, and from the wall to the surroundings) as a single overall process (Fig. 18.1). In this case, the rate of heat transfer is written as:

$$Q_{cond} = h A (T_{bed\ outer\ surface} - T_{surroundings}), \tag{18.6}$$

where h is the "overall heat transfer coefficient". The temperatures are self-explanatory. On the other hand, if the bioreactor wall is treated as a different sub-system, then for transfer from the bed to the inner surface of the wall we write:

$$Q_{cond1} = h_1 A_1 (T_{bed\ outer\ surface} - T_{wall\ inner\ surface}), \tag{18.7}$$

where h_1 is the heat transfer coefficient between the bed and the inner surface of the wall and A_1 is the area of contact between the bed and wall.

For transfer across the bioreactor wall we can write:

$$Q_{cond2} = h_2 A_2 (T_{wall\ inner\ surface} - T_{wall\ outer\ surface}), \tag{18.8}$$

where h_2 is the heat transfer coefficient for transfer within the material of the bioreactor wall and A_2 is the area of the wall.

In order to describe transfer from the wall outer surface to the surroundings (Q_{cond3}) we would use an term of similar form, but describing convective heat transfer from a surface to a cooling fluid (see Eq. (18.10) in Sect. 18.4.1).

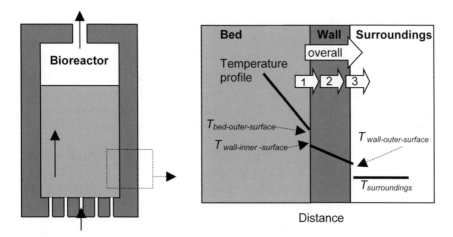

Fig. 18.1. Conductive heat transfer across the bioreactor wall, highlighting that it can be treated as consisting of three individual steps or simply as one overall process. Steps: (1) Heat transfer from the outer surface of the substrate bed to the inner surface of the bioreactor wall; (2) Conduction across the bioreactor wall; (3) Convective heat removal from the outer surface of the bioreactor wall to a well-mixed cooling fluid (air or water)

18.3.2 Conduction Within a Phase

Conduction will also occur within a phase, such as the substrate bed, the headspace gas, or even the bioreactor wall, although the significance of the contribution that it makes to overall heat removal will depend on the presence of other heat removal mechanisms such as convection and evaporation. Conduction will be the dominant mechanism within static beds without forced aeration (Group I bioreactors), that is, within the bed within tray bioreactors. In other bioreactors its contribution to heat removal may be relatively minor.

The rate of transfer of heat by conduction within a static phase (Q_{cond}, J h^{-1}) is determined by:

- the temperature gradient in the phase (dT/dx, °C m^{-1});
- the thermal conductivity of the phase (k, J m^{-1} h^{-1} °C^{-1}). This is a property of the material that characterizes how easily it conducts heat, and which will be significantly affected by its composition. In the case of beds of solid particles, it depends on the bed water content, being higher with higher water contents. Note that the bed may be treated as a single pseudo-homogenous phase in which the thermal conductivity is calculated as a weighted average of the thermal conductivities of the solid phase and the inter-particle gas phase;
- the area across which heat transfer is being considered (A, m^2). Note that this area term may be cancelled out in the final equation after it is rearranged.

Therefore the term for conductive heat transfer within a phase is given by:

$$Q_{cond} = -kA\frac{dT}{dz}.$$ (18.9)

Depending on the design and operation of an aerated bed, conduction within the bed can occur: (1) co-linearly with the air flow (in which case the transfer by conduction will be in the opposite direction to the air flow); (2) normal to the air flow; or (3) in both the co-linear and normal directions (Fig. 18.2). In other words, an energy balance may contain a term that includes dT/dz, a term that includes dT/dx, or two terms, one including dT/dz and the other including dT/dz.

Once there is a temperature gradient, conductive heat transfer will occur. Conversely, if conductive cooling is the only heat transfer mechanism in the bed (i.e., in the case of a static unaerated bed) and the surface is being cooled by heat transfer to the surroundings, then temperature gradients will arise in the bed. As shown in Fig. 18.2, conduction occurs "down" the temperature gradient, hence the minus sign on the right hand side of Eq. (18.9). In other words, the flux of heat is positive in the direction in which the temperature gradient is negative.

During the rearrangements made in simplifying the energy balance for a static bed, Eq. (18.9) is often divided by the volume of the bioreactor (volume being given by an axial distance, z, multiplied by a cross-sectional area, A). This has two consequences: firstly, the area term cancels out and, secondly, the axial distance (z) that is left over combines with the term dz to make the derivative a second-order derivative. That is, the conductive term will often appear as "kd^2T/dz^2".

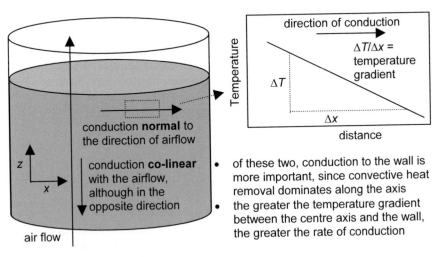

Fig. 18.2. Conductive heat transfer within aerated static beds. Note that conduction normal to the direction of air flow can be promoted by the presence of water jackets. It is important in thin beds but in wide beds its contribution to overall heat removal may be small. The graph on the upper right shows how conduction occurs down a temperature gradient. For the case where the temperature gradient is uniform, $dT/dx = \Delta T/\Delta x$

18.4 Convection

Convective cooling, that is, cooling by transfer of heat to a moving fluid, which then transports the heat away due to bulk flow, occurs in various situations in SSF bioreactors that we might like to describe within bioreactor models:

- at the bioreactor wall, the removal of heat to flowing water in a water jacket, or to flowing air, which might either be forcefully agitated or be undergoing natural convection;
- at a bed surface in which there is a cross-flow of air;
- within a forcefully aerated bed, in which heat is removed from the solid phase to the flowing air phase between the particles and then removed from that location by the flow of air through the bed.

18.4.1 Convection at the Bioreactor Wall

The rate of heat removal by convection (Q_{conv}, J h^{-1}) at a surface in contact with a fluid depends on (Fig. 18.3):

- the coefficient of convective heat transfer (h, J m^{-2} h^{-1} °C^{-1}). This depends on the velocity of the fluid flow because there is a layer of stagnant fluid at the solid surface, and heat transfer through this stagnant layer is limited to conduction. The thickness of the stagnant layer decreases as the flow velocity of the bulk fluid increases; this decreases the resistance to heat transfer and therefore increases the coefficient;
- the area of contact between the surface and the fluid (A, m^2);
- the difference in temperature between the surface and the bulk fluid (°C).

That is, for the case where heat is transferred from the outer surface of the bioreactor wall to cooling water in a cooling jacket, we would write:

$$Q_{conv} = h A (T_{wall\ outer\ surface} - T_{water}). \tag{18.10}$$

This equation applies if we can assume that the fluid is well mixed and can therefore be represented by a single temperature. The equation will be more complicated if we want to describe how the temperature of a fluid increases as it flows in a unidirectional manner past the surface.

To increase heat removal from the bioreactor wall, it is necessary to increase one or more of the three terms. The heat transfer coefficient can often be increased by increasing the velocity of fluid flow, while the area of contact can be increased by using projections on the wall or a bioreactor geometry that increases the overall wall surface area (for a given bioreactor volume). The driving force for heat transfer (i.e., the temperature difference) can be increased by cooling the water before it is passed through the water jacket.

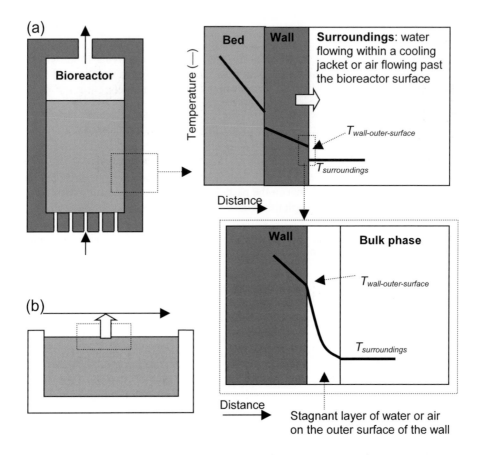

Fig. 18.3. Convective heat transfer from a surface to a well-mixed flowing phase. **(a)** The example shown here is for heat transfer from the bioreactor wall to surrounding air or the water in a water jacket. **(b)** Similar considerations apply for the transfer of heat from the surface of a bed to a passing gas phase

18.4.2 Convective Heat Removal from Solids to Air

The rate of heat removal from the solid phase to the gas phase by convection (Q_{conv}, J h^{-1}) depends on (Fig. 18.4):

- the coefficient for heat transfer between the solid particles and the air phase (h, J m^{-2} h^{-1} °C^{-1}), the value of which depends on the velocity of the air flow;
- the superficial area of contact between the solids and the air phase (A, m^2);
- the difference in temperature between the solids and the air phase (°C).

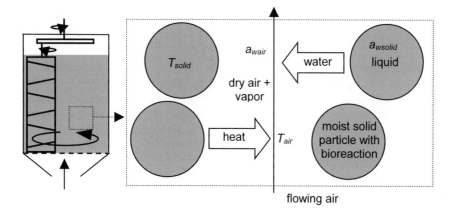

Fig. 18.4. Heat and mass transfer between the solid and gas phases in the case where the solid and gas phases are treated as separate phases

To describe solid-to-gas heat transfer we therefore write:

$$Q_{conv} = h A (T_{solid} - T_{air}).\qquad(18.11)$$

Note that the area of contact between the solid and gas phases can be difficult to measure and therefore the product "$h A$" is often expressed and determined as a global heat transfer coefficient that combines the two quantities ("hA", J h^{-1} °C^{-1}). It may even be expressed as the overall coefficient per m^3 of bed volume (i.e., with units of J h^{-1} °C^{-1} m^{-3}-bed).

The amount of heat removed from the solids by convective cooling can be increased by increasing the air flow rate or decreasing the air temperature at the air inlet. Either of these strategies should increase the average temperature difference between the air and solid phases. Also, the higher air flow rate will increase the value of the heat transfer coefficient.

At times the solids and air are assumed to be in thermal equilibrium (this is the assumption of a pseudo-homogeneous bed). Note that this does not necessarily mean that the bed has the same temperature at all positions. It means that the solid particles at any particular position within the bed are at the same temperature as the gas phase at that position. Therefore a single temperature variable can be used to represent the temperature at a given position in the bed. In this case it is not necessary to write an equation describing solids-to-air heat transfer, as this is subsumed in the term that describes the heat removal associated with the flow of gas through the bed.

18.4.3 Convective Heat Removal Due to Air Flow Through the Bed

The rate of heat removal by flow of the air through the bed (Q_{conv}, J h^{-1}) depends on (Fig. 18.5):

- the mass flux of dry air (G, kg-dry-air m^{-2} h^{-1}), which is given by the superficial velocity of the air (V_Z, m h^{-1}) multiplied by the density of the air (ρ_{air}, kg-dry-air m^{-3}). Of course, the superficial velocity itself is simply equal to the volumetric flow rate (m^3-dry-air h^{-1}) divided by the total cross sectional area of the bed (note that this is the total area, not the area occupied by the void spaces);
- the cross-sectional area of the bed (A_b, m^2);
- the heat capacity of the air (C_{Pair}, J kg-dry-air^{-1} °C^{-1});
- the difference between the air temperatures at two different locations (°C).

Applied over the whole bed (i.e., in a balance that considers the difference between the air inlet and the air outlet), the rate of heat removal by convection (J h^{-1}) would be given by:

$$Q_{conv} = G_{air} C_{Pair} A_b (T_{outlet} - T_{inlet}), \qquad (18.12)$$

where in this case A_b is the cross sectional area of the bioreactor.

However, in static beds, in which the temperature is a function of height within the bed, it is often of more interest to write an equation that allows the calculation of the temperature as a function of height. In this case, the balance equation is initially written over a thin layer of the bed. Within this equation the convection term will appear as:

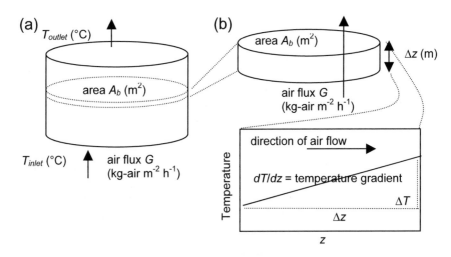

Fig. 18.5. Illustration of the various parts of the expression for the removal of sensible energy in the air stream. (a) In an overall energy balance over a bioreactor; (b) In a balance written over thin layer of the bed

$$Q_{conv} = G_{air} C_{Pair} A_b \frac{dT}{dz} \Delta z , \qquad (18.13)$$

since the temperature difference (°C) between the inlet and outlet of this thin layer is simply the temperature gradient (dT/dz, °C m^{-1}) multiplied by the thickness of the thin layer (Δz, m).

Typically the energy balance equation will be divided through by the volume of the thin layer during later rearrangements, such that in the final equation this term will appear containing neither A_b nor Δz. Note also that if the temperature at the outlet of the thin layer is higher than the temperature at the inlet of the thin layer, then convection will be reducing the sensible energy of the thin layer, and therefore this term will be preceded by a negative sign if it appears on the right hand side of an equation such as Eq. (18.1). In fact, it is often put on the left hand side of the balance equation.

Note also that it is often convenient to use the same term to express the contribution of the water vapor to the removal of sensible energy. Given the humidity (H, kg-water kg-dry-air^{-1}) and the heat capacity of the vapor (C_{Pvapor}, J kg-vapor^{-1} °C^{-1}), the term would simply become:

$$Q_{conv} = G(C_{Pair} + HC_{Pvapor}) A_b \frac{dT}{dz} \Delta z . \qquad (18.14)$$

Again, A_b and Δz may be cancelled out in the manipulations that are made to arrive at the final equation in the bioreactor model.

Note that convective cooling by the forced aeration of a static bed in which there is continual heat liberation by the growth process will cause temperature gradients in the bed. This phenomenon was explained in Fig. 4.3.

18.5 Evaporation

Evaporation can be important in various instances within SSF bioreactors:

- at the surface of a bed exposed to the air (for example, the surface of a tray);
- between the air and solid phases in a forcefully-aerated bed.

The equations used to describe evaporation in the various circumstances will have many similarities with the equations used to describe heat transfer, as will become apparent in the subsections below. Note that the diffusion of liquid water or water vapor is not described here, since bioreactor models typically assume that it is negligible. If it were to be included in a model, the diffusion term would have a mathematical form similar to Eq. (18.9).

18.5.1 Evaporation from the Solids to the Air Phase

The rate of evaporation from the solids to the gas phase within the bed depends on (Fig. 18.6):

- the difference between the water activity that the solid actually has (a_{wsolid}, dimensionless) and the water activity that it would have if it were in equilibrium with the gas phase (a_{wsolid}^*);
- the area of contact between the solid and gas phases (A, m^2);
- the mass transfer coefficient (k_w), which is the mass of water transferred per unit of time per unit of area per unit of driving force. Since the driving force is expressed in terms of water activity, which is dimensionless, the units of k_w are simply kg-H$_2$O m^{-2} h^{-1}.

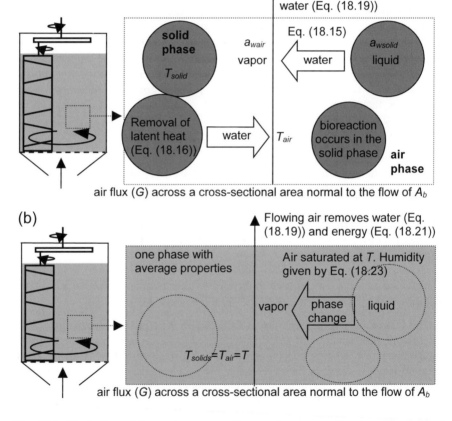

Fig. 18.6. Illustration of the various ways of expressing evaporative loss of water and evaporative heat removal (convective heat removal is not considered here). **(a)** In the case in which the solid and gas phases are treated as separate phases; **(b)** In the case in which the solid and gas phases are treated as a single pseudo-homogeneous phase

We can therefore write the local rate of evaporation R_{evap} (kg-H_2O h^{-1}) as:

$$R_{evap} = k_w A(a_{wsolid} - a_{wsolid}*).$$

(18.15)

As in the case of convective heat removal, it is common to combine the mass transfer coefficient and the area to obtain an overall transfer coefficient ("kA").

The local rate of heat removal from the solid phase by evaporation (Q_{evap}, J h^{-1}) is given by:

$$Q_{evap} = \lambda k_w A(a_{wsolid} - a_{wsolid}*),$$

(18.16)

where λ is the enthalpy of vaporization of water (J kg-H_2O^{-1}).

Note that an isotherm can be used to order to write the driving force in terms of the water content. In this case the driving force for evaporation is the difference between the water content that the solid actually has (W, kg-water kg-dry solid^{-1}) and the water content that it would have it were in equilibrium with the gas phase (W_{sat}, kg-water kg-dry solid^{-1}). This affects the units used in the mass transfer co-efficient. Chapter 22 will describe how the equation is written in this case.

At times the solids and air are assumed to be in moisture equilibrium, or, in other words, the air phase is assumed to be saturated with water at the temperature of the solids (this is the assumption of a pseudo-homogeneous bed). This has the consequence that the humidity at a particular position can be expressed as a function of the temperature at that position. In this case, it is not necessary to write equations describing solids-to-air water transfer and evaporative heat transfer, as these are subsumed within the terms that describe the water and heat removal associated with the flow of gas through the bed, as explained in the next section.

18.5.2 Water Removal Due to Air Flow Through the Bed

The flow of moist air through the bed typically leads to the removal of water from the bed (see Fig. 4.3). The overall rate of water removal (kg-water h^{-1}) from the bed is:

$$R_{conv} = G A_b(H_{outlet} - H_{inlet}),$$

(18.17)

where H_{inlet} and H_{outlet} are the humidities (kg-water kg-dry-air^{-1}) at the air inlet and outlet, respectively. G (kg-dry-air m^{-2} h^{-1}) and A_b (m^2) are as described in Sect. 18.4.3.

For an overall energy balance on a bioreactor, the rate of heat removal due to evaporation will then be:

$$Q_{conv} = \lambda G A_b(H_{outlet} - H_{inlet}).$$

(18.18)

However, as before, for beds that are not well mixed, it is often of more interest to write balances that allow the calculation of the temperature and humidity as functions of the position within the bed. In this case, a balance equation is written over a thin layer of the bed. The approach is different depending on whether the solids and gas phases are treated as different phases or are lumped together and treated as a pseudo-homogenous phase.

18.5.2.1 Solids and Gas Treated as Separate Phases

If the solids and gas are treated as separate phases, then the convective flow term within the mass balance equation for water will appear as:

$$R_{conv} = G \, A_b \, \frac{dH}{dz} \, \Delta z \,, \tag{18.19}$$

since the humidity difference (kg-water kg-dry-air^{-1}) between the inlet and outlet of this thin layer is simply the humidity gradient (dH/dz, kg-water kg-dry-air^{-1} m^{-1}) multiplied by the thickness of the thin layer (Δz, m).

Typically during the manipulations of the water balance equation, it will be divided through by the volume of the thin layer, such that the term will appear without containing A_b and Δz. Note also that if the humidity at the outlet of the thin layer is higher than the humidity at the inlet of the thin layer, then the flow of air will be reducing the humidity of the thin layer, and therefore this term will be preceded by a negative sign if it appears on the right hand side of an equation such as Eq. (18.3). In fact, it is often put on the left hand side of the balance equation.

Note that evaporation removes energy from the solids and not from the air phase. Energy removal from the solids phase, which does not flow, has already been taken into account by Eq. (18.16). Therefore the energy balance on the air phase will not contain a term of the form of Eq. (18.19) multiplied by the enthalpy of evaporation.

18.5.2.2 Solids and Gas Treated as a Pseudo-Homogeneous Phase

When the assumption is made that the air is always saturated at the temperature of the solids (i.e., the assumption of a pseudo-homogeneous bed), the rate of evaporation (R_{evap}, kg-H$_2$O h^{-1}) is still written in the form of Eq. (18.19). In this case the rate of evaporative heat removal is given by

$$Q_{evap} = \lambda \, G \, A_b \, \frac{dH_{sat}}{dz} \, \Delta z \,. \tag{18.20}$$

This is not inconsistent with Sect. 18.5.2.1, since Eq. (18.16) is not used when the assumption of a pseudo-homogeneous bed is made. Further, even though evaporation removes the energy from the solids and not the gas, this makes no difference since the solids and gas are assumed to equilibrate immediately to the same temperature. The Antoine equation can be used to calculate the saturation humidity (H_{sat}) as a function of temperature, so it is useful to apply the chain rule of differentiation to cause the term "dH_{sat}/dT" to appear explicitly in the equation:

$$\frac{dH_{sat}}{dz} = \frac{dH_{sat}}{dT} \frac{dT}{dz} \,. \tag{18.21}$$

Substituting Eq. (18.21) into Eq. (18.20) gives:

$$Q_{evap} = \lambda\, G_{air} A_b \frac{dH_{sat}}{dT} \frac{dT}{dz} \Delta z .$$ (18.22)

An equation relating dH_{sat}/dT to the temperature is developed in Sect. 19.4.1.

18.6 Conclusions

This chapter has identified the forms of various terms that may appear within the balance/transport sub-model of a bioreactor model. Several of these will appear in energy and mass balances in the mathematical models of bioreactors presented in Chaps. 22 to 25. These equations contain various parameters that it will be necessary to determine before the model can be solved. Chapters 19 and 20 describe how these and other necessary parameters can be determined.

Further Reading

Selected examples from the SSF literature in which models are developed that describe the various heat and mass transfer phenomena described in this chapter. These models therefore contain terms similar to those shown in this chapter

Mitchell DA, von Meien OF (2000) Mathematical modeling as a tool to investigate the design and operation of the Zymotis packed-bed bioreactor for solid-state fermentation. Biotechnol Bioeng 68:127–135

Mitchell DA, Tongta A, Stuart DM, Krieger N (2002) The potential for establishment of axial temperature profiles during solid-state fermentation in rotating drum bioreactors. Biotechnol Bioeng 80:114–122

Oostra J, Tramper J, Rinzema A (2000) Model-based bioreactor selection for large-scale solid-state cultivation of *Coniothyrium minitans* spores on oats. Enz Microbial Technol 27:652–663

Rajagopalan S, Modak JM (1994) Heat and mass transfer simulation studies for solid-state fermentation processes. Chem Eng Sci 49:2187–2193

Saucedo-Castañeda G, Gutierrez-Rojas M, Bacquet G, Raimbault M, Viniegra-Gonzalez G (1990) Heat transfer simulation in solid substrate fermentation. Biotechnol Bioeng 35:802–808

Stuart DM, Mitchell DA (2003) Mathematical model of heat transfer during solid-state fermentation in well-mixed rotating drum bioreactors. J Chem Technol Biotechnol 78:1180–1192

Weber FJ, Oostra J, Tramper J, Rinzema A (2002) Validation of a model for process development and scale-up of packed-bed solid-state bioreactors. Biotechnol Bioeng 77:381–393

Basic and easy to read introductions to the modeling of heat and mass transfer in bioprocessing

Van Den Akker HEA, Heijnen JJ, Leach CK, Mudde RF (1992) Bioprocess Technology, Modelling and Transport Phenomena. Butterworth Heinemann, Oxford.

Johnson AT (1998) Biological process engineering: an analogical approach to fluid flow, heat transfer, and mass transfer applied to biological systems. John Wiley & Sons, New York.

Introductions to heat and mass transfer phenomena (not in the context of bioprocesses)

Thomson WJ (2000) Introduction to transport phenomena. Prentice Hall, Upper Saddle River

Brodkey RS, Hershey HC (2003) Transport Phenomena: A Unified Approach, vols. 1 and 2. Brodkey Publishing, Columbus, Ohio

Overview of approaches to modeling heat and mass transfer in SSF bioreactors

Mitchell DA, von Meien OF, Krieger N (2003) Recent developments in modeling of solid-state fermentation: heat and mass transfer in bioreactors. Biochem Eng J 13:137–147

19 Substrate, Air, and Thermodynamic Parameters for SSF Bioreactor Models

David A. Mitchell, Oscar F. von Meien, Luiz F.L. Luz Jr, and Marin Berovič

19.1 Introduction

Chapter 18 presented mathematical expressions to describe the various macroscale heat and mass transfer processes that are important in SSF bioreactors. These expressions contain various parameters. In order to solve mathematical models of SSF bioreactors, it is necessary to supply values for these parameters. This chapter gives recommendations for how these parameters can be estimated, by experimental or other means. However, it does not give detailed instructions for how to carry out such determinations. It also addresses other parameters that become important in the bioreactor design and optimization process. The determination of transfer coefficients is addressed in Chap. 19.

19.2 Substrate Properties

Substrate properties can be quite important in affecting how an SSF bioreactor performs. Some of these properties need to be determined to be included in bioreactor models. Others may not appear in models, but can influence bioreactor performance, and therefore it is necessary to think about these during the bioreactor development process. As yet, there has been relatively little effort to characterize these properties quantitatively for SSF systems, therefore much of this section will be qualitative.

Substrate properties related to the intra-particle structure are not included here. They might be important in influencing the resistance to diffusion of enzymes and nutrients and to hyphal penetration. However, as yet bioreactor models tend to ignore intra-particle phenomena.

19.2.1 Particle Size and Shape

The size and the shape of the prepared substrate particles influence the accessibility of nutrients to the organism. The greater the size of the smallest particle dimension is, the greater is the average depth of the nutrients within the particle (Fig. 19.1(a)). It may be difficult for the organism to utilize the nutrients located in the interior of the particle, especially polymers. This will affect both the rate and final amount of growth that occurs during the fermentation. In fact, the maximum biomass content is an important parameter of the logistic equation (see Eq. (14.6) in Table 14.1 and Eq. (16.3) in Table 16.1). It might be determined by the surface area available for growth, which in turn is determined by the size and shape of the particles. In other words, if the biomass covers the surfaces of all the particles at the maximum biomass packing density (biomass per cm^2), further growth will not occur, even if there are still nutrients within the substrate particle. However, note

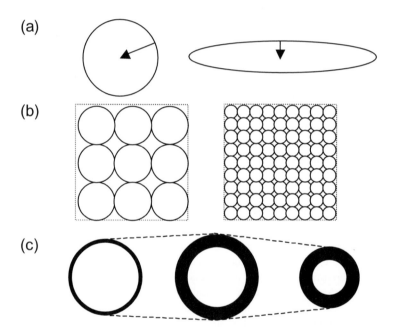

Fig. 19.1. Some important considerations about particle size and shape. **(a)** The effect of particle shape on nutrient accessibility. For two substrate particles of the same overall volume, the more spherical the shape then the greater the average depth from the surface and the lower the surface area to volume ratio. **(b)** Effect of particle size on porosity and pressure drop, illustrated with regularly-packed spherical particles. For the two different particle sizes, the percentage of the total volume occupied by the void spaces is identical. However, for forced aeration at the same superficial velocity, the pressure drop across the bed will be greater for the bed on the right, that is, with the smaller particles. **(c)** The overall particle size can increase early during the fermentation and then decrease during the latter stages. This is a result of two phenomena, the consumption of residual substrate and the expansion of the biofilm

that in models of SSF bioreactors the particle size does not appear within the kinetic equation, which is usually simply an empirical description of observed growth curves. Experiments should be done in the laboratory with different particle sizes in order to determine the optimum particle size from the point of view of growth kinetics.

The particle size will also influence the packing within the bed and therefore the aeration of the bed. Comparing two beds of different particle sizes but the same porosity (void fraction), it will be more difficult to force air through the bed of smaller particles (the phenomenon of pressure drop) (Fig. 19.1(b)). However, on the other hand, the air may tend to follow preferential routes in a bed of larger particles (the phenomenon of channeling). Therefore studies to determine the optimal particle size for the process should also be done in the production-scale bioreactor, if possible.

The particle diameter may appear within various correlations that are used in modeling, for example, in correlations for heat and mass transfer coefficients that appear in the model. Therefore it may be necessary to determine the particle diameter experimentally. This may not be simple for irregularly-shaped particles. For non-spherical but regular particles, it is possible to use the equivalent radius, defined as the volume of the particle divided by its surface area.

Determination of the particle size will be further complicated if not all particles are identical. The homogeneity of particle size and shape will depend on the source of the particles and the manner in which they were prepared. For example, grains might be expected to be more homogeneous than substrates like chopped straw or rasped tubers. For heterogeneous substrates, it will be necessary to determine the particle size distribution. For particles with a shape that is reasonably close to spherical, this can be done by passing a sample through a number of graded sieves. Of course, sieves can also be used to select a particular range of particle sizes for use in the fermentation.

As mentioned in Chap. 2, the particle size can change during the process. The overall particle size, that is, the overall diameter of the biomass layer and the residual substrate, can first increase and then later decrease (Fig. 19.1(c)). However, as yet there is little quantitative data available and, although some models have been proposed to describe particle size reduction during the fermentation (Rajagopalan et al. 1997), such models are typically not incorporated into bioreactor models.

19.2.2 Particle Density

It may be useful to know the density of the prepared substrate particles since, as described in Sect. 19.2.4, this parameter can be used in an estimation of the bed porosity. However, substrate particle density is not necessarily an easy parameter to determine. If the prepared particles do not absorb water and therefore do not swell quickly when put into contact with excess water, one method may be to flood a sample of particles with water within a container such as a measuring cylinder. The substrate particle density can then be calculated as:

$$\rho_s = \frac{(m_{total} - m_{container}) - m_w}{V_{total} - m_w / \rho_w} = \frac{m_s}{V_s}, \tag{19.1}$$

where m_{total} is the mass of the system after flooding (g), $m_{container}$ is the mass of the empty container (g), m_w is the mass of the water added to flood the bed (g), V_{total} is the total volume of the flooded bed (L), ρ_w is the density of water (g L^{-1}), and m_s is the mass of the substrate particles (g). The advantage of using this system of units is that it will give the correct numerical value of the substrate density in SI (since g L^{-1} is equivalent to kg m^{-3}).

19.2.3 Bed Packing Density

It is often necessary to know the apparent density of the bed, that is, the bed packing density (ρ_b) (Fig. 19.2). This relates the bed mass with the volume that the bed will occupy. It can be determined experimentally by packing the substrate, prepared in a manner identical to preparation for the fermentation, to fill a container of known volume (V, liters) and mass, and reweighing. The difference between the masses of the container when it is packed with substrate and when it is empty is the packed mass of the bed (m_p, g). Of course, the packing process must also be identical to that which is used in the fermentation. The packing density is then calculated as:

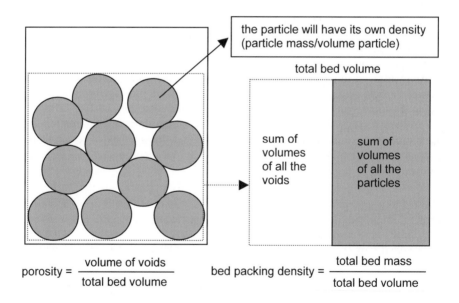

Fig. 19.2. Physical meaning of two important bed properties: the porosity (void fraction) and bed packing density

$$\rho_b = \frac{m_p}{V},$$
(19.2)

where ρ_b has the units of g L^{-1}. Once again, the advantage of using this system of units is that it will give the correct numerical value of the packing density in SI (since g L^{-1} is equivalent to kg m^{-3}). Note that the packing density can also be the calculated if both the porosity of the bed (ε, m^3-voids m^{-3}-bed) and the density of the substrate particles themselves (ρ_s, kg m^{-3}) are known:

$$\rho_b = \rho_s(1-\varepsilon) + \rho_a\varepsilon,$$
(19.3)

where ρ_a is the density of air. Typically the second term on the right hand side will make only a minor contribution, since the air density is typically two to three orders of magnitude smaller than the density of the substrate particles. However, it is not a simple matter to determine the porosity and, as shown in the next section, it would be more common to use this equation to give an estimate of the porosity.

The value of ρ_b will change during the fermentation, for a variety of reasons:

- agitation will affect bed structure;
- the growth of hyphae between particles will affect bed structure;
- the growth process will typically affect particle size and density;
- changes in the moisture content of the bed will affect particle size and density.

Little work has been done to characterize such changes during SSF processes. It is often assumed that the total bed volume does not change. This might occur if the major nutrient is not a structural polymer of the substrate particle, such that the particle size changes little during the fermentation, even as nutrients are converted into CO_2. However, if both particle size and density change, the packing density will change in a complex way; these changes must be measured experimentally.

Weber et al. (1999) studied the effect of moisture content on a parameter related to the bed density, namely the specific packed volume on a dry basis (V_P, m^3 kg-dry-matter^{-1}). They prepared a known mass of moist solid substrate (m_b, g), packed it in a manner identical to the packing used for the fermentation, measured the volume occupied by this moist bed (V_m, L), dried it, and then measured the volume occupied by the dried bed (V_d, L). The shrinkage factor (S, m^3-dry-bed m^{-3}-moist-bed) can be calculated directly as the ratio of the two volumes (i.e., V_d/V_m) since the amount of dry matter in the moist and dry samples is identical. Shrinkage was negligible or minor for solid supports designed to be impregnated with a nutrient solution, such as hemp ($S = 1.0$), bagasse ($S = 1.0$), and perlite ($S = 0.9$). In the case of oats the shrinkage was quite significant ($S = 0.55$). Of course, in an actual fermentation the bed would not be allowed to dry out completely, but Oostra et al. (2000) did show that the value of V_P of oat particles can fall by as much as 30% (from about 0.0020 to about 0.0015 m^3 kg-dry-matter^{-1}) as the moisture content falls from 1.1 kg-water kg-dry-matter^{-1} to 0.57 kg-water kg-dry-matter^{-1}, a fall of this magnitude being expected during a packed-bed fermentation with this substrate for the production of spores of the biocontrol fungus *Coniothyrium minitans*.

19.2.4 Porosity (Void Fraction)

The way that the substrate bed as a whole packs is important in determining the effectiveness of aeration. The packing will affect the size and continuity of the inter-particle spaces, and it is through these inter-particle spaces that O_2 is made accessible to the organism at the particle surface. These effects are characterized by the porosity (ε), that is, the fraction of the total bed volume that is comprised by the void spaces (Fig. 19.2):

$$\varepsilon = \frac{V_a}{V_b} = \frac{V_b - V_s}{V_b} = 1 - \frac{V_s}{V_b},$$ (19.4)

where V_b is the total volume occupied by the bed (m³), V_a is the volume within the bed occupied by air (m³), and V_s is the volume within the bed occupied by the substrate particles (m³).

The smaller that particles are, the smaller will be the size of the inter-particle spaces. However, note that the overall porosity does not change significantly, especially for spherical particles, for which the porosity is independent of particle size. Of course, the smaller size of the particles causes larger pressure drops when air is being forced through the substrate mass. This is due to the larger overall surface area of solids that is present, which causes the static gas film that occurs on solid surfaces to occupy a greater proportion of the void volume.

Uniformly sized spheres cannot pack in such a way as to exclude air, even when they are tightly packed (Fig. 19.3(a)). For solid spheres, the porosity can be predicted reasonably easily depending on the way the substrate was packed. However, substrate particles might deform within the bed due to the overlying weight of the bed or due to agitation, even if they were originally spherical. With irregularly sized particles, smaller particles can tend to fill the inter-particle spaces that would otherwise be vacant (Fig. 19.3(b)). This can happen when the substrate releases fines during movement and handling of the dry substrate. For irregularly sized and shaped solids it is not possible to predict the porosity with any accuracy with simple equations and it must be measured experimentally. Particles with large flat surfaces will tend to lie with the flat surfaces touching, excluding O_2 and therefore greatly limiting the amount of growth (Fig. 19.3(c)).

The bed porosity typically appears as a key parameter within bioreactor models. However, it is not necessarily an easy thing to measure, especially in SSF processes. If the density of the solid particle is known, it may be possible to estimate the porosity from the bed packing density. A container of known volume is filled with a substrate bed in the same manner in which a bioreactor would be packed, and this is reweighed to give the bed weight (m_b, kg). Writing the volume terms in Eq. (19.4) as a mass divided by a density and assuming that the mass of air in the bed (m_a, kg) makes a negligible contribution, it is possible to arrive at an equation for the porosity in terms of the overall bed density (ρ_b, kg m⁻³) and the substrate particle density (ρ_s, kg m⁻³):

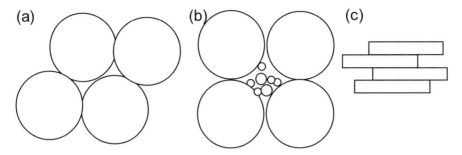

Fig. 19.3. Effect of particle size and shape on the porosity of the bed. **(a)** Regular spheres cannot pack in such a manner as to exclude void spaces even when they are packed in the most tightly packed conformation possible. **(b)** However, if spheres are not uniform in size, then smaller spheres can pack within the void spaces between larger spheres. **(c)** Particles with flat surfaces can tend to lie side-by-side, excluding air

$$\varepsilon = 1 - \frac{V_s}{V_b} = 1 - \frac{\left(\dfrac{m_s}{\rho_s}\right)}{\left(\dfrac{m_b}{\rho_b}\right)} = 1 - \frac{\dfrac{m_s}{\rho_s}}{\dfrac{m_b}{\rho_b}} \approx 1 - \frac{\dfrac{m_s}{\rho_s}}{\dfrac{m_s}{\rho_b}} = 1 - \frac{\rho_b}{\rho_s}. \tag{19.5}$$

Weber et al. (1999) quotes various values of the porosity ε for various different substrates: 0.31 for oats, 0.48 for hemp impregnated with a nutrient solution, 0.47 for impregnated bagasse, and 0.41 for impregnated perlite.

Note that the porosity of the bed is not constant during the fermentation, since the microorganism will tend to fill the inter-particle spaces. This is especially true in static beds during the growth of a mycelial organism, where aerial hyphae extend into the void spaces. Although this has received some attention (Auria et al. 1995), the phenomenon has not been sufficiently studied to incorporate these changes in porosity into bioreactor models. In any case, changes in porosity due to microbial growth will not be such a problem if the bed is agitated, since the movement of particles will tend to squash hyphae onto the surface, and rip apart any hyphae that do manage to span between particles during periods of static operation. In fact, in some cases intermittent agitation is used not to aid in heat transfer, but rather to restore the porosity of the bed and therefore reduce the pressure drop through the bed. Porosity can also change as the overall particle size and shape change due to consumption of dry matter.

19.2.5 Water Activity of the Solids

The water activity of the solids is a key parameter in bioreactor models for two reasons. Firstly, microbial growth depends on the water activity of the solids (see Sect. 16.3.2) and, secondly, the driving force for evaporation is the difference between the water activity of the solid phase and the water activity that it would have if it were in equilibrium with the gas phase (see Sect. 18.5.1). Within a bioreactor there are many processes that affect the water content of the substrate and these changes will affect the water activity. Therefore it is necessary to have an equation relating the water content and water activity of the solids, or in other words, an equation describing the isotherm of the solids.

Typical isotherms for the types of solid materials used as substrates in SSF processes are shown in Fig. 19.4. For each particular substrate it will be necessary to determine the isotherm experimentally. A simple experimental method for doing this involves placing samples of the prepared but uninoculated substrate in several hermetically sealed containers, each container having a volume of a saturated salt solution (Fig. 19.4(a)). The containers are placed in a temperature-controlled incubator. Each sample is allowed to equilibrate with its salt solution. Once equilibrated, the fresh weights of the samples are determined, then they are dried and their dry weights are determined. This allows the construction of an isotherm, a plot of the water content of the sample at equilibrium against the water activity of the salt solution with which it was equilibrated (Fig. 19.4(b)). An empirical equation can then be fitted to the isotherm. For example, Nagel et al. (2001b) fitted the following equation to give the isotherm of autoclaved wheat at 35°C as a function of its moisture content (W, kg-water kg-dry-solids^{-1}):

$$a_{ws} = -2.917 + \frac{3.919}{1 + (W/0.0344)^{-1.861}} \, . \tag{19.7}$$

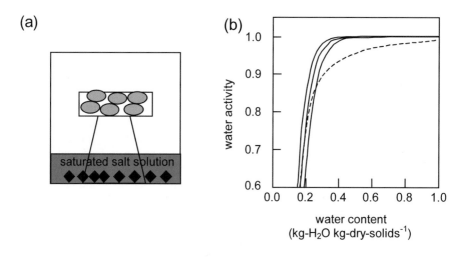

Fig. 19.4. Moisture isotherms. (a) Hermetically sealed container in which the substrate is allowed to equilibrate with a saturated salt solution. "♦" represents a salt crystal. (b) Isotherms of some substrates used in SSF processes: (- - -) Desorption isotherm of autoclaved wheat grains at 35°C as determined by Nagel et al. (2001b), reproduced with kind permission from John Wiley & Sons, Inc.; (—) Isotherm of corn as described by Eq. (19.8) at 20°C (lower curve), 35°C (middle curve), and 50°C (upper curve) (Calçada 1998)

Likewise, Calçada (1998) fitted the following equation to give the water activity of corn (a_{ws}) as a function of the moisture content (W, kg-water kg-dry-solids^{-1}) and the temperature:

$$a_{ws} = \left(1 - \exp\left(-W^{(1.275-0.0029T_s)}\exp\left(2.9 + 0.004T_s\right)\right)\right)^{1/0.32}. \tag{19.8}$$

Note that this equation can be rearranged to be explicit in the solids water content and can be used to calculate the water content necessary to give a desired water activity of the corn at a given temperature:

$$W = \left[\frac{\ln\left(1 - a_{ws}^{0.32}\right)}{-\exp\left(2.9 + 0.004T_s\right)}\right]^{\frac{1}{(1.275-0.0029T_s)}}. \tag{19.9}$$

Microbial growth on the substrate could potentially change the isotherm significantly. That is, for the same water content, uninoculated substrate and fermented substrate could have quite different water activities. However, this point has received relatively little attention. In the development of bioreactor models it has been assumed that the fermenting solids have the same isotherm as the uninoculated substrate. However, Nagel et al. (2001b) did develop a mathematical model in which the water in the particle was segregated into intracellular and extracellular water, and such an approach could be incorporated into bioreactor models.

19.3 Air Density

It will be necessary to know the air density in order to convert between masses and volumes of air. For example, in a bioreactor model, the aeration rate may be input into the model in terms of the volumetric flow rate (m^3-air s^{-1}) while the heat capacity used may contain units of mass (i.e., J kg-air^{-1} °C^{-1}). In this case, in order to calculate the amount of energy stored within a given volume of air for a given temperature rise, it is necessary to first multiply the air volume by its density to calculate the mass of air, before multiplying the mass of air by the heat capacity and the temperature rise.

Air density is a function of temperature and pressure. At low pressures ideal gas behavior can be assumed:

$$P = \frac{n}{V}RT_K, \tag{19.10}$$

where P is the pressure (Pa), n is the number of moles (mol), V is the volume (m^3), T_K is the temperature (K), and R is the universal gas constant (8.314 J mol^{-1} K^{-1}). The number of moles can be replaced with the mass of the gas (m_g, kg) divided by its molecular weight (M_g, kg mol^{-1}). This leads to a term in which the mass of gas is divided by the volume, and this combination can be replaced by the gas density (ρ_g, kg m^{-3}):

$$P = \frac{n}{V} RT_K = \frac{\left(\dfrac{m_g}{M_g}\right)}{V} RT_K = \frac{m_g}{V} \frac{1}{M_g} RT_K = \rho_g \frac{1}{M_g} RT_K. \tag{19.11}$$

The gas density can then be isolated on the left hand side of the equation:

$$\rho_g = \frac{PM_g}{RT_K} = \frac{PM_g}{R(T+273)}, \tag{19.12}$$

where the symbol T represents the temperature in °C. Since the gas is moist air and therefore has several components, it is necessary to use a mole-weighted average molecular weight for M_g. For a given humidity (H, kg-vapor kg-dry-air^{-1}) it is possible to show that:

$$M_g = \frac{1+H}{\left(\dfrac{1}{0.0288} + \dfrac{H}{0.018}\right)}, \tag{19.13}$$

where the value of 0.0288 represents the average molecular weight of dry air (in kg mol^{-1}, treating it as 79 mol% N_2 and 21 mol% O_2) and the value 0.018 represents the molecular weight of water (in kg mol^{-1}). Therefore, it is possible to write

$$\rho_g = \frac{P}{R(T+273)} \frac{1+H}{\left(\dfrac{1}{0.0288} + \dfrac{H}{0.018}\right)}. \tag{19.14}$$

At times properties are related to the mass of dry air and to the mass of water separately. If the dry air density (ρ_a, kg-dry-air m^{-3})) is defined as the mass of dry air that is held within a unit volume of a humid gas sample, it can be calculated as:

$$\rho_a = \frac{\rho_g}{1+H} = \frac{P}{R(T+273)} \frac{1}{\left(\dfrac{1}{0.0288} + \dfrac{H}{0.018}\right)}. \tag{19.15}$$

19.4 Thermodynamic Properties

This section considers those parameters not involved in transfer processes within or between phases but which affect system performance, such as heat capacities and phase equilibria.

19.4.1 Saturation Humidity

The saturation humidity may need to be known in modeling in order to calculate the water activity of the gas phase within an SSF bioreactor (a_{wg}), which is defined as:

$$a_{wg} = \frac{P_w}{P_{sat}},$$

(19.16)

where P_w is the vapor pressure (Pa) of water within the gas phase at a particular temperature and P_{sat} is the saturation vapor pressure (Pa) of water at the same temperature.

The Antoine equation gives the saturation vapor pressure of water as a function of temperature (T, °C) (Reid et al. 1977):

$$P_{sat} = d \exp\left(a - \frac{b}{T+c}\right).$$

(19.17)

For water vapor the constants are $a=18.3036$, $b=3816.44$, $c=227.02$, and $d=133.322$. The humidity at saturation (kg-vapor kg-dry-air^{-1}) is then given by:

$$H_{sat} = 0.62413\frac{P_{sat}}{P - P_{sat}} = \frac{0.62413}{\dfrac{P}{P_{sat}} - 1},$$

(19.18)

where 0.62413 represents the ratio of the molecular weight of water to the average molecular weight of dry air and P is the total pressure (Pa). This allows us to write an equation for the saturation humidity in terms of the temperature:

$$H_{sat} = \frac{0.62413}{\dfrac{P}{d \exp\left(a - \dfrac{b}{T+c}\right)} - 1}.$$

(19.19)

Note that this equation can be differentiated to obtain an expression for the quantity dH_{sat}/dT, which appeared in Eq. (18.22). At a given total pressure (P, Pa):

$$\frac{dH_{sat}}{dT} = \frac{0.62413bP}{(T+c)^2\left(\dfrac{P}{d \exp\left(a - \dfrac{b}{T+c}\right)} - 1\right)^2 d \exp\left(a - \dfrac{b}{T+c}\right)}.$$

(19.20)

19.4.2 Heat Capacity of the Substrate Bed

The heat capacity of the substrate bed will appear in energy balance equations since it relates the amount of energy stored within the bed to the temperature of the bed. Typically the model will be written in terms of the heat capacity at constant pressure, even though at times the heat capacity at constant volume is the more correct term. For a bed of solid particles there will be little difference between these two heat capacities.

The heat capacity of the bed depends on the heat capacities of its various components, namely the dry solid, the liquid water, the dry air, and the water vapor. However, it is not simply the arithmetic mean of these heat capacities. It must be calculated as a "mass-weighted average".

Firstly, the heat capacity of the moist solids (C_{Ps}, J kg-wet-solids^{-1} °C^{-1}) is a weighted average of the heat capacities of the dry solids (C_{Pd}, J kg-dry-solids^{-1} °C^{-1}) and the liquid water (C_{Pw}, J kg-water^{-1} °C^{-1}):

$$C_{Ps} = \frac{C_{Pd} + W C_{Pw}}{1 + W}, \tag{19.21}$$

where W is the water content of the solids on a dry basis (kg-water kg-dry-solids^{-1}). Likewise, the heat capacity of the gas phase is a weighted average of the heat capacities of the dry air (C_{Pa}, J kg-dry-air^{-1} °C^{-1}) and the water vapor (C_{Pw}, J kg-vapor^{-1} °C^{-1}):

$$C_{Pg} = \frac{C_{Pa} + H C_{Pv}}{1 + H}, \tag{19.22}$$

where H is the air humidity (kg-vapor kg-dry-air^{-1}). The overall heat capacity of the bed (C_{Pb}, J kg-bed^{-1} °C^{-1}) is a weighted average of these two heat capacities.

$$C_{Pb} = \frac{m_s C_{Ps} + m_g C_{Pg}}{m_b}, \tag{19.23}$$

where m_b is the overall mass of the bed (kg), equal to the sum of m_s and m_g.

The heat capacities of liquid water, dry air, and water vapor can be found in various books (see the further reading section at the end of the chapter). All three are functions of temperature. The heat capacity of liquid water (J kg^{-1} °C^{-1}) is given by (Himmelblau 1982):

$$C_{Pw} = 1015.56 + 26.206(T + 273) - 0.0743117(T + 273)^2 + 7.2946 \times 10^{-5}(T + 273)^3. \tag{19.24}$$

The heat capacity of dry air (J kg^{-1} °C^{-1}) is given by (Himmelblau 1982):

$$C_{Pa} = 997.9 + 0.143T - 0.00011T^2 - 6.776 \times 10^{-8}T^3. \tag{19.25}$$

The heat capacity of water vapor is given by (Himmelblau 1982):

$$C_{Pv} = 1857 + 0.382T - 0.0004221T^2 - 1.994 \times 10^{-7}T^3. \tag{19.26}$$

In all three cases the temperature is in °C. However, the influence of temperature is likely to be small over the temperature range experienced during a fermentation, and therefore heat capacities determined for a temperature in the middle of the expected temperature range can be used. Reasonable values are 1006 J kg^{-1} °C^{-1} for C_{Pa}, 1880 J kg^{-1} °C^{-1} for C_{Pv}, and 4184 J kg^{-1} °C^{-1} for C_{Pw}.

Of the various heat capacities, that of the dry solids is the most difficult to obtain. You can use the following equation to estimate the heat capacity if you know the composition of the substrate and have available literature data for the heat capacities of the various components:

$$C_{Pd} = \sum (w_i C_{Pi}),$$ (19.27)

where w_i is the fraction of the total mass of the substrate (or "mass fraction") contributed by component number "i" and C_{Pi} is the heat capacity of that component. Since the components present in solid substrates used in SSF will typically be similar to the components of foodstuffs, books that tabulate the heat capacities of food components may be used (e.g., Sweat 1986; Hallstrom et al. 1988).

It is also possible to estimate C_{Pb} experimentally. In this case it would be necessary to transfer a known quantity of energy into the bed and then measure the temperature rise. It would be best to do this in a bomb calorimeter. Although this will give C_{Vb}, the difference between C_{Vb} and C_{Pb} will be negligible. Schutyser et al. (2004) determined the heat capacity of oats experimentally as 2300 J kg-dry-solids^{-1} °C^{-1}.

Note that the heat capacity of the bed can also change as the microorganism grows, given that the microbial biomass has a different composition and water content than the substrate particle itself. Little attention has been paid to this in bioreactor models.

19.4.3 Enthalpy of Vaporization of Water

This is an important property because of the importance of evaporation as a heat removal mechanism. It describes the enthalpy change in the process:

$$H_2O \text{ (liquid)} \rightarrow H_2O \text{ (vapor)}$$

The enthalpy of vaporization of water (λ, J kg-H$_2$O^{-1}) depends on the temperature, but over the range of temperatures that might be expected in a fermentation, this variation is not large. For example, the enthalpy of vaporization of water is 2438.4 kJ kg^{-1} at 27°C and 2389.8 kJ kg^{-1} at 47°C (Himmelblau 1982), which represents a decrease of only 2%. An average of 2414 kJ kg^{-1} would be appropriate for most situations.

Further Reading

A mathematical model that recognized that fermented substrate may not have the same isotherm as uninoculated substrate
Nagel FJI, Tramper J, Bakker MSN, Rinzema A (2001) Model for on-line moisture-content control during solid-state fermentation. Biotechnol Bioeng 72:231–243

The heat capacities of liquid water, dry air, and water vapor can be found in various books, such as
Himmelblau DM (1982) Basic principles and calculations in chemical engineering, 5th edn. Prentice Hall, Englewood Cliffs
Himmelblau DM (1996) Basic principles and calculations in chemical engineering, 6th edn. Prentice Hall, Upper Saddle River

Books that give properties of foodstuffs and agricultural products, which are similar to the properties of substrates for SSF
Mohsenin NN (1980) Thermal properties of foods and agricultural materials. Gordon and Breach, New York
Rao MA, Rizvi SSH (1995) Engineering properties of foods, 2nd edn. Marcel Dekker, New York
Rahman S (1995) Food properties handbook. CRC Press, Boca Raton, Florida

20 Estimation of Transfer Coefficients for SSF Bioreactors

David A. Mitchell, Oscar F. von Meien, Luiz F.L. Luz Jr, and Marin Berovič

20.1 Introduction

Chapter 18 presented equations for various heat and mass transfer processes. These contain various heat and mass transfer coefficients for which it is important to have reasonable values if the model is to make acceptably accurate predictions about bioreactor performance. Relatively little attention has been paid to experimental determination of transfer coefficients in SSF systems, so various of the works reported here represent data borrowed from non-SSF systems.

Note that several of the studies reported here have determined transfer coefficients on the basis of "per second", whereas other chapters have considered rates on the basis of "per hour". This presents no difficulties, since it is a simple matter to convert a value "per second" into a value "per hour", by multiplying by the conversion factor "3600 s h^{-1}".

20.2 Thermal Conductivities of Substrate Beds

The thermal conductivity appears in equations describing conduction within the bed (see Eq. (18.9) in Sect. 18.3.2). Little attention has been given to direct experimental determination of thermal conductivities of solid beds, with many workers simply using values tabulated for foodstuffs. Lai et al. (1989) describe a simple system that can be used to determine the thermal diffusivity of the bed (α_b, $\text{m}^2 \text{ h}^{-1}$) (Fig. 20.1), from which the thermal conductivity of the bed (k_b, $\text{J h}^{-1} \text{ m}^{-1}$ $^\circ\text{C}^{-1}$) can be calculated by the following equation:

$$k_b = \alpha_b \rho_b C_{Pb},\qquad(20.1)$$

where C_{Pb} is the heat capacity of the bed ($\text{J kg-bed}^{-1} \, ^\circ\text{C}^{-1}$) and ρ_b is the bed density ($\text{kg m}^{-3}\text{-bed}$). Over a 10-day fermentation of sorghum mash, the thermal diffusivity of the bed varied between $3.8\times10^{-4} \text{ m}^2 \text{ h}^{-1}$ and $4.0\times10^{-4} \text{ m}^2 \text{ h}^{-1}$, indicating that the thermal diffusivity itself does not change significantly as a result of microbial

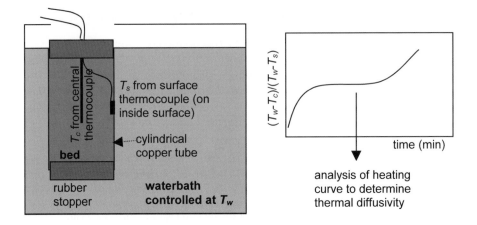

Fig. 20.1 Apparatus that can be used for the determination of the thermal diffusivity of a solid substrate (adapted from Lai et al. 1989). For the determination of thermal diffusivity, the "diffusivity cell" was initially equilibrated at 20°C and then quickly transferred to a 46°C waterbath

growth. However, it will be sensitive to any changes in the bed packing density and bed moisture content, as shown by the work of Costa et al. (1998) who determined the thermal conductivity of defatted rice bran over a range of packing densities and moisture contents, and derived the following equation:

$$k_b = 47.508 + 0.0115\rho_b + 0.1295M - 6.0737\ln\rho_b - 5.5555\ln M, \qquad (20.2)$$

where k_b is in J s^{-1} m^{-1} °C^{-1}, ρ_b is the bed packing density (kg m^{-3}), and M is the percentage moisture content (wet basis). Under the various conditions tested, the value of k_b varied between 0.1166 and 2.6551 J s^{-1} m^{-1} °C^{-1}. Oostra et al. (2000) determined the thermal conductivity of moisturized oats, with a water content of 1.1 kg-water kg-dry-matter^{-1}, as 0.1 J s^{-1} m^{-1} °C^{-1}.

20.3 Heat Transfer Coefficients Involving the Wall

If the bioreactor wall is not recognized explicitly as a separate phase in the model, then the heat transfer to the surroundings will be described as a direct transfer from the outer surface to the bed to the surroundings (Eq. (18.6) in Sect. 18.3.1). In this case, the various resistances, that is, for transfer from the bed to the wall, transfer across the wall and transfer from the wall to the surroundings, will be lumped together into an overall heat transfer coefficient h_{ov}. In cases where the bioreactor wall is recognized as a separate system, then separate heat transfer coefficients may be needed for terms describing heat transfer from the bed to the wall, from the headspace air to the wall and from the wall to the surroundings.

20.3.1. Bed-to-Wall Heat Transfer Coefficients

The efficiency of the heat transfer from the bed to the bioreactor wall will depend on whether the bed is agitated or static. Further, in an agitated bed, the heat transfer coefficient is likely to depend on the type of agitation.

For rotating drum bioreactors, Stuart and Mitchell (2003) used literature data for non-SSF applications of rotating drum bioreactors to estimate the bed to wall heat transfer coefficient (h_{bw}, J s^{-1} m^{-2} °C^{-1}) as:

$$h_{bw} = \frac{77100}{D} \, , \qquad (20.3)$$

for the drum diameter (D) in meters.

The major source of resistance in transfer from the bed to the wall resides in the bed itself, so the heat transfer coefficient of the bed (h_b, J s^{-1} m^{-2} °C^{-1}) can be used as an approximation of the bed-to-wall heat transfer coefficient (Oostra et al. 2000). The value of h_b will depend on whether the bed is mixed or not, with h_b increasing as the intensity of mixing increases (Oostra et al. 2000). They estimated h_b as:

$$h_b = 2\sqrt{\frac{k_b \rho_b C_{Pb}}{\pi \, t_c}} \, , \qquad (20.4)$$

where k_b is the thermal conductivity of the bed (W m^{-1} °C^{-1}), ρ_b is the bed density (kg m^{-3}), C_{Pb} is the bed heat capacity (J kg-bed °C^{-1}), and t_c is the time of contact between the solid particles and the wall (s), which is inversely proportional to the rotational speed of the agitator. In a static bed, a value of h_b of 40 J s^{-1} m^{-2} °C^{-1} was determined experimentally for moist oats (Schutyser et al. 2004).

However, it is often not a simple matter to determine h_b and, instead of doing this, the overall heat transfer coefficient for transfer from the bed to the outside is often determined experimentally (see Sect. 20.3.4).

20.3.2 Wall-to-Headspace Heat Transfer Coefficients

In most bioreactors heat transfer between the wall and the headspace gases is not described within mathematical models because the air has already left the bed and will leave the bioreactor without any further interaction with the bed. The situation is different in rotating drum bioreactors, in which the headspace air interacts with the bed as it travels along the drum (see Sect. 20.5). This headspace air also interacts with the drum wall. Once again, this transfer has received little attention in SSF bioreactors.

Stuart and Mitchell (2003) derived the following equation for the wall-to-headspace heat transfer coefficient (h_{wg}, J s^{-1} m^{-2} °C^{-1}) from the studies undertaken in a rotary kiln by Tscheng and Watkinson (1979):

$$h_{wg} = \frac{0.1055}{D(D^{1.5}S)^{0.292}} \left[F\left(\frac{1}{0.029} + \frac{H_{in}}{0.018}\right)\frac{D}{(D^2L)^{0.667}} \right]^{0.575}, \qquad (20.5)$$

where D is the drum diameter (m), L is the drum length (m), S is the fraction of the critical speed (see Sect. 8.4.1), F is the inlet air flow rate (kg-dry-air s^{-1}), and H_{in} is the inlet air humidity (kg-vapor kg-dry-air^{-1}).

A much simpler equation, given as a general approximation for linear flow of air past a surface, is (Geankoplis 1993):

$$h_{wg} = 14.278 \left(\frac{F}{A_g}\right)^{0.8}, \qquad (20.6)$$

where F is the inlet air flow rate (kg-dry-air s^{-1}) and A_g is the cross-sectional area of the headspace gases normal to the direction of gas flow (m^2).

20.3.3 Wall-to-Surroundings Heat Transfer Coefficients

The wall-to-surroundings heat transfer coefficient (h_{wsurr} J s^{-1} m^{-2} °C^{-1}) will vary markedly, depending on whether the bioreactor wall is surrounded by air or by the water in a water jacket, and also by the flow of this cooling fluid. In the case of air, the air may be blown forcefully past the bioreactor (forced convection) or not. In the latter case, flow will be due to natural circulation, with heat being removed by "natural convection".

Oostra et al. (2000) quote a value of 500 J s^{-1} m^{-2} °C^{-1} for h_{wsurr} for a water-jacketed stainless-steel bioreactor. In the absence of a water jacket, correlations for vertical walls in air can be used (Churchill and Chu 1975). For rotating drum bioreactors, in which the outside wall of the bioreactor is in motion, Stuart (1996) estimated that h_{wsurr} would be of the order of 5 J s^{-2} m^{-2} °C^{-1}, based on correlations provided for heat transfer from rotating cylinders by Kays and Bjorklund (1958), assuming a 20°C difference between the temperatures of the drum wall and the surrounding air.

20.3.4 Overall Heat Transfer Coefficients

Often the bioreactor wall is not recognized as a separate subsystem and an overall heat transfer coefficient from the outside of the bed to the surroundings (h_{ov}, J s^{-1} m^{-2} °C^{-1}) is used.

The overall heat transfer coefficient can be estimated from the law of resistances in series if the heat transfer properties of the various system components are known (Oostra et al. 2000):

$$h_{ov} = \frac{1}{\dfrac{1}{h_b} + \dfrac{L_{wall}}{k_{wall}} + \dfrac{1}{h_{ext}}} ,$$
(20.7)

where h_b is the heat transfer coefficient of the bed ($J\ s^{-1}\ m^{-2}\ °C^{-1}$), h_{ext} is the heat transfer coefficient at the outer surface ($J\ s^{-1}\ m^{-2}\ °C^{-1}$), k_{wall} is the thermal conductivity of the wall ($J\ s^{-1}\ m^{-1}\ °C^{-1}$), and L_{wall} is the thickness of the wall (m). For a water-jacketed stainless-steel bioreactor, Oostra et al. (2000) quote a typical value of 80 $J\ s^{-1}\ m^{-1}\ °C^{-1}$ for k_{wall}. The two heat transfer coefficients h_b and h_{ext} were considered in Sects. 20.3.1 and 20.3.3, respectively.

Nagel et al. (2001a) report values of $h_{ov}A$ of 6 to 8.5 $J\ s^{-1}\ °C^{-1}$ for a glass-walled 35 L bioreactor around which a flat plastic hose containing cooling water was wrapped. Using the diameter of 30 cm and length of 50 cm, the curved wall of the cylinder has an area of 0.47 m^2. This gives an overall heat transfer coefficient of the order of 15 $J\ s^{-1}\ m^{-2}\ °C^{-1}$. Possibly values of this order of magnitude can be expected in glass-walled laboratory bioreactors, although of course the exact value will depend on the thickness of glass used. Further, greater values of h_{ov} would be expected for a proper water jacket, as the wall of the plastic hose must have represented an additional resistance to heat transfer. Nagel et al. (2001a) also report that, for a stainless-steel water-jacketed industrial solids mixer, adapted for use as an SSF bioreactor, the overall heat transfer coefficient was 100 $J\ s^{-1}\ m^{-2}\ °C^{-1}$. Such a variation in the value of the overall heat transfer coefficient is not unexpected due to the different materials used in the laboratory-scale bioreactors (glass) and industrial-scale bioreactors (stainless steel).

20.4 Solids-to-Air Heat and Mass Transfer Coefficients Within Beds

Little attention has been given in the SSF literature to the effectiveness of heat and mass transfer between the solids and gas phases within a bed of moist solid particles. However, studies of drying of agricultural products have led to equations that can possibly be adapted for SSF systems.

In their two-phase model of heat and mass transfer in a packed-bed bioreactor, von Meien and Mitchell (2002) used correlations that had been determined for the drying of corn. The coefficient that they used for convective heat transfer between the solid and gas phases, which appears in (Eq. (18.11) in Sect. 18.4.2), is given by (Calado 1993):

$$ha = 44209.85\left(\frac{G(T_g + 273)}{0.0075P}\right)^{0.6011} ,$$
(20.8)

where "ha" has the units of $J\ s^{-1}\ m^{-3}\ °C^{-1}$, G is the air flux passing through the bed (kg-air $m^{-2}\ s^{-1}$), T_g is its temperature (°C), and P is the pressure (Pa). Note that the units of "*ha*" are different from those indicated for *h* in Eq. (18.11) and that "*ha*" multiplied by a driving force gives the overall transfer rate per cubic meter of bed volume. This set of units is used for "*ha*" since the area for heat transfer in the bed is impossible to measure accurately, and therefore the effect of area is combined with the underlying heat transfer coefficient as a single variable. Conversely, overall bed volume is simple to measure and, when the bed is packed in a certain way, each unit volume of the bed will have a certain interfacial area. As a result, even though the area across which heat transfer occurs is not known explicitly, it is in fact taken into account. Potentially, an equation such as Eq. (20.8) could include the packing density of the bed as a variable, although the dependence of "*ha*" on the packing density would need to be determined empirically.

In a similar manner, the coefficient for water mass transfer between the solid and gas phases, which appears in Eq. (18.15) in Sect. 18.5.1, is given by an empirical equation determined for the drying of corn (Mancini 1996):

$$Ka = (7.304 - 0.0177(T_g + 273))W - 2.202 + 0.00618(T_g + 273),$$ (20.9)

where T_g is the gas temperature (°C), W is the solids water content (kg-water kg-dry-solids^{-1}), and "*Ka*" has units of kg-H$_2$O s^{-1} m^{-3} (kg-H$_2$O kg-dry-solids^{-1}). The units within the parentheses are the units of the driving force, such that kg-H$_2$O cancels out, giving the units of "*Ka*" as kg-dry-solids s^{-1} m^{-3}. As with "*ha*", "*Ka*" is related to the overall transfer rate per cubic meter of bed volume.

Schutyser et al. (2004) report a mass transfer coefficient for water of 0.00492 m s^{-1} (presumably these units are kg-H$_2$O s^{-1} m^{-2} (kg-H$_2$O m^{-3})$^{-1}$). However, such a value only becomes useful if an estimate is available of the superficial area for mass transfer within the bed.

20.5 Bed-to-Headspace Transfer Coefficients

Bed-to-headspace heat and mass transfer is of crucial importance in rotating drums while in bioreactors with forced aeration it is typically ignored since the air that leaves the bed is assumed to leave at the temperature of the bed and to not interact any further with the bed before leaving the bioreactor. Note that the rate of transfer depends on the driving force, and the driving force may vary in a complex way with position due to the complex flow patterns that occur with the drum headspace (see Sect. 8.4.2). Further, these flow patterns and the resulting value of the transfer coefficient may depend on how the drum is operated, especially the rotational rate and the aeration rate. Therefore it is not an easy matter to determine the bed-to-headspace transfer coefficients.

Assuming that the headspace of their rotating drum bioreactor was well mixed, Stuart and Mitchell (2003) used the work of Tscheng and Watkinson (1979) to estimate the bed to headspace heat transfer coefficient (h_{bg}, J s^{-1} m^{-2} °C^{-1}):

$$h_{bg} = \frac{2.351\left(D^{1.5}S\right)^{0.104}}{D\left[F\left(\frac{1}{0.029} + \frac{H_{in}}{0.018}\right)\frac{D}{(D^2 L)^{0.667}}\right]^{0.535}},$$ (20.10)

where D is the drum diameter (m), L is the drum length (m), S is the fraction of the critical speed (see Sect. 8.4.1), F is the inlet air flow rate (kg-dry-air s^{-1}), and H_{in} is the inlet air humidity (kg-vapor kg-dry-air^{-1}).

For plug flow of the air, potentially Eq. (20.6) can be used to estimate h_{bg}. On the other hand, the work of Hardin et al. (2002) gives an insight into what is necessary in order to determine bed-to-headspace transfer coefficients in a rotating drum bioreactor in which neither the well-mixed nor plug-flow regimes occur. They monitored the outlet humidity of a 200-L drum with a bed of moist wheat bran under various different operating conditions (Fig. 20.2). They then used a mathematical model of the flow patterns (which had been developed on the basis of residence time distribution studies to describe the flow pattern shown in Fig. 8.15) in order deduce the value of the transfer coefficient, "ka", which is a lumped overall transfer coefficient (i.e., the area term was not determined separately). Note that they defined "ka" in such a way as to have units of s^{-1}, since it was used in combination with dimensionless concentrations of water vapor in the headspace air. The dimensionless concentration of water vapor was calculated from the humidity (H, kg-H$_2$O kg-dry-air^{-1}) as:

$$c_{air} = \frac{H_{sat} - H_{air}}{H_{sat} - H_{inlet}}.$$ (20.11)

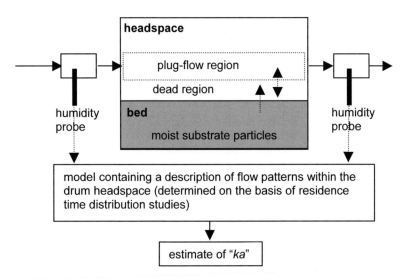

Fig. 20.2. Experimental set-up used by Hardin et al. (2002) for investigating mass transfer between the bead and the headspace in a 200-L rotating drum bioreactor

The term for overall water transfer from the bed to the dead volume in the head-space (R_w, s^{-1}) would then be written as:

$$R_w = ka(c_{bed} - c_{air}),$$ (20.12)

where c_{bed} is the dimensionless saturation water vapor concentration at the temperature of the bran.

The results of Hardin et al. (2002) are summarized in Table 20.1. Note that the experimentally determined values of "ka" vary over 4-fold, with the rotational rate of the drum having a strong influence. For example, for experiments done at the same aeration rate of 155 L min^{-1} and the same percentage filling of 30%, the value of "ka" varied from 0.0472 s^{-1} at 0.9 rpm to 0.2024 s^{-1} at 9 rpm.

In order to use these results in a model of a drum, that is, to calculate an appropriate value for "ka" for any particular combination of operating conditions, it is necessary to combine the various operating conditions, namely the % filling, the aeration rate, and the rotational rate, into a single variable. This is done by defining an effective Peclet number (Pe_{eff}), given by Hardin et al. (2002) as:

Table 20.1 Experimentally determined values of the bed-to-headspace water mass transfer coefficient, as determined by Hardin et al. (2002) for a variety of operating conditions. Adapted from Hardin et al. (2002) with kind permission of Elsevier

Rotational speed (rpm)	Air Rate (L min^{-1})	Percentage Filling (%)	Pe_{eff} (calculated)	ka (s^{-1})
0.9	155	30	21.9	0.0472
1.8	155	30	32.9	0.0504
3.6	155	30	49.1	0.1323
5.4	155	30	61.7	0.1373
7.2	155	30	72.5	0.1252
9.0	155	30	82.3	0.2024
0.9	155	22.5	19.8	0.0538
1.8	155	22.5	29.5	0.0880
3.6	155	22.5	43.7	0.1337
5.4	155	22.5	54.9	0.1400
7.2	155	22.5	63.9	0.1742
9.0	155	22.5	72.2	0.1141
0.9	155	15	17.3	0.0449
1.8	155	15	25.6	0.0564
3.6	155	15	37.5	0.0746
5.4	155	15	46.5	0.0923
7.2	155	15	54.1	0.1164
9.0	155	15	60.6	0.1179
Group of studies that showed that the aeration rate has a relatively small effect				
5.4	92.6	30	61.3	0.1423
5.4	118	30	61.8	0.1286
5.4	137	30	61.8	0.1746
5.4	155	30	61.8	0.1799
5.4	174	30	61.9	0.1692

$$Pe_{eff} = \frac{u_p d}{\delta},$$

(20.13)

where u_p is the average particle velocity in the moving layer (m s^{-1}), d is the particle diameter (m), and δ is the diffusivity of water vapor in air (m^2 s^{-1}). The particle diameter can be measured experimentally and the diffusivity of water vapor in air can be obtained from a reference book, such as McCabe et al. (1985). In order to calculate the average particle velocity in the moving layer, it is necessary to know (see Fig. 20.3):

- N, the rotational speed (revolutions per second), determined by the operator;
- γ the dynamic angle of repose of the solids (degrees), determined experimentally. This was 37° in the system of Hardin et al. (2002);
- D, the drum diameter (m), determined by the drum design;
- h, the maximum height of the bed (m), which will be a function of the fractional filling of the drum. It can be calculated according to geometric principles or simply measured experimentally.

Firstly, two secondary variables, K and s need to be calculated. K can be estimated from the following equation (Savage 1979):

$$K = \frac{2}{3}\left(\frac{g \sin \gamma}{C_v d^2}\right)^{0.5}\left(\frac{3}{5} - \frac{12}{11}f + \frac{18}{17}f^2 - \frac{12}{23}f^3 + \frac{3}{29}f^4\right),$$

(20.14)

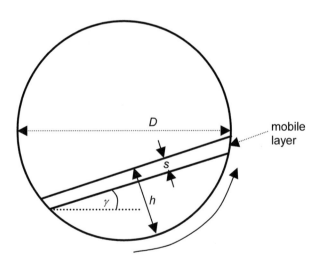

Fig. 20.3. Detail of the drum in the tumbling flow regime, showing the nomenclature used in the calculations of the mobile layer thickness, "s". From Hardin et al. (2002) with kind permission of Elsevier

where f is a dimensionless factor of porosity, equal to 0.8 for most materials. C_V is a dimensionless constant associated with the bed viscosity, equal to 0.6 for most materials and g is gravitational acceleration (9.81 m s^{-2}). Once K has been calculated, the thickness of the mobile layer of solids (s, m) can be calculated by solving the following equation, which is quadratic in s (Blumberg and Schlünder, 1996):

$$\pi N\left(D(h-s)-(h-s)^2\right)- Ks^{2.5} + 2\pi Ns(0.5D-h+s) = 0 . \tag{20.15}$$

With both K and s it is possible to calculate the average particle velocity in the mobile layer. This is done using an equation of Blumberg and Schlünder (1996)

$$u_P = - Ks^{3/2} + 2\pi N(D/2-h+s). \tag{20.16}$$

The effective Peclet number can now be calculated. Hardin et al. (2002) did this and plotted the experimentally-determined value of "ka" against the calculated value of Pe_{eff} (Fig. 20.4). For their drum they obtained the relationship:

$$ka = 2.32\times10^{-3}Pe_{eff}. \tag{20.17}$$

It would be necessary to undertake a broader study to investigate whether this correlation is generally valid for all rotating drum bioreactors. Such a study would need to involve a number of different rotating drum bioreactors of different length to diameter ratios and with different air inlet and outlet positions and designs. Further, as part of this study, it would be necessary to determine the particular air flow patterns within the headspace of each bioreactor. This is not a small task.

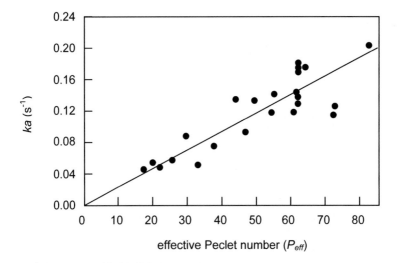

Fig. 20.4. Experimental "ka" values plotted against the calculated value of the Peclet number (Hardin et al. 2002). The line of best fit is forced through the origin, giving Eq. (20.17). Adapted from Hardin et al. (2002) with kind permission of Elsevier

20.6 Conclusions

Transfer coefficients have received relatively little attention within SSF bioreactors. At the moment the available information is not sufficient to allow the proposal of general correlations and therefore it will be necessary to determine the coefficients experimentally for each particular bioreactor. In the absence of experimentally determined correlations, correlations for non-SSF systems can be used, although it must be realized that doing this may bring inaccuracies in the model predictions.

Further Reading

General theory of heat transfer, explaining the significance of heat transfer coefficents
Incropera FP, DeWitt DP (1996) Introduction to heat transfer. John Wiley & Sons, New York

Estimation of transfer coefficients in SSF
Andre G, Moo-Young M, Robinson CW (1981) Improved method for the dynamic measurement of mass transfer coefficient for application to solid-substrate fermentation. Biotechnol Bioeng 23:1611–1622
Zhao H, Zhang X, Zhou X, Li Z (2002) Use of the glucose oxidase system for estimation of oxygen transfer rate in a solid-state bioreactor. Enzyme Microb Technol 30:843–846

Thermal conductivity in composting systems
Van Ginkel JT, Van Haneghem IA, Raats PAC (2002) Physical properties of composting material: Gas permeability oxygen diffusion coefficient and thermal conductivity. Biosystems Eng 81:113–125

Selected examples of models of SSF bioreactors in which heat transfer coefficients were estimated or taken from the literature
Costa JAV, Alegre RM, Hasan SDM (1998) Packing density and thermal conductivity determination for rice bran solid-state fermentation. Biotechnol Techniques 12:747–750
Mitchell DA, von Meien OF (2000) Mathematical modeling as a tool to investigate the design and operation of the Zymotis packed-bed bioreactor for solid-state fermentation. Biotechnol Bioeng 68:127–135
Mitchell DA, Tongta A, Stuart DM, Krieger N (2002) The potential for establishment of axial temperature profiles during solid-state fermentation in rotating drum bioreactors. Biotechnol Bioeng 80:114–122
Nagel FJI, Tramper J, Bakker MSN, Rinzema A (2001a) Temperature control in a continuously mixed bioreactor for solid state fermentation. Biotechnol Bioeng 72:219–230
Oostra J, Tramper J, Rinzema A (2000) Model-based bioreactor selection for large-scale solid-state cultivation of *Coniothyrium minitans* spores on oats. Enzyme Microb Technol 27:652–663

Rajagopalan S, Modak JM (1994) Heat and mass transfer simulation studies for solid-state fermentation processes. Chemical Engineering Science 49:2187–2193

Sangsurasak P, Mitchell DA (1998) Validation of a model describing two-dimensional heat transfer during solid-state fermentation in packed bed bioreactors. Biotechnol Bioeng 60:739–749

Schutyser MAI, Briels WJ, Boom RM, Rinzema A (2004) Combined discrete particle and continuum model predicting solid-state fermentation in a drum fermenter. Biotechnol Bioeng 86: 405–413

Schutyser MAI, Weber FJ, Briels WJ, Rinzema A, Boom RM (2003) Heat and water transfer in a rotating drum containing solid substrate particles. Biotechnol Bioeng 82:552–563

Stuart DM, Mitchell DA (2003) Mathematical model of heat transfer during solid-state fermentation in well-mixed rotating drum bioreactors. J Chem Technol Biotechnol 78:1180–1192

von Meien OF, Mitchell DA (2002) A two-phase model for water and heat transfer within an intermittently-mixed solid-state fermentation bioreactor with forced aeration. Biotechnol Bioeng 79:416–428

Weber FJ, Oostra J, Tramper J, Rinzema A (2002) Validation of a model for process development and scale-up of packed-bed solid-state bioreactors. Biotechnol Bioeng 77:381–393

21 Bioreactor Modeling Case Studies: Overview

David A. Mitchell

21.1 What Can the Models Be Used to Do?

Chapters 22 to 25 present case studies intended to give insights into how mathematical models can be useful tools in the design of SSF bioreactors and the optimization of their performance. Several of the models are available in the form of programs (see Appendix for details). Note that details of how to run the programs and interpret the output files are also given in the appendix. You will obtain a better understanding of these chapters if you:

- use the supplied programs to simulate performance of each bioreactor under design and operating conditions that are different from those presented in the various case studies;
- use a spreadsheeting or graphing program to plot the results;
- inspect and interpret the results;
- return to the model and further explore the predicted performance, changing the design and operating variables on the basis of the analysis of the results generated in the previous simulations.

In fact, you can use the models in two different manners:

- You can use them directly as tools in bioreactor design without trying to understand why a certain combination of design and operating variables is optimal. For example, you might want to build a certain type of bioreactor with a bed volume of 1 m^3. You can explore how various design variables (bioreactor length and height) and operating variables (e.g., aeration rate) affect the performance of such a bioreactor, seeking to find the combination that gives the most growth in the least time (i.e., the highest productivity). You can change the variables by trial-and-error or be more systematic, using a strategy in which you vary the variables one-by-one in order to search for the optimum combination. It is also possible to use more sophisticated means to search for the optimum combination, namely by incorporating the bioreactor model as part of the objective function within an optimization program. This program will find the optimum combination of design and operating variables using powerful search algorithms;

- You can use them as tools to increase your understanding of how certain types of bioreactors might be expected to operate under a range of different design and operating conditions and sizes, and to explore the question as to why they would be expected to operate in that manner. For instance, various of the models presented in the case studies give detailed predictions about temperature gradients within subsystems and about temperatures and moisture contents of different subsystems. These can be plotted and analyzed, in an effort to understand how the various phenomena are interacting with each other. For example, in a system in which the model describes spatial gradients and recognizes the substrate particles and inter-particle air in the bed as separate systems, you can plot both the air and solids temperatures as functions of height. This will give an idea of how close the bed is to equilibrium: The solids and air might not be in thermal equilibrium near the air inlet, but soon after the air inlet they may have almost the same temperature, this being maintained until the top of the bed. Further, knowing these temperatures as a function of height may help you to understand why one part of the bed dries out more quickly than another.

Obviously, there are no limits on the range of simulations that can be done. Chapters 22 to 25 do not aim to give in-depth demonstrations of all of the possibilities. Rather, they present relatively brief studies into the question of design and operation of large-scale bioreactors. If you wish more detail about how the models can be used, you should consult the original papers that are cited, in which the model predictions are explored in greater depth. However, you should not restrict yourself simply to what has already been done. You are encouraged to use the models to explore the various bioreactors further.

21.2 Limitations of the Models

Of course, the models provided have various limitations that mean that, although they are certainly valuable tools for increasing the understanding of the principles of bioreactor operation, they are not yet the powerful and flexible tools that we would like to have available in the bioreactor design process. For example:

- The models only simulate performance in terms of growth. They do not give any analysis of the economic consequences of certain operating modes. They do not even simulate performance in terms of product formation.
- The models do not impose practical limits on operating variables. For example, the packed-bed model does not limit the superficial air velocity to remain at values below the velocity that would fluidize the substrate particles in the bed.
- The flexibility in changing the kinetics of growth is greatly limited. It is possible to specify the optimum temperature for growth and the value of the specific growth rate constant at this temperature, but not the form of the curve describing the specific growth rate constant as a function of temperature. It is only possible to choose from two different types of dependence of growth on water

activity (the "*Rhizopus*-type" has optimum growth at a water activity of 1.0, while the "*Aspergillus*-type" has an optimum water activity of 0.95).

- Various properties of the substrate are pre-determined and cannot be changed (e.g., the absorption/desorption isotherm).
- Although it is possible to change the size and often the length to diameter ratio of the bioreactor, it is not possible to change the shape (e.g., you cannot change from a circular cross-section to a rectangular cross-section).

Obviously, much more needs to be done in order to improve models as tools in the bioreactor design process. We hope that the current book plays a role in stimulating the continued improvement of models.

21.3 The Amount of Detail Provided about Model Development

Complete deductions of the model equations and all of the thinking used in the model development process are not presented in the case studies. You will need to consult the original papers if you want more detail than that which is provided,

Details are not given about how the various equations were originally written and manipulated in order to arrive at the equations that are presented. Chapters 12 to 20 have given the general principles for the deduction of the equations. In fact, in order to be able to write the correct equations and solve them, it is necessary to have skills in several areas:

- mass and energy balancing;
- heat and mass transfer phenomena;
- differential and integral calculus;
- numerical methods for solving differential equations;
- programming in a computer language such as FORTRAN

We have assumed that the majority of our readers do not in fact have such skills. Of course, this is not a problem since, if you want to develop or modify a mathematical model of an SSF bioreactor and you do not have the necessary skills yourself, you can interact with people who do, whom we can refer to as "modelers". The aim of this book (which has guided the detail in the previous chapters and will guide the detail presented in these modeling case study chapters) is to give you sufficient understanding to allow you to interact effectively with these modelers. For example:

- you will understand something about what modeling can and cannot do, thereby having realistic expectations about what benefits a final model can bring (i.e., after reading this book you will be better able to discuss Step 1 of the modeling process with the modeler, see Sect. 12.4.1);

- you will understand something about heat and mass transfer and kinetic phe- nomena, improving your ability to communicate with the modeler and under- stand what he or she says;
- you will be able to recognize the mathematical forms of expressions used to de- scribe various heat and mass transfer phenomena and kinetic phenomena (even if you do not know how to derive the correct form of the expression for a par- ticular situation). In other words, looking at an equation within the model, you will have an idea about which phenomena it describes, and how. This eases the interaction with the modeler.

Further, a modeler with whom you interact may have the necessary engineer- ing, mathematical, and programming skills but may not be familiar with SSF sys- tems. Such a person should also read this book, with the aim of understanding the general features of the various bioreactors and the principles of growth kinetics in SSF systems. With this information, the modeler will be better placed to partici- pate in decisions about what level of detail to use to describe the system and the processes occurring within and between the phases in the system.

21.4 The Order of the Case Studies

In the chapters that follow, the order of presentation will not follow the order of classification in Groups I to IV presented in Chap. 3.3. Rather, the "mathemati- cally simpler" models will be presented first. The first two case study chapters deal with bioreactors within which each of the subsystems is assumed to be well mixed, namely the continuously-mixed, forcefully-aerated bioreactor from Group IVa (Chap. 22) and the rotating-drum bioreactor from Group III (Chap. 23). The models of these bioreactors involve ordinary differential equations. The other two chapters deal with models in which there are gradients within the substrate bed, namely, packed-bed bioreactors from Group II (Chap. 24) and intermittently- mixed, forcefully-aerated bioreactors from Group IVb (Chap. 25). The models of these bioreactors involve partial differential equations. A case study is not pre- sented for tray bioreactors (Group I).

The balance on the overall mass of dry solids (i.e., the sum of dry biomass and dry residual substrate) is necessary since not all the consumed substrate is converted into biomass; a proportion is lost in the form of CO_2. This is Eq. (16.11), although without the maintenance term:

$$\frac{dM}{dt} = \left(1 - \frac{1}{Y_{XS}}\right)\frac{dX}{dt}. \tag{22.4}$$

In the gas phase water balance presented in Fig. 22.2, all terms have units of $kg\text{-}H_2O\ h^{-1}$ and:

- the left hand side represents the temporal variation in the amount of water vapor in the air phase within the bed;
- the first term on the right hand side represents the entry of water vapor with the inlet air and the leaving of water vapor with the outlet air;
- the second term on the right hand side represents the water exchange between the solid and gas phases.

In the gas phase energy balance presented in Fig. 22.2 all terms have units of $J\ h^{-1}$ and:

- the left hand side represents the temporal variation in the sensible energy of the dry air and water vapor in the air within the bed;
- the first term on the right hand side represents the sensible energy of the dry air entering and leaving the bed in the process air stream;
- the second term on the right hand side represents the sensible energy of the water vapor entering and leaving the bed in the process air stream;
- the third term on the right hand side represents the sensible heat exchange between the solid phase and the gas phase.
- the fourth term on the right hand side represents the sensible heat exchange between the gas phase and the bioreactor wall, using the void fraction (ε) as an estimate of the fraction of the total wall area in contact with the gas phase;

In the solid phase water balance presented in Fig. 22.2 all terms have units of $kg\text{-}H_2O\ h^{-1}$ and:

- the left hand side represents the temporal variation in the water content of the solids phase;
- the first term on the right hand side represents metabolic water production;
- the second term on the right hand side represents the exchange of water between the solid and gas phases.

In the solid phase energy balance presented in Fig. 22.2 all terms have units of $J\ h^{-1}$ and:

- The left hand side represents the temporal variation of the sensible energy within the solids phase;

- the first term on the right hand side represents the liberation of waste metabolic heat in the growth process
- the second term on the right hand side represents sensible energy exchange between the solids and the gas phase;
- the third term on the right hand side represents the removal of energy from the solid as the latent heat of evaporation or addition of energy to the solid due to condensation, depending on the direction of water transfer;
- the fourth term on the right hand side represents the sensible energy exchange with the bioreactor wall, using $(1-\varepsilon)$ as an estimate of the fraction of the total wall area in contact with the solids phase;

In the energy balance over the bioreactor wall presented in Fig. 22.2 all terms have units of J h^{-1} and:

- The left hand side represents the temporal variation of the sensible energy within the wall;
- the first term on the right hand side represents sensible energy exchange between the wall and the gas phase within the bed, using the void fraction (ε) as an estimate of the fraction of the total wall area in contact with the gas phase;
- the second term on the right hand side represents sensible energy exchange between the wall and the solids phase within the bed, using $(1-\varepsilon)$ as an estimate of the fraction of the total wall area in contact with the solids phase;
- the third term on the right hand side represents sensible energy exchange between the wall and the water in the water jacket.

Several of the assumptions that this model makes are:

- the bed volume does not change during the fermentation. The effect of consumption of dry matter is to decrease the packing density of the bed.
- the fermenting solids have the same isotherm as the substrate itself.
- only the side walls of the bioreactor are available for heat transfer to the cooling water in the jacket. Further, the whole of this side wall has the same temperature.
- The bed porosity (ε) does not change during the fermentation. Further, in choosing a value for ε, it is assumed that the porosity is higher than that for a normal packed bed due to the continuous mixing action.
- there is no maintenance metabolism.
- there is no microbial death. The sole effect of high temperatures is to limit the growth rate.

Further, no special attempt is made to describe the deleterious effects of mixing on the growth of the organism. The value of the optimal specific growth rate constant (μ_{opt}) used in the model therefore should be an experimental value obtained in a continuously-mixed system.

22.2.2 Values of Parameters and Variables

Tables 22.1 and 22.2 show the values used in the base case simulation for the various parameters and variables in the model.

The coefficients for heat transfer from (1) the gas and the solid phase to the bioreactor wall and (2) the bioreactor wall to the water in the water jacket were chosen as 30 W m^{-2} °C^{-1} in order to give an overall coefficient for heat transfer from the bed to the water in the water jacket (calculated on the basis of the law of resistances in series) of the order of magnitude of 15 W m^{-2} °C^{-1}, a value calculated from data provided by Nagel et al. (2001a) for a glass-walled laboratory-scale bioreactor (see Sect. 20.3.1).

Note that the side wall area (A, m^2) and the bed volume (V_{bed}, m^3) are calculated on geometric principles for an upright cylinder of circular cross-section.

The mass balance part of the model calculates the water content of the solids (W, kg-H$_2$O kg-dry-solids^{-1}), whereas in the growth kinetic part of the model the specific growth rate constant is expressed as a function of the water activity of the solids phase and not its water content (see Eq. (22.2)). The isotherm determined for corn by Calçada (1998) is used to convert the water content into the corresponding water activity (see Eqs. (19.8) and (19.9) in Sect. 19.2.5).

This isotherm is also used in the calculation of the evaporation term. The driving force for evaporation is the difference between the water content of the solids (W, kg-H$_2$O kg-dry-solids^{-1}), given by Eq. (19.9), and the water content that the solids would have if they were in equilibrium with the gas in the headspace phase (W_{sat}). To calculate W_{sat}, Eq. (19.9) is again used, but with the gas phase water activity and temperature, giving:

$$W_{sat} = \left[\frac{\ln\left(1 - a_{wg}^{0.32}\right)}{-\exp\left(2.9 + 0.004 T_g\right)} \right]^{\frac{1}{(1.275 - 0.0029 T_g)}}. \qquad (22.5)$$

The coefficients for heat transfer and water mass transfer between the solid and gas phases were those determined for corn (see Eqs. (20.8) and (20.9)). Note that, since these transfer coefficients were determined for a packed-bed and, considering that solids/gas transfer is potentially more efficient in a mixed bed, the model allows for manipulation of these transfer coefficients, through the input variable "fold", which is used to multiply the values calculated in Eqs. (20.8) and (20.9).

A simple control scheme is incorporated to control the temperature of the water in the water jacket.

$$T_w = T_{setpoint} - J(T_s - T_{setpoint}), \qquad (22.6)$$

where J is the proportional gain. In other words, the program calculates the temperature difference between the solids and the set point temperature. It then sets the cooling water temperature so that the difference between the cooling water temperature and the set point temperature is J-fold greater, but in the opposite direction, such that the cooling water will warm the bed if the bed temperature is below the set point and cool the bed if the bed temperature is above the set point.

Table 22.1. Values used for the base case simulation of those parameters and variables that can be changed in the accompanying model of a well-mixed forcefully-aerated bioreactor

Symbol	Significance	Base case value and units[a]
Design and operating variables and initial values of state variables		
H_B	Height of the bed in the bioreactor	0.2 m
D	Diameter of the bioreactor	0.15 m
L	Thickness of the bioreactor wall	5 mm
B	Total mass of bioreactor wall	calculated as πDHL kg (L in m)
vvm	Volumes of air per bed volume per minute	1 vvm
a_{wg}^*	Outlet gas water humidity set point[b]	0.87
fold	Fold increase in the solids-to-gas heat and mass transfer coefficients	1
T_{sys}	Initial systems temperature[c]	35°C
T_{in}	Temperature of the inlet air	35°C
$T_{setpoint}$	Set point (cooling water temperature control)	30°C
J	Gain (cooling water temperature control)	0 (T_w constant at $T_{setpoint}$)
a_{wgo}	Initial water activity of the gas phase[d]	0.99
a_{wgin}	Water activity of the inlet air[e]	0.99
a_{wso}	Initial water activity of the solids phase[f]	0.99
Microbial parameters (can be varied in the input file for the model)		
b_o	Initial biomass content[g]	0.001 kg-biomass kg-IDS^{-1e}
b_m	Maximum possible biomass content[g]	0.25 kg- biomass kg-IDS^{-1e}
μ_{opt}	Specific growth rate constant at T = T_{opt}	0.236 h^{-1}
T_{opt}	Optimum temperature for growth	38°C
Y_{XS}	Yield of biomass from dry substrate	0.5 kg-biomass kg-dry-substrate^{-1}
Type	Type of relation of growth with a_{ws}	*Aspergillus*-type (see Fig. 22.3(b))
Other parameters		
ε	Effective bed porosity during mixing[h]	0.5 (m^3-voids m^{-3}-total)
ρ_S	Density of dry solid particles[h]	450 kg m^{-3}
h_{gb}	Gas/wall heat transfer coefficient	30 W m^{-2} °C^{-1}
h_{sb}	Solids/wall heat transfer coefficient	30 W m^{-2} °C^{-1}
h_{bw}	Wall/surroundings heat transfer coefficient	30 W m^{-2} °C^{-1}

[a] The program converts all input variables and parameters to a consistent set of units. Note that where "biomass" appears within the units, this represents dry biomass.
[b] The addition of water is triggered when the outlet gas relative humidity falls below this set point.
[c] Used as initial temperature of the solids (T_{so}), the gas phase (T_{go}), and the bioreactor wall (T_{bo}).
[d] Used, in conjunction with T_{go}, to calculate the initial gas phase air humidity, H (kg-H$_2$O kg-dry-air^{-1}).
[e] Used, in conjunction with T_{in}, to calculate the inlet air humidity, H_{in} (kg-H$_2$O kg-dry-air^{-1})
[f] Used, in conjunction with T_{so}, to calculate the zero time water content of the solid phase (W_o, kg-H$_2$O kg-dry-solids^{-1}).
[g] Used to calculate X_o and X_m (kg). Note that IDS = initial dry solids.
[h] The initial mass of dry substrate S_o is calculated as $(1-\varepsilon)\rho_S V_{bed}$ This value is used with X_o to determine the initial mass of dry solids in the solids phase (M_o).

Table 22.2. Values used for the base case simulation of those parameters and variables that cannot be changed in the model of a well-mixed forcefully-aerated bioreactor

Symbol	Significance	Base case value and units[a]
Microbial parameters		
Y_{QX}	Yield of metabolic heat from growth	8.366×10^6 J kg-biomass^{-1}
Y_{WX}	Yield of metabolic water from growth	0.3 kg-water kg-biomass^{-1}
Other parameters and constants		
C_{pg}	Heat capacity of dry air	1006 J kg^{-1} °C^{-1}
C_{pv}	Heat capacity of water vapor	1880 J kg^{-1} °C^{-1}
C_{pw}	Heat capacity of liquid water	4184 J kg^{-1} °C^{-1}
C_{pb}	Heat capacity of the bioreactor wall	420 J kg^{-1} °C^{-1}
C_{pm}	Heat capacity of the dry matter	1000 J kg^{-1} °C^{-1}
R	Universal gas constant	8.314 J mol^{-1} °C^{-1}
P	Overall pressure within the bioreactor	760 mm Hg[b]
λ	Enthalpy of vaporization of water	2.414×10^6 J kg-water^{-1}
ρ_a	Density of the air phase	1.14 kg-dry-air m^{-3} [c]
ρ_b	Density of the bioreactor wall	7820 kg m^{-3}

[a] The program converts all variables and parameters to a consistent set of units. Note that where "biomass" appears within the units, this represents dry biomass.
[b] Needed for the calculation of the air water activity.
[c] Used in the calculation of the headspace gas mass (G, kg) and to calculate the mass flow rate of air (F_{in}, kg-dry-air s^{-1}) from the value input as vvm.

As shown in Fig. 22.1(a), it is also assumed that the relative humidity of the off-gases is monitored. When the water activity (i.e., the relative humidity divided by 100) falls below a set point value ($a_{wg}*$), then sufficient water is added to the bed to bring the water activity of the solids back to the initial value. Note that it is assumed that this water is added at the temperature of the bed.

22.3 Insights the Model Gives into the Operation of Well-Mixed Bioreactors

22.3.1 Insights into Operation at Laboratory Scale

The base case simulation is for a small bioreactor of 3.5 L volume, in which the water in the water jacket is maintained constant at the optimum temperature for growth. This bioreactor is of a size comparable to the rocking drum bioreactor of Barstow et al. (1988) and Ryoo et al. (1991). It begins with approximately 800 g of dry substrate. Table 22.3 shows the values used for the key design and operating variables in this and other simulations done in the case study.

Growth is sub-optimal (Fig. 22.4(a)) since, even at this small scale, high solids temperatures will be reached if special efforts are not made to cool the bed (Fig. 22.4(b)). Note that, since the air fed to the bed is almost saturated, the solids do not dry out during the fermentation. The relative importance of temperature and water activity in controlling the growth rate are most easily seem by plotting the fractional specific growth rates, μ_{FT} and μ_{WT} (Fig. 22.4(c)). At the time of peak heat generation, the value of μ_{FT} falls to values around 0.5.

Growth is good for a bioreactor operated under the same conditions but with control of the temperature in the cooling jacket with a value of J of 2 (Fig. 22.4(d)) because the solids temperature is controlled within reasonable limits (Fig. 22.4(e)), that is, the solids temperature is maintained within conditions for which μ_{FT} is over 0.9 (Fig. 22.4(f)).

Figure 22.5 shows simulations for a bioreactor of 0.32 m bed height and 0.3 m diameter, which starts with 5 kg of dry solids, the same amount of substrate as Nagel et al. (2001a) used in their bioreactor. Since Nagel et al. (2001a) humidified their inlet air, but did not manage to saturate it, the inlet air water activity is set to 0.9. The simulation is not intended to describe the results of Nagel et al. (2001a) directly, but it is interesting to compare the results.

Growth is poor for a fermentation done without any control of the temperature of the cooling water (Fig. 22.5(a)) because temperatures as high as 47.5°C are reached in the bed (Fig. 22.5(b)). Note that the use of dry air causes the bed to dry out sufficiently to trigger the addition of water, which occurs at 36 h. This alleviates the water-limitation of growth, which becomes quite severe at this time (Fig. 22.5 (c)).

Table 22.3. Design and operating variables changed in the various explorations of performance of a well-mixed forcefully-aerated bioreactor[a]

Figure	Column in Figure	$H_B \times D$ (m × m)	M_o (kg)	a_{wgin}	J	vvm	Run
22.4	left	0.2 × 0.15	0.797	0.99	0	1	1
22.4	right	0.2 × 0.15	0.797	0.99	2	1	2
22.5	left	0.32 × 0.3	5.1	0.90	0	1	3
22.5	right	0.32 × 0.3	5.1	0.90	30	1	4
22.6	left	1.0 × 1.0	177.1	0.99	0	1	5
22.6	right	1.0 × 1.0	177.1	0.99	2	1	6
22.7	left	2.0 × 2.0	1416.5	0.50	0	1	7
22.7	right	2.0 × 2.0	1416.5	0.99	3	1	8
22.9	left	2.0 × 2.0	1416.5	0.50	-1	1	9
22.9	right	2.0 × 2.0	1416.5	0.50	-1	3	10

[a] Other conditions are as given in Tables 22.1 and 22.2.

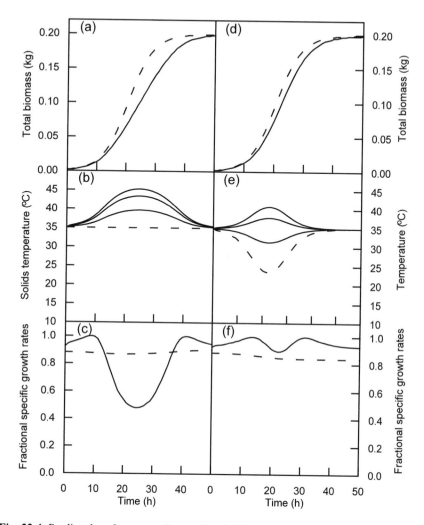

Fig. 22.4. Predicted performance of a small-scale bioreactor, 0.2 m high by 0.15 m diameter. In the left hand column the temperature of cooling water in the water jacket is maintained constant (Run 1) whereas in the right hand column it is controlled according to Eq. (22.6) with J=2 (Run 2). **(a)** and **(d)** Predicted growth (—) compared to that which would be achieved if optimal conditions were maintained throughout the fermentation (- - -); **(b)** and **(e)** Temperatures of the solids phase (top solid curve), the gas phase (middle solid curve), the bioreactor wall (bottom solid curve), and the cooling water (dashed curve); **(c)** and **(f)** Relative limitations of growth by temperature and water: (—) fractional specific growth rate based on temperature, μ_{FT}; (- - -) fractional specific growth rate based on water activity, μ_{WT}

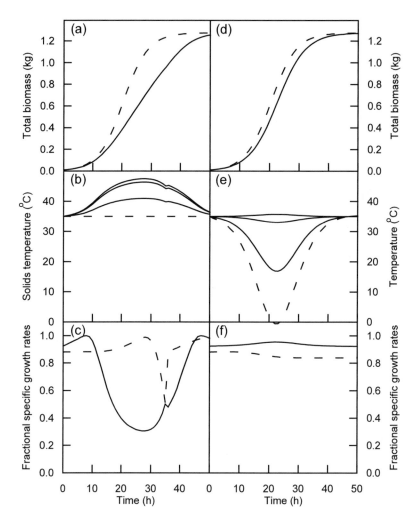

Fig. 22.5. Predicted performance of a bioreactor 0.32 m high by 0.3 m diameter. In the left hand column the temperature of cooling water in the water jacket is maintained constant (Run 3) whereas in the right hand column it is controlled according to Eq. (22.6) with J=30 (Run 4). **(a)** and **(d)** Predicted growth (—) compared to that which would be achieved if optimal conditions were maintained throughout the fermentation (- - -); **(b)** and **(e)** Temperatures of the solids phase (top solid curve), the gas phase (middle solid curve), the bioreactor wall (bottom solid curve), and the cooling water (dashed curve); **(c)** and **(f)** Relative limitations of growth by temperature and water: (—) fractional specific growth rate based on temperature, μ_{FT}; (- - -) fractional specific growth rate based on water activity, μ_{WT}

Since Nagel et al. (2001a) also investigated wall cooling, a simulation was done with a value of J of 30. Growth is much better (Fig. 22.5(a)) because this strategy manages to control the temperature of the solids, which does not exceed 35.7°C (Fig. 22.5(b)). This leads to water temperatures as low as –0.5°C (requiring the addition of antifreeze to the cooling water), and to wall temperatures as low as 16.9°C. Such low cooling-water temperatures are not impossible to obtain in the laboratory, but might be too expensive at large scale. Note that these predictions agree in general terms with those of Nagel et al. (2001a), who had to control the wall temperature at values as low as 18°C in order to keep the bed temperature around 35°C.

22.3.2 Insights into Operation at Large Scale

Simulations were done for larger scale bioreactors. For these simulations the coefficients for heat transfer from the gas and the solid phase to the bioreactor wall and from the bioreactor wall to the water in the water jacket are chosen as 200 W m^{-2} °C^{-1} in order to give an overall coefficient for heat transfer from the bed to the water in the water jacket (calculated on the basis of the law of resistances in series) of the order of magnitude of 100 W m^{-2} °C^{-1}, a value determined experimentally for a water-jacketed industrial solids mixer adapted as an SSF bioreactor (Nagel et al. 2001a) (see Sect. 19.5.2).

Figure 22.6 shows simulations for a bed of 1 m diameter and 1 m height, which contains an initial substrate loading of 177 kg. When the temperature in the water jacket is held constant (J=0) growth is poor (Fig 22.6(a)), despite the higher heat transfer coefficients at the wall, because undesirably high solids temperatures are still reached, peaking at 44.9°C (Fig 22.6(b)). On the other hand, with a value of J of 2, growth is good (Fig. 22.6(d)) because the solids temperature does not exceed 40.4°C (Fig 22.6(e)). In this case, the minimum temperature of the cooling water is 24.2°C, which is quite reasonable and may be possible to achieve without refrigeration.

Figure 22.7 shows simulations for a bed of 2 m diameter and 2 m height, which contains an initial substrate loading of 1417 kg. In Figs 22.7(a) to (c) the temperature in the water jacket is held constant (J=0), but in order to promote evaporation, the water activity of the inlet air is set at 0.5. However, undesirably high temperatures of 46.1°C are reached (Fig 22.7(b)). Note that water is added at 23 and 35 h.

In Figs 22.7(d) to (f) near-saturated air is used (a_{wgin}=0.99) along with a value of J of 3. The temperature does not exceed 42.0°C (Fig 22.7(e)). In this case the minimum temperature of the cooling water is 14.1°C. Reasonable temperature control might also be achieved by using higher aeration rates. In this case the best strategy will depend on the comparative operating costs of higher aeration rates versus refrigeration of a cooling jacket.

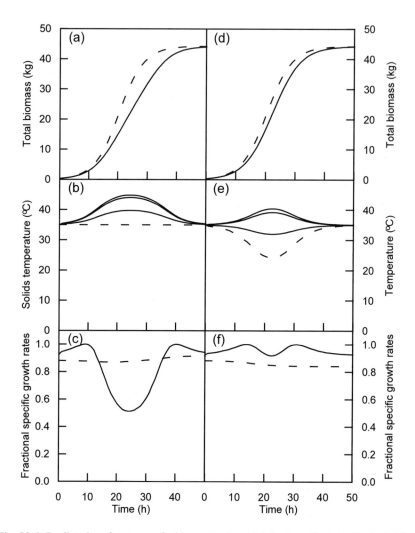

Fig. 22.6. Predicted performance of a bioreactor 1 m high by 1 m diameter. In the left hand column the temperature of cooling water in the water jacket is maintained constant (Run 5) whereas in the right hand column it is controlled according to Eq. (22.6) with J=2 (Run 6). **(a)** and **(d)** Predicted growth (—) compared to that which would be achieved if optimal conditions were maintained throughout the fermentation (- - -); **(b)** and **(e)** Temperatures of the solids phase (top solid curve), the gas phase (middle solid curve), the bioreactor wall (bottom solid curve), and the cooling water (dashed curve); **(c)** and **(f)** Relative limitations of growth by temperature and water: (—) fractional specific growth rate based on temperature, μ_{FT}; (- - -) fractional specific growth rate based on water activity, μ_{WT}

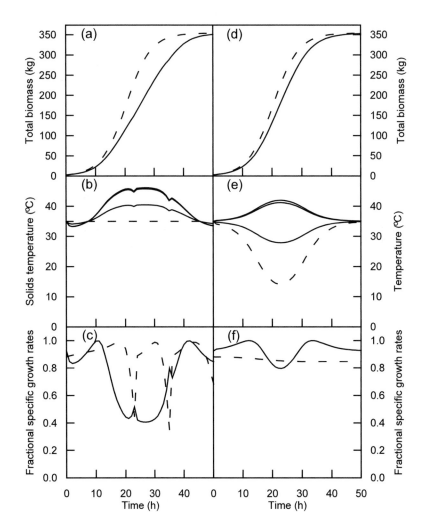

Fig. 22.7. Predicted performance of a bioreactor 2 m high by 2 m diameter. In the left hand column the temperature of cooling water in the water jacket is maintained constant (Run 7) whereas in the right hand column it is controlled according to Eq. (22.6) with $J=3$ (Run 8). **(a)** and **(d)** Predicted growth (—) compared to that which would be achieved if optimal conditions were maintained throughout the fermentation (- - -); **(b)** and **(e)** Temperatures of the solids phase (top solid curve), the gas phase (middle solid curve), the bioreactor wall (bottom solid curve), and the cooling water (dashed curve); **(c)** and **(f)** Relative limitations of growth by temperature and water: (—) fractional specific growth rate based on temperature, μ_{FT}; (- - -) fractional specific growth rate based on water activity, μ_{WT}

22.3.3 Effect of Scale and Operation on Contributions to Cooling of the Solids

It is interesting to explore the relative importance of the various mechanisms for removal of heat from the solids, and to explore how this relative importance varies under different operating conditions.

The operating conditions of runs 1, 3, 5, and 7 are such that wall cooling is not maximized, while the operating conditions of runs 2, 4, 6, and 8 maximize wall cooling and minimize evaporation. This is of course reflected in Fig. 22.8, which plots the contributions to cooling for these runs. Table 22.4 compares the contributions in percentage terms at the time of peak heat production.

An interesting question arises. If a water jacket is not practical on a large-scale bioreactor and heat removal through the bioreactor wall is negligible, how should the bioreactor be operated? This question becomes increasingly relevant as the scale of the bioreactor increases since it might be expected that it would become increasingly difficult to mix the solids. If good mixing is not assured, then a static layer of solids near the wall may be overly cooled by a water jacket and would provide an insulating layer preventing the cooling water from removing heat from a mixed region in the center.

Table 22.4. Comparison of the predicted contributions of the various mechanisms to the cooling of the solids at the time of peak heat production, for bioreactors at different scales and under different operating conditions[a]

Run	Time (h)[b]	Growth (W)[c]	% to wall[d]	% sensible to gas[d]	% evaporation[d]
1	24	17.61	45	34	21
2	22	23.748	56	39	5
3	26	95.83	31	32	37
4	22	156.8	54	46	0
5	24	4058.0	40	38	22
6	22	5354.3	49	44	7
7	25	29556	24	28	47
8	22	39974	44	44	12
9	31	20798	0	12	88
10	24	38232	0	11	87

[a] The key conditions that were changed between runs are given in Table 22.3. Other conditions are as given in Tables 22.1 and 22.2.
[b] Time of peak heat production.
[c] Rate of heat liberation by the growth reaction in Watts (the first term on the right hand side of the solid phase energy balance in Fig. 22.2).
[d] These values were calculated on the basis of the values of the remaining terms on the right hand side of the solid phase energy balance in Fig. 22.2, each of which has the units of Watts. The % of the heat transferred to the wall was calculated using the fourth term, the % of heat removed as sensible energy to the gas phase was calculated using the second term and the % of heat removed as the latent heat associated with evaporation of water from solids was calculate using the fourth term.

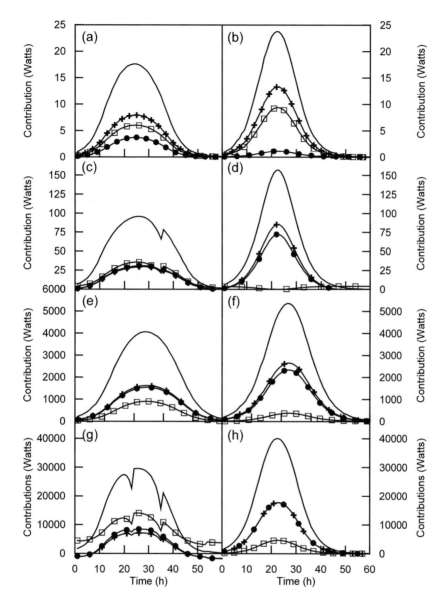

Fig. 22.8. Predicted rates of metabolic heat generation and heat removal by different mechanisms for the simulations shown in Figs. 22.4 to 22.7. **(a)** Run 1; **(b)** Run 2; **(c)** Run 3; **(d)** Run 4; **(e)** Run 5; **(f)** Run 6; **(g)** Run 7; **(h)** Run 8. Key: (——) Rate of metabolic heat production; (+) Rate of sensible heat removal from the solids to the bioreactor wall; (□) Rate of sensible heat removal from the solids to the gas phase; (●) Rate of evaporative heat removal

In fact, negligible heat removal through the bioreactor wall is easily simulated by putting the proportional gain, J, equal to -1. This makes the cooling water temperature equal to the bed temperature. A simulation was done for the 2 m height by 2 m diameter bioreactor, using the conditions listed for run 9 in Table 22.3. Under these conditions growth is very poor (Fig. 22.9(a)) due to very high solids temperatures (Fig. 22.9(b)). Both water and temperature restrictions are significant (Fig. 22.9(c)), although the temperature restrictions have the greater effect. Evaporation, promoted by the use of 50% relative humidity air at the air inlet, removes almost 90% of the metabolic heat liberated by the microorganism (Fig. 22.9(d) and run 9 in Table 22.4). Quite reasonable performance can be obtained by increasing the air flow to the bioreactor (from 1 to 3 vvm) (see run 10 in Table 22.3) as shown by Figs 22.9(e), (f), (g), and (h). As with the previous simulation, evaporation is responsible for almost 90% of the overall heat removal, but note that the overall metabolic heat generation rate is almost double due to the better growth (Table 22.4).

22.4 Conclusions on the Operation of Well-Mixed Bioreactors

Based on the simulations presented in this chapter, it appears that, if we can in fact keep the bed well mixed, then continuously-mixed, forcefully-aerated bioreactors can be operated well at scales containing over a ton of dry substrate. In fact, if we increased the porosity and the solids-to-gas heat transfer coefficient in order to simulate an air-solid fluidized bed, then we would predict reasonable operation at even much larger scales. Of course, the model does not take into account any practical difficulties that there might be in operating air-solid fluidized beds at large scale.

The ability to provide a well mixed bed at large scale will depend on the effectiveness of the mixing method. Poor mixing will lead to poorer bioreactor performance than that predicted by the simulations done in this chapter. Further, the practicality of continuous mixing depends strongly on the susceptibility of the process organism to shear damage.

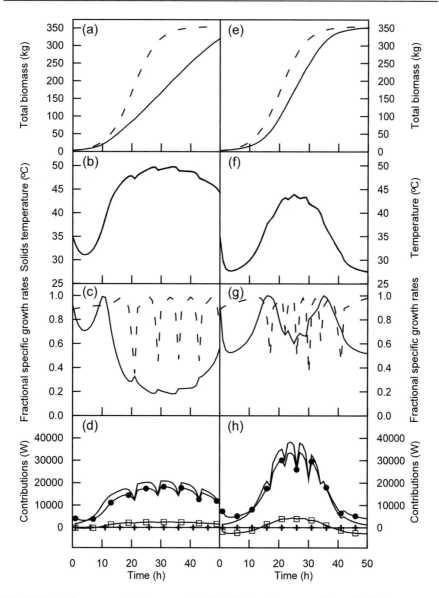

Fig. 22.9. Performance of a bioreactor 2 m high by 2 m diameter with negligible heat transfer through the wall. The aeration rate is 1 vvm for the left hand column (Run 9) and 3 vvm for the right hand column (Run 10). **(a)** and **(e)** Growth (—), compared to that which would be achieved with $\mu = 0.236$ h^{-1} throughout the fermentation (- - -); **(b)** and **(f)** Bed temperature. **(c)** and **(g)** Relative limitations of growth: (—) μ_{FT}; (- - -) μ_{WT}; **(d)** and **(h)** Rates of heat production and removal: (—) Metabolic heat production; (+) Sensible heat removal from the solids to the bioreactor wall; (□) Sensible heat removal from the solids to the gas phase; (●) Evaporative heat removal

Further Reading

Models of well-mixed bioreactors presented in the literature

dos Santos MM, da Rosa AS, Dal'Boit S, Mitchell DA, Krieger N (2004) Thermal denaturation: Is solid-state fermentation really a good technology for the production of enzymes? Bioresource Technology 93:261–268

Nagel FJI, Tramper J, Bakker MSN, Rinzema A (2001b) Model for on-line moisture-
content control during solid-state fermentation. Biotechnol Bioeng 72:231–243

23 A Model of a Rotating-Drum Bioreactor

David A. Mitchell, Deidre M. Stuart, and Nadia Krieger

23.1 Introduction

This chapter presents a case study to show how modeling work can provide insights into how to best design and operate well-mixed rotating and stirred drum bioreactors. Recently, Schutyser et al. (2001, 2002, 2003c) have developed more sophisticated models for rotating-drum bioreactors. These models describe the movement of individual particles during drum operation. Although they are potentially quite powerful tools for exploring bioreactor behavior, they are much more complex to set up, and solution times can be as long as several days.

23.2 A Model of a Well-Mixed Rotating-Drum Bioreactor

The system modeled is a rotating-drum bioreactor, as shown in Fig. 23.1. It is divided into three subsystems, the substrate bed, the headspace, and the bioreactor wall. Each is assumed to be well mixed. In other words, each system is represented by a single value for each state variable. Note that this means that the drum wall has a single temperature, uniform across the whole drum, from inside to outside and from the section of the wall in contact with the substrate bed to the section of the wall in contact with the headspace.

23.2.1 Synopsis of the Mathematical Model and its Solution

Figure 23.2 summarizes the model, which is a modified version of that of Stuart and Mitchell (2003). Water and energy balances are done over the substrate bed and headspace gases and an energy balance is written over the bioreactor wall.

Many assumptions are made about the system, the most important ones being:

- Although the solid is consumed during the process, this affects only the density of the bed and not the volume that it occupies, which remains constant at the original value;

- The total amount of dry gas in the headspace is constant;
- The gas flow rates at the air inlet and air outlet are the same (that is, the exchange of O_2 for CO_2 does not cause any difference between the mass of dry gas entering and leaving).

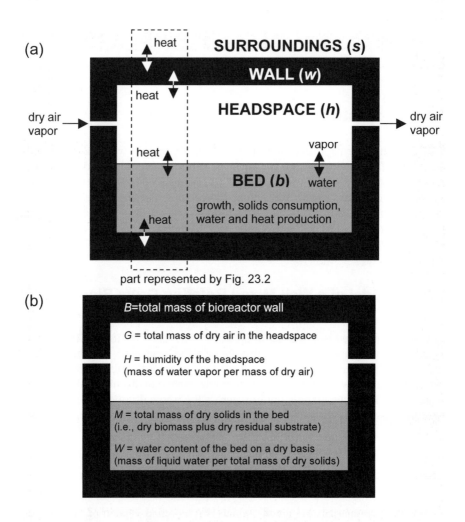

Fig. 23.1. (a) The rotating-drum system as modeled in the case study. The letters in parentheses represent the subscripts that are used in the model equations to denote the various subsystems within the model equations. **(b)** The meaning of the symbols related to the masses and water contents of the various subsystems

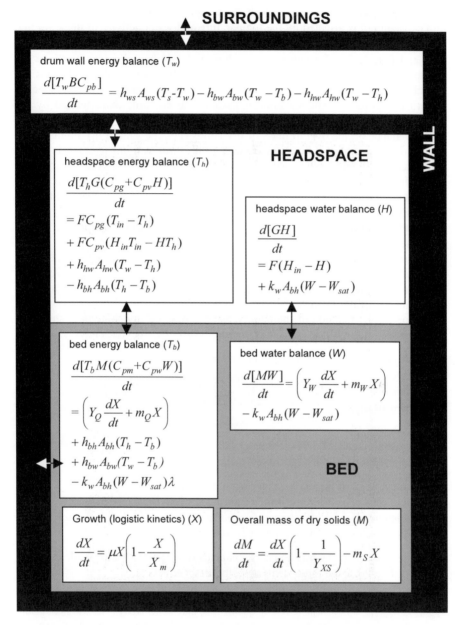

Fig. 23.2. Summary of the model of a well-mixed rotating drum, which is a slightly simplified version of the model of Stuart and Mitchell (2003). In each text box, the variable shown in parentheses after the heading is the variable that is isolated in the differential term on the left hand side of the differential equation before the equation set is solved. Subscripts: h = headspace, b = bed, w = wall of bioreactor, s = surroundings, in = inlet, sat = saturation

Growth occurs according to the logistic equation, where the specific growth rate constant (μ) is affected by the temperature and water activity of the solid in a manner identical to that described by Eqs. (22.1), (22.2), and (22.3) (see Chap. 22.2 and Fig. 22.3). The value of μ_{opt} chosen should be valid for growth of the process organism in a continuously-mixed system since the model itself does not include any equations to describe the effect of the rotation rate on the specific growth rate constant. The substrate is assumed to be corn, characterized by the isotherm given by Eqs. (19.8) and (19.9).

The balance on the overall mass of dry solids (i.e., the sum of dry biomass and dry residual substrate), shown in the lower right of the bed in Fig. 23.2, is necessary since not all the consumed substrate is converted into biomass; a proportion is lost in the form of CO_2. This is Eq. (16.11), which was deduced in Sect. 16.2.

The mass balance on water in the substrate bed, shown in the upper right of the bed in Fig. 23.2, has terms on the right hand side to describe, respectively, metabolic water production due to growth and maintenance and the evaporation of water to the headspace.

The mass balance on water in the headspace, shown on the right of the headspace in Fig. 23.2, has terms on the right hand side to describe, respectively, the entry and leaving of water with the gas flow through the headspace and the evaporation of water from the bed.

On the right hand side of the energy balance over the substrate bed, shown in the upper left of the bed in Fig. 23.2, the four terms describe, respectively:

- metabolic heat production due to growth and maintenance;
- sensible heat transfer between the bed and the headspace gases;
- sensible energy transfer between the bed and the bioreactor wall;
- removal of energy from the bed by evaporation of water into the headspace gases.

On the right hand side of the energy balance over the headspace gases, shown on the left of the headspace in Fig. 23.2, the four terms describe, respectively:

- sensible energy of the dry air entering and leaving the headspace;
- sensible energy of the water vapor entering and leaving the headspace;
- sensible energy transfer between the headspace gases and the bioreactor wall;
- sensible heat transfer between the bed and the headspace gases.

On the right hand side of the energy balance over the bioreactor wall, shown at the top of Fig. 23.2, the three terms describe, respectively:

- sensible energy transfer between the wall and the surroundings of the bioreactor;
- sensible energy transfer between the bed and the bioreactor wall;
- sensible energy transfer between the headspace gases and the bioreactor wall.

Within these energy balance equations, there are several heat and mass transfer coefficients. The bed-to-wall heat transfer coefficient (h_{bw}) is calculated using Eq. (20.3). Before the other heat transfer coefficients are calculated, the value of the air flow rate in vvm (volumes of air per total bioreactor volume per minute) is used to calculate the air flow rate, F (kg-dry-air s^{-1}). This and the cross sectional area of the headspace normal to the gas flow (A_g, m^2) are used to calculate the two heat transfer coefficients involving the headspace gases (W m^{-2} °C^{-1}), namely the bed-to-headspace coefficient (h_{bh}) and the headspace-to-wall coefficient (h_{hw}), according to the relationship given by Geankoplis (1993) (see Eq. (20.6)). The psychrometric ratio is then used to calculate the bed-to-headspace mass transfer coefficient:

$$k_w = \frac{h_{bh}}{6342}.$$

(23.1)

Note that the denominator also contains a conversion factor due to the units used for the driving force in the term that describes evaporation. This equation uses h_{bh} in W m^{-2} °C^{-1} and gives k_w in kg-H$_2$O s^{-1} m^{-2} (kg-H$_2$O kg-dry-solids^{-1})$^{-1}$. Note that the units within the parentheses are the units of the driving force. After simplification, the units of k_w are kg-dry-solids s^{-1} m^{-2}. Note that the driving force for evaporation is the difference between the water content of the solids (W, kg-H$_2$O kg-dry-solids^{-1}) and the water content that the solids would have if they were in equilibrium with the gas in the headspace phase (W_{sat}). W_{sat} is calculated using Eq. (22.5). In this case, the equation uses the headspace temperature and water activity, the latter calculated as explained in Sect. 19.4.1 (see Eq. (19.16)).

The values of the transfer coefficients h_{bh} and k_w calculated as described in the previous paragraph are for a drum that is not mixed. Mixing should increase bed-to-headspace heat and mass transfer. However, there is not sufficient information available to incorporate this mechanistically within the equation. Therefore, in this model, h_{bh} and k_w are simply multiplied by an empirical factor "n", which represents the fold-increase in transfer rates due to mixing.

The model incorporates two simple control schemes, a control of the inlet air humidity and a control of the bed water activity (Fig. 23.3). Control of the inlet air humidity is desirable since, although the use of dry air to promote evaporation is an effective cooling strategy, if dry air is used at the beginning of the fermentation when the rate of metabolic heat production is low, then the bed temperature can fall to values low enough to retard early growth. Therefore the temperature of the bed is monitored hourly. If it is less than the optimum temperature for growth (38°C) then high-humidity air is fed to the bioreactor, while if it exceeds this temperature then low-humidity air is supplied.

The promotion of evaporation by supplying low-humidity air could potentially dry the bed to water activities low enough to restrict growth. Therefore it is assumed that samples are removed from the bed every hour and their water activity rapidly determined. If the water activity falls below a set point, then sufficient water is added to the bed to bring the water activity back to the initial value. Note that it is assumed that the added water is at the temperature of the solids and therefore does not affect the bed temperature.

The model equations are ordinary differential equations, since time is the only independent variable. They are solved using the Runge-Kutta numerical integration algorithm.

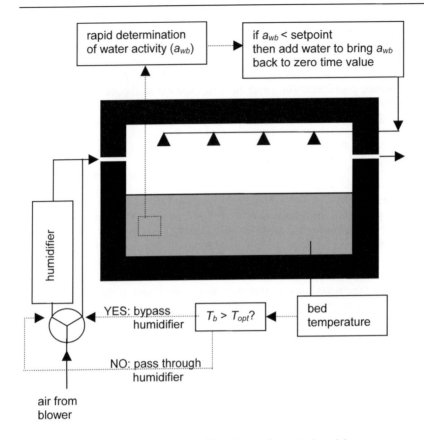

Fig. 23.3. Control schemes incorporated into the mathematical model

23.2.2 Predictions about Operation at Laboratory Scale

Figure 23.4 shows the kind of information that can be obtained from the model. In this case the data from Table 23.1 and 23.2 were used, with $n = 10$. The output can be plotted so as to see temporal variations in:

- The growth of biomass and decrease in the overall mass of dry solids (Fig. 23.4(a)).
- The value of the specific growth rate parameter (μ) in the logistic equation and the relative effects of temperature and water activity on the value of this parameter (Fig. 23.4(b)). In this case the temperature has the greater influence on the value of this parameter.
- The driving force for evaporation (Fig. 23.4(c)).
- The degree of saturation of the headspace (Fig. 23.4(d)). In this case the headspace remains saturated throughout the fermentation.

Table 23.1. Values used for the base case simulation for those parameters and variables that <u>can</u> be changed in the accompanying model of a well-mixed rotating-drum bioreactor

Symbol	Significance	Base case value and units[a]
Design and operating variables (can be varied in the input file for the model)		
D	Diameter of the bioreactor	0.8 m
L	Length of the bioreactor	0.2 m
%fill	Percentage of the drum volume occupied by the solid bed	0.3 (used to calculate the initial dry mass of solids in the bed M_o)
T_s	Temperature of the surroundings	30°C
vvm	Rate of air flow (dry basis)	0.01 m^3-air (m^3-bioreactor)$^{-1}$ min^{-1}[b]
T_{in}	Temperature of the inlet air	30°C
a_{win1}	Water activity of the inlet air when $T \le T_{opt}$	0.15[c]
a_{win2}	Water activity of the inlet air when $T > T_{opt}$	0.15
a_{wSP}	Set point bed water activity (below which the addition of water to the bed is triggered)	0.9
Initial values (can be varied in the input file for the model)		
T_{bo}	Initial bed temperature	30°C
T_{wo}	Initial temperature of the bioreactor body	30°C
T_{ho}	Initial temperature of the headspace gases	30°C
a_{wbo}	Initial water activity of the solids	0.99[d]
Microbial parameters (can be varied in the input file for the model)		
b_o	Initial biomass concentration	0.001 kg-biomass kg-IDS^{-1}[e]
b_m	Maximum possible biomass concentration	0.23 kg-biomass kg-IDS^{-1}[e]
μ_{opt}	Specific growth rate constant at $T = T_{opt}$	0.236 h^{-1}
T_{opt}	Optimum temperature for growth	38°C
Type	Type of relation of growth with a_w	*Aspergillus*-type (see Fig. 22.3(b))
Y_Q	Yield of metabolic heat from growth	8.366 x 10^6 J kg-biomass^{-1}
m_Q	Maintenance coefficient for metabolic heat	0 J s^{-1} kg-biomass^{-1}
Y_{XS}	Yield of biomass from dry substrate	0.5 kg-biomass kg-dry-substrate^{-1}
m_S	Maintenance coefficient for substrate	0 kg-dry-substrate s^{-1} kg-biomass^{-1}
Y_W	Yield of metabolic water from growth	0.3 kg-water kg-biomass^{-1}
m_W	Maintenance coefficient for water	0 kg-water s^{-1} kg-biomass^{-1}
n	Fold-increase in transfer coefficients	10 or 1

[a] The program converts all input variables and parameters to a consistent set of units. Note that where "biomass" is mentioned within the units, this represents dry biomass.
[b] Used to calculate the inlet air flow rate (F, kg-dry-air min^{-1}).
[c] Used, in conjunction with T_{in}, to calculate the inlet air humidity, H_{in} (kg-water kg-dry-air^{-1}). This calculated value of H_{in} is used as the zero time value, H_{ino}.
[d] Used in Eq. (19.9) to calculate the initial water content of the bed (W_o, kg-H$_2$O kg-dry-solids^{-1}).
[e] Used to calculate X_o and X_m (kg). IDS = initial dry solids.

Table 23.2. Values used for the base case simulation of those parameters and variables that cannot be changed in the accompanying model of a well-mixed rotating-drum bioreactor

Symbol	Significance	Base case value and units[a]
Parameters related to the effect of temperature on growth		
A_1	Constant in the equation describing $\mu=f(T)$	8.31×10^{11}
A_2	Constant in the equation describing $\mu=f(T)$	70225 J mol^{-1}
A_3	Constant in the equation describing $\mu=f(T)$	1.3×10^{47}
A_4	Constant in the equation describing $\mu=f(T)$	283356 J mol^{-1}
Other parameters and constants		
C_{pg}	Heat capacity of dry air	1006 J kg^{-1} °C^{-1}
C_{pv}	Heat capacity of water vapor	1880 J kg^{-1} °C^{-1}
C_{pw}	Heat capacity of liquid water	4184 J kg^{-1} °C^{-1}
C_{pb}	Heat capacity of the bioreactor wall	420 J kg^{-1} °C^{-1}
C_{pm}	Heat capacity of the dry matter	1000 J kg^{-1} °C^{-1}
R	Universal gas constant	8.314 J mol^{-1} °C^{-1}
h_{bw}	Bed/wall heat transfer coefficient	W m^{-2} °C^{-1} (Eq. (20.3))
h_{ws}	Wall/surroundings heat transfer coefficient	5 W m^{-2} °C^{-1}
A_{bw}	Bed/wall contact area	m^2 (geometric principles)
A_{bh}	Bed/headspace contact area	m^2 (geometric principles)
A_{hw}	Headspace/wall contact area	m^2 (geometric principles)
A_{ws}	Wall/surroundings contact area	m^2 (geometric principles)
k_w	Mass transfer coefficient for evaporation	kg-dry-solids s^{-1} m^{-2} (Eq. (23.1))
W_{sat}	Value of W the substrate bed would have if it were in equilibrium with the headspace	kg-water kg-dry-solids^{-1} (Eq. (19.9))
ρ_b	Overall density of the solid bed (wet basis)	387 kg-wet-solids m^{-3}
P	Overall pressure within the bioreactor	760 mm Hg^b
λ	Enthalpy of vaporization of water	2.414 x 10^6 J kg-water^{-1}
ρ_a	Density of the air phase	1.14 kg-dry-air m^{-3} ^{c}

[a] The program converts all variables and parameters to a consistent set of units. Note that where "biomass" is mentioned within the units, this represents dry biomass.
[b] Needed for the calculation of the air water activity.
[c] Used in the calculation of the headspace gas mass (G, kg) and to calculate the mass flow rate of air (F, kg-dry-air s^{-1}) from the value input as vvm.

The base case simulation conditions are similar to those used by Stuart et al. (1999). A simulation done with $n = 1$ is similar to their static fermentation while a simulation done with $n = 10$ is similar to their rolled fermentations. The predictions of the model about the magnitude and timing of the peak bed temperatures and about headspace gas temperatures at the time of the peak bed temperature are similar to experimental results obtained by Stuart et al. (1999) (Table 23.3).

The model predicts that, during static operation (i.e., with $n = 1$), the wall temperature will be several degrees higher than the headspace temperature, such that the headspace air receives energy not only by direct transfer of heat from the bed, but also by the more indirect route of bed to wall to headspace (Fig. 23.5(a)). For agitated operation (i.e., with $n = 10$), the model predicts that the headspace temperature will be slightly higher than the drum wall temperature (Fig. 23.5(c)).

Table 23.3. Comparison of experimental and predicted peak bed temperatures and of the experimental and predicted headspace temperatures at the time of the peak bed temperature

	Experimental (time reached) Stuart et al. (1999)	Model prediction (time reached)
static operation ($n = 1$ in model)		
peak bed temperature	47.2°C (28 h)	47.5°C (32 h)
headspace temperature	38.6°C (28 h)	37.8°C (32 h)
rotating at 5 rpm ($n = 10$ in model)		
peak bed temperature	48.8°C (36 h)	47.4°C (32 h)
headspace temperature	44.0°C (36 h)	43.8°C (32 h)

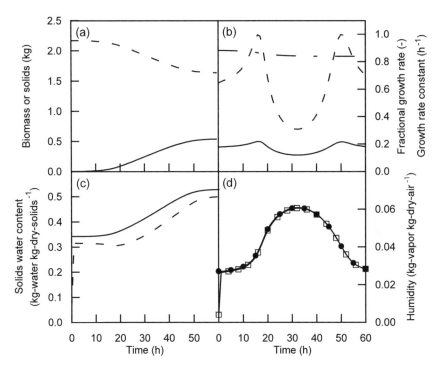

Fig. 23.4. Some of the outputs of the model (other outputs are shown in Fig. 23.5). A simulation was done with the case study model using the data in Table 23.1, with $n = 10$. **(a)** Plots of (- - -) total solids and (—) total biomass; **(b)** Fractional specific growth rates on the basis of (- - -) temperature and (— - —) water activity and (—) the resulting value of the specific growth rate constant, μ; **(c)** Water content of the solids: (—) actual water content of the solids, (- - -) water content that the solids would have to have to be in equilibrium with the gas phase; **(d)** Humidity of the gas phase: (—□—) actual humidity of the gas phase, (—●—) saturation humidity of the gas phase

The model can be used to explore the relative contributions of the various heat removal mechanisms (Figs. 23.5(b) and 23.5(d)). One of the output files gives the values of the four terms on the right hand side of the energy balance equation for the substrate bed (see the equation on the upper left of the bed within Fig. 23.2). Heat removal to the bioreactor wall is the major contributor under the conditions of the base case simulation, this holding for both $n = 1$ and $n = 10$. For $n = 10$, at the time of peak heat production (32 h) there is a metabolic heat generation rate of 37.9 W within the bed. Of this, 0.1 W is being removed by convection to the head-space, 0.7 W by evaporation, and 37.1 W by conduction to the bioreactor wall. The contribution of evaporation is low because of the low air flow rate. In fact, evaporation is so slow that the substrate does not lose enough water to make the addition of further water necessary, even during 100 h of operation.

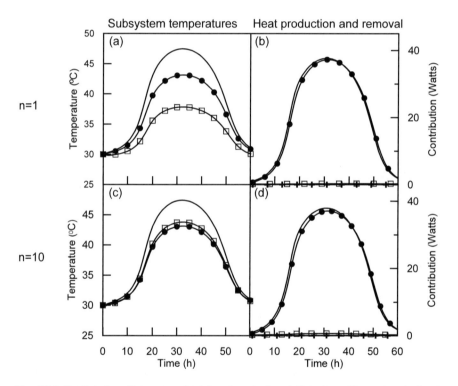

Fig. 23.5. Predicted performance of a laboratory-scale rotating-drum bioreactor similar to that of Stuart et al. (1999). The parameters are as shown in Table 23.1 except that two different values of n were used, in order to simulate static ($n = 1$) and rotated fermentations ($n = 10$). In the left hand column are the predicted temperatures of the (———) bed, (—□—) headspace, and (—●—) bioreactor wall. In the right hand column are the various contributions to heat removal. Key: (———) rate of production of waste metabolic heat; (—●—) rate of heat removal to the bioreactor wall; (—□—) rate of heat removal by evaporation; (—+—) rate of heat removal by sensible energy transfer to the headspace gases

23.2.3 Scale-up of Well-Mixed Rotating-Drum Bioreactors

Explorations with the model give insights into how to operate well-mixed rotating-drum bioreactors, both at laboratory scale and at large scale.

The model can be used to show that the air flow rates of around 0.01 vvm that were used by Stuart et al. (1999) and that were used in the base case simulation in the previous section, were simply insufficient. For good performance, they should have used flow rates of around 1 vvm. Figure 23.6(a)) shows a simulation done using the base case values (with $n = 10$) but with an aeration rate of 1 vvm. Note that at this higher flow rate, the substrate dries out during the fermentation. Therefore the fermentation is started with a high inlet air water activity of 0.99, and the inlet air water activity is only decreased to 0.15 when the bed temperature exceeds 38°C. Note also that the addition of water is triggered due to the drying out of the substrate bed, with water being added at 29 h and at 35 h.

For this simulation, the predicted peak bed temperature is 42.4°C at 27 h (Fig. 23.6(a)). At this time the overall metabolic heat generation rate is 62.8 W. The heat removal rates are 4.5 W by convection to the headspace, 27.0 W by conduction to the bioreactor wall and 31.1 W by evaporation (Fig. 23.7(a)). Therefore the increase in the air flow rate to 1 vvm allows evaporation to contribute around 50% of the heat removal, whereas at 0.01 vvm it was contributing only 2%.

In order for the bed temperature not to exceed 40°C, aeration rates higher than 1 vvm are needed. Figures 23.6(b) and 23.7(b) show the predicted results for a fermentation undertaken under the same conditions but with an aeration rate of 10 vvm. Due to the effectiveness of the cooling, the inlet air temperature is set at 35°C. In this case the predicted maximum temperature is 39.0°C at 29 h.

The model can also be used to explore scale-up of rotating-drum bioreactors. The simulations are done with $n = 10$ in order to simulate mixed operation. The inlet air temperature is maintained at 30°C. In the simulations described here, the aim is to prevent the bed temperature from exceeding 40°C. Readers can use the accompanying program to explore the predictions in greater depth.

Several bioreactors of ever-increasing size are compared. For each increase in scale the diameter and length are both doubled. The volumes and initial bed weights of these bioreactors are shown in Table 23.4. Note that the "large-scale" rotating drum has a bed capacity equal to the large-scale rotating drum used in *koji* production that was mentioned in Chap. 8, which has a capacity for 1500 kg of cooked substrate (Sato and Sudo 1999).

Table 23.4. Comparison of drum sizes and initial bed weights in the simulations done to investigate the effect of increase in scale on the performance of a rotating-drum bioreactor

Scale of bioreactor	Diameter (m)	Length (m)	Volume (L)	Initial wet bed weight (kg)
Laboratory (Sect. 23.2.2)	0.2	0.8	25	2.9
Small-pilot	0.4	1.6	200	23.3
Large-pilot	0.8	3.2	1600	187
Large-scale	1.6	6.4	12800	1494

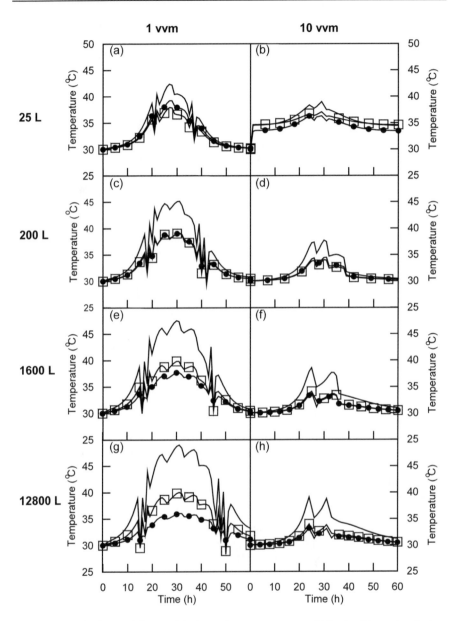

Fig. 23.6. Predictions of the model about the temperatures within the bioreactor during scale-up at a constant aeration rate in terms of vvm. The graphs on the left are for increasingly larger scales, each aerated at 1 vvm. The graphs on the right are for increasingly larger scales, each aerated at 10 vvm. In all cases $n = 10$. Key: (———) bed temperature; (—□—) headspace temperature; (—●—) bioreactor wall temperature

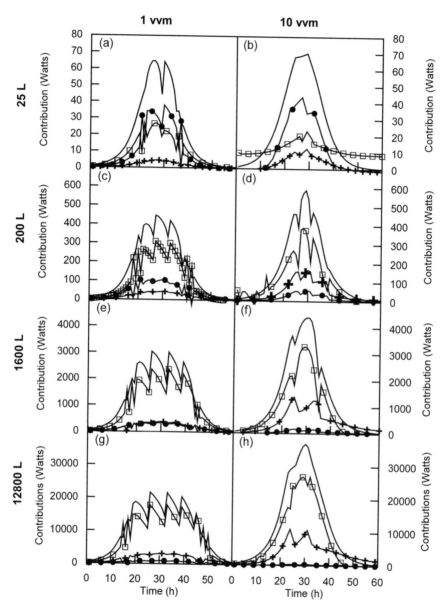

Fig. 23.7. The importance of evaporative cooling in large-scale rotating-drum bioreactors. The predictions about the contributions to heat removal are made for scale up at a constant aeration rate in terms of vvm, as explained in Fig. 23.6. Key: (——) rate of production of waste metabolic heat; (—●—) rate of heat removal to the bioreactor wall (—□—) rate of heat removal by evaporation; (—+—) rate of heat removal by sensible energy transfer to the headspace gases

The temperature profiles plotted in Fig. 23.6 show that as scale increases at a constant aeration rate of 1 vvm, the peak temperature reached in the bed increases, peaking near 50°C in the largest bioreactor. On the other hand, as scale increases with a constant aeration rate of 10 vvm, the maximum bed temperature stays below 40°C. It is clear that high aeration rates, of the order of magnitude of 10 vvm, will be required for adequate temperature control in large-scale rotating-drum bioreactors. An analysis of the contributions to cooling (Fig. 23.7) shows that heat removal to the drum wall removes a significant proportion of the metabolic heat in the 25 L bioreactor but becomes less and less significant as scale increases. At large scale the majority of the metabolic heat is removed by evaporation.

23.3 What Modeling Work Says about Rotating-Drum Bioreactors Without Axial Mixing

Mitchell et al. (2002a) developed a model to describe the operation of rotating-drum bioreactors that have large length to diameter ratios, for example, similar to the 11 m long by 1 m diameter rotating-drum bioreactors reported by Ziffer (1998), and that have end-to-end aeration. In such bioreactors, axial mixing may make a relatively small contribution, and in the model of Mitchell et al. (2002a) there was no axial mixing in either the substrate bed or the air phase.

In this model the bioreactor wall is not recognized as a separate subsystem (Fig. 23.8). A water balance is not done because it is assumed that water is periodically added to the bed in order to maintain a sufficiently high water activity so as not to limit growth. The model equations are partial differential equations, since both time and axial distance are independent variables. The partial differential equations are converted into ordinary differential equations by orthogonal collocation.

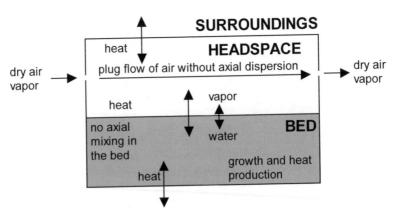

Fig. 23.8. The rotating drum without axial mixing, as modeled by Mitchell et al. (2002a)

Significant axial temperature profiles can be expected, both in the substrate bed and in the headspace, even at small scale (Fig. 23.9). Although detailed studies have not been undertaken, air temperature differences between the air inlet and air outlet can be as high as 4°C, over a bioreactor length of 0.85 m (Mitchell et al. 2002a). The model predicts axial temperature gradients of this magnitude.

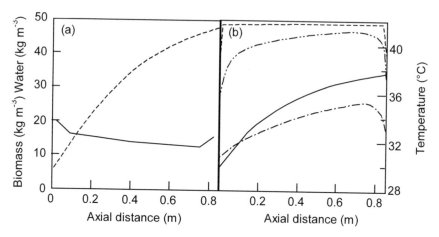

Fig. 23.9. Predictions made by the model of a rotating drum without axial mixing, for the growth of *Aspergillus oryzae* on wheat bran in a 0.85 m long drum (Mitchell et al. 2002a). **(a)** Axial profiles for (- - -) concentration of water vapor in the headspace; (——) biomass within the substrate bed; **(b)** Axial substrate temperature profiles at (—·—) 5 h, (—··—) 10 h, and (– – –) 20 h and (——) headspace gas temperature profile at 20 h. Reproduced from Mitchell et al. (2002a) with kind permission from John Wiley & Sons, Inc.

23.4 Conclusions on the Design and Operation of Rotating-Drum Bioreactors

When operated in batch mode, rotating- and stirred-drum bioreactors should be designed to promote axial homogeneity. Axial mixing can be promoted by having an inclined axis and angled lifters that push the substrate along the drum (Fig. 8.11). However, it can also be achieved to some degree by ensuring that the headspace gases are well mixed, since convection and evaporation to the headspace gases will be major pathways for heat removal at large scale, and a uniform temperature within the headspace will tend to promote uniform rates of heat removal and therefore a uniform temperature along the substrate bed. However, it is not practical to insert a fan within the headspace. The other option is to introduce and remove air along the whole axial length (Fig. 22.10).

Due to the importance of evaporative cooling at large scale, it will be necessary to add water to the substrate during the fermentation. This will require internal piping with spray nozzles (Fig. 22.10).

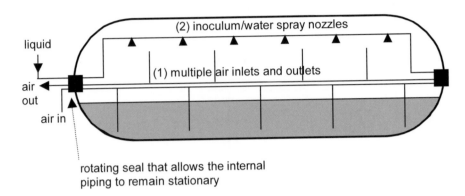

Fig. 23.10. Design features for large-scale rotating-drum bioreactors, including (1) multiple air inlets and outlets for promoting homogeneity within the headspace and (2) nozzles to al-low the addition of water, which will be necessary at large scale due to the dependence on evaporative cooling. If the bioreactor is baffled, then it will be necessary to leave an appro-priate clearance between the baffles and the air and water lines

The model predictions suggest that the substrate-to-headspace heat and mass transfer coefficients are important. This is an area that has received only relatively little attention. Based on the predictions of the model, the "major contributor to heat removal" is likely to change quite significantly as scale is increased. Conduc-tion through the bioreactor wall, which makes a large contribution to heat removal at small scales, is insufficient to remove the heat at large scales. This of course is due to the fact that the ratio of the surface area of the drum to the substrate bed volume decreases with scale (if geometric similarity is maintained). To try to maximize the heat transfer through the drum wall at large scales, you might con-sider either (1) including a water jacket, although this greatly complicates the de-sign and increases power requirements for rotation; or (2) increasing the L to D ra-tio in order to minimize the reduction of the ratio of the bed-wall contact area to bed volume that occurs with increase in scale when geometric similarity is main-tained. Of course there will be practical limits as to how long and thin the reactor can be.

Further reading

Fast-solving models of rotating-drum bioreactors
Stuart DM, Mitchell DA (2003) Mathematical model of heat transfer during solid-state fermentation in well-mixed rotating drum bioreactors. J Chem Technol Biotechnol 78:1180–1192
Mitchell DA, Tongta A, Stuart DM, Krieger N (2002a) The potential for establishment of axial temperature profiles during solid-state fermentation in rotating drum bioreactors. Biotechnol Bioeng 80:114–122

24 Models of Packed-Bed Bioreactors

David A. Mitchell, Penjit Srinophakun, Oscar F. von Meien, Luiz F.L. Luz Jr, and Nadia Krieger

24.1 Introduction

This chapter provides two case studies to show how modeling work can provide insights into how to design and operate traditional packed-beds and Zymotis-type packed-beds. Various other mathematical modeling case studies have been undertaken with packed-bed bioreactors:

- Saucedo-Castaneda et al. (1990). A model of a thin packed-bed in which axial convection is not taken into account, rather only horizontal conduction through the walls.
- Gutierrez-Rojas et al. (1995). A mathematical model for solid-state fermentation of mycelial fungi on inert support.
- Hasan et al. (1998). Heat transfer simulation of solid-state fermentation in a packed-bed bioreactor.
- Oostra et al. (2000). A model of a packed-bed bioreactor was used as part of the process of bioreactor selection.
- Weber et al. (1999). A simplified material and energy balance approach for process development and scale-up of *Coniothyrium minitans* conidia production by solid-state cultivation in a packed-bed reactor.
- Weber et al. (2002). Validation of a model for process development and scale-up of packed-bed solid-state bioreactor.

The aim of this chapter is not to review these models. Readers with a deeper interest in modeling of packed-beds are recommended to find the original papers.

24.2 A Model of a Traditional Packed-Bed Bioreactor

The system modeled is a traditional packed-bed bioreactor, as shown in Fig. 24.1. It is assumed that the bioreactor is sufficiently wide such that heat transfer to the side walls is negligible, and therefore only heat transfer in the axial direction is included in the equations.

Note that this model is relatively limited. It is aimed only at predicting temperatures within the bed. It does not aim to describe what happens with water in the bed. As such, it is only useful for processes in which the substrate can undergo large decreases in water content with only minor changes in water activity, such that growth is never water-limited. In any case, such substrates must be used if strict packed-bed operation is to be used, that is, with absolutely no mixing events. Such a substrate is nutrient impregnated hemp, used by Weber et al. (1999).

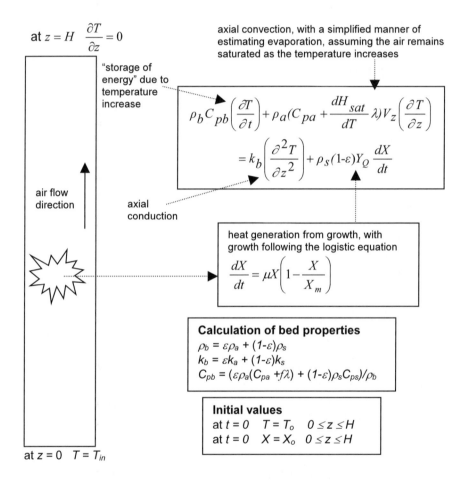

Fig. 24.1. Summary of the model used in the first packed-bed case study. The subscripts for the parameters ρ, k, and C_p are "s" for substrate particle, "a" for air, and "b" for the weighted average calculated for the bed

24.2.1 Synopsis of the Mathematical Model and its Solution

The model is based on the work of Sangsurasak (Sangsurasak and Mitchell 1995, 1998). It is almost identical to the version used by Mitchell et al. (1999), although the equation describing the effect of temperature on growth has been changed. The full model is not reproduced here; Fig. 24.1 summarizes its main features. Growth occurs according to the logistic equation, where the specific growth rate constant (μ) is affected by the temperature of the solid in a manner identical to that described by Eq. (22.1) (see Sect. 22.2). The energy balance takes into account axial convection, conduction, and evaporation and the production of metabolic heat. These two differential equations allow the biomass and temperature to be predicted as functions of time and space. The fact that the model is so simple means that it has many implicit assumptions and simplifications. Amongst these, some of the most important are:

- Growth depends only on biomass density and temperature. The bed does not dry out sufficiently during the fermentation to limit the growth;
- The bed is treated as a single pseudo-homogeneous phase that has the average properties of the gas and solid phases;
- Biomass does not move in space;
- Growth does not affect the void fraction;
- The substrate bed properties do not change with temperature or during consumption of substrate and production of biomass;
- Flow phenomena arising from increased pressure drop are not important;
- The air is always in thermal and moisture equilibrium with the solid (i.e., as the air heats up as it passes through the bed, water evaporates from the solid to maintain saturation of the air).

Table 24.1. gives the values of parameters, initial values of state variables and values of operating variables used in the base-case simulation. Note that some parameters are considered as constants even though they are not truly constant. For example the heat capacity of the air will change as the air passes through the bed, due to the increase in water content.

The mathematical model summarized in Fig. 24.1 contains partial differential equations. This model is solved by application of orthogonal collocation to convert each partial differential equation into a set of ordinary differential equations, which can then be integrated numerically. The principles of orthogonal collocation are beyond the scope of this book.

Note that the differential term dH_{sat}/dT in the energy balance is given by Eq. (19.20) (see Sect. 19.4.1). It has been used to replace the constant value of 0.00246 kg-H_2O kg-dry-air $°C^{-1}$ used by Mitchell et al. (1999).

Table 24.1. Values of the parameters and variables used for the base case simulation

Symbol	Significance	Base case value and units[a,b]
Design and operating variables (can be varied in the input file for the model)		
T_o	Initial bed temperature*	38°C
T_{in}	Inlet air temperature*	38°C
K	Control factor for the inlet air temperature	1 (see Eq. (24.1))
V_Z	Superficial velocity of the air*	5 cm s^{-1}
H	Height of the bioreactor*	1.0 m
Microbial parameters that you can vary		
X_o	Initial biomass concentration	0.001 kg-biomass kg-substrate^{-1}
X_m	Maximum possible biomass concentration	0.125 kg-biomass kg-substrate^{-1}
μ_{opt}	Maximum possible value of μ (at T=T$_{opt}$)	0.236 h^{-1}
T_{opt}	Optimum temperature for growth	38°C
Parameters related to the effect of temperature on growth (cannot be varied)		
A_1	Constant in the equation describing $\mu=f(T)$	8.31x10^{11}
A_2	Constant in the equation describing $\mu=f(T)$	70225 J mol^{-1}
A_3	Constant in the equation describing $\mu=f(T)$	1.3x10^{47}
A_4	Constant in the equation describing $\mu=f(T)$	283356 J mol^{-1}
Other parameters and constants (cannot be varied)		
C_{pa}	Heat capacity of the air	1180 J kg^{-1} °C^{-1}
C_{ps}	Heat capacity of the substrate particles	2500 J kg^{-1} °C^{-1}
k_a	Thermal conductivity of the air phase	0.0206 W m^{-1} °C^{-1}
k_s	Thermal conductivity of the substrate	0.3 W m^{-1} °C^{-1}
R	Universal gas constant	8.314 J mol^{-1} °C^{-1}
Y_Q	Yield of metabolic heat from growth	8.366 x 10^6 J kg-dry-biomass^{-1}
ε	Void fraction in the bed	0.35
λ	Enthalpy of vaporization of water	2.414 x 10^6 J kg-water^{-1}
ρ_a	Density of the air phase	1.14 kg-dry-air m^{-3}
ρ_s	Density of the solid particles*	700 kg-substrate m^{-3}

[a] Note that those values highlighted with an asterisk must be supplied in the input file. The remaining values are already in the program and cannot be changed.
[b] The program converts all variables and parameters to a consistent set of units.

24.2.2 Base-Case Predictions

Figures 24.2 and 24.3 represent the type of information that such a model provides. Both axial and temporal temperature variations occur (Fig. 24.2). The axial temperature profile is a result of convective cooling (see Fig. 4.3), with the steepness of this profile depending on the rate of growth and the superficial air velocity.

Note that the temperature variations in time are relatively slow compared to the temperature variations along the length of the bed. For example, in Fig. 24.2(b), the difference in temperature between 10 and 11 h is much smaller than the temperature difference between the inlet and outlet of the bed. This has implications for understanding the behavior of intermittently stirred beds (see Chap. 25).

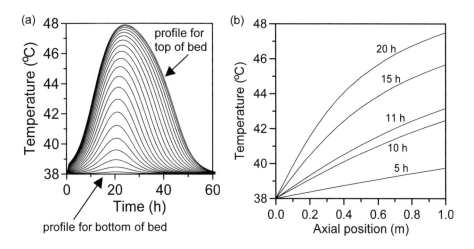

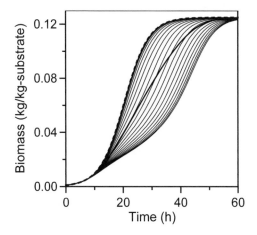

Fig. 24.2 (a) Typical output of the model, showing axial and temporal temperature profiles. Each curve represents the temperature profile against time at a particular location in the bioreactor. Moving upwards, the curves are for greater and greater heights in the bed. **(b)** The same data as shown in Fig. 24.2(a), but plotted against axial distance at several different times, in order to highlight the spatial temperature gradients. After 21 h the axial temperature profile begins to decrease again

Fig. 24.3. Predicted growth profiles at different heights within the bed. The lines at the top of the graph represent the bottom of the bed. The lines at the bottom represent the top of the bed, which is much hotter and therefore causes growth to be slower. The dashed line at the left represents logistic growth with $\mu=\mu_{opt}$ throughout the growth phase. The darker solid line in the middle of the lighter lines represents the average biomass concentration in the bed

Note the similarity of the general appearance of Fig. 24.2(a) to the experimental results obtained by Weber et al. (2002) in Fig. 7.6(a). Of course, the temperatures and times are different because the organism simulated by the model is quite different from that which they used. One feature that is common to the two graphs is the lack of symmetry around the peak, that is, the decrease in temperature takes slightly longer than the initial rise in temperature.

As a result of the spatial temperature profiles, growth will occur at different rates in the different regions of the bed (Fig. 24.3).

24.2.3 Insights that Modeling Has Given into Optimal Design and Operation of Traditional Packed-Beds

Section 7.2 showed that the design and operational variables that can be manipulated for traditional packed-bed bioreactors are the inlet air temperature and flow rate, the presence of a water jacket and the temperature of the water in this jacket, and the height and width of the bioreactor. The model can be used to investigate the effect of some of these design and operating variables on bioreactor performance. It cannot describe the effect of the bioreactor width or of a water jacket since it does not describe heat removal by conduction normal to the direction of air flow.

In the simulations presented in this section, in which only the effect of temperature on growth is considered, the aim is to minimize temperature gradients in order to maintain the average temperature as close as possible to the optimum temperature for growth and product formation. The effects of bed height, aeration rate, and air temperature are interrelated, but, in the subsections that follow, one-at-a-time changes will be made in order to make the contribution of each individual variable clear.

Effect of inlet air temperature. The inlet air temperature can be reduced below the optimum for growth in order to combat the temperature rise that occurs within the bed. However, it is important not to maintain the air temperature at a constant low value during the fermentation. During the initial stages of the fermentation the air temperature must remain near the optimum in order not to retard the initial growth. Therefore, in the simulations shown in Fig. 24.4, a simple temperature control scheme was included in the model:

$$T_{in} = T_{opt} - K(T_{ou} - T_{opt}),\qquad(24.1)$$

where K is a factor that determines by how much the temperature of the inlet air (T_{in}) is decreased for a given rise in the outlet air temperature (T_{out}) above the optimum temperature for growth (T_{opt}).

This strategy causes the temperature in the bed to vary around the optimum for growth (Fig. 24.4(a)). At the time of maximum heat production, the axial temperature profile is actually steeper than for aeration with the inlet air at T_{opt}: Fig. 24.4(a) shows a difference of almost 15°C between the air inlet and outlet, compared to 10°C in Fig. 24.2(a). However, the maximum deviation from the optimum temperature for growth (38°C) is only 7.5°C, and the average deviation from T_{opt} is also smaller, because the axial gradient straddles the optimum temperature. This leads to corresponding predictions of better growth (Fig. 24.4(b)).

Effect of inlet airflow rate. Doubling the airflow rate (i.e., increasing the superficial velocity of the air from 0.05 to 0.1 m s^{-1}) in the absence of any control of the air temperature, the performance of the bioreactor is predicted to improve significantly. Increasing the air flow rate decreases the gradient of the axial temperature profile: the highest temperature reached decreases from 48°C (Fig. 24.2(a)) to 45°C (Fig. 24.5(a)) and, as a result, the growth profiles in the different regions of the bed are closer to the optimum profile (Fig. 24.5(b)).

No work has been done to investigate the upper limits on the superficial velocities that can be used in packed-bed bioreactors, and this model does not take pressure drops into account. Obviously, the higher the airflow rate, the greater the operating cost, not only because more air must be supplied, but also because the pressure drop is greater. Therefore an economic optimum will need to be found between improved packed-bed performance and increasing operating costs. The best strategy might be to increase the air flow rate only during the period of peak heat production. In this case fluidization of the particles in the bed will not be a problem, because the microorganism will bind the particles together before high air flow rates are used. However, it is possible for the pressure drop to be sufficiently high that the whole knitted bed is ejected from the bioreactor!

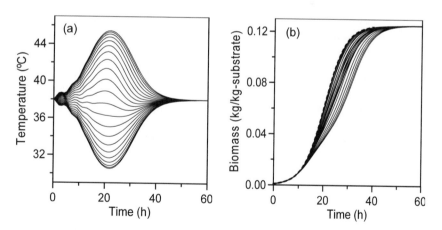

Fig. 24.4. (a) Predicted temperatures within the bed for a bioreactor in which the inlet air temperature (bottom-most line) is reduced in response to the temperature at the bed outlet (topmost line) according to Eq. (24.1). The other operating conditions are as in Table 24.1. Each curve represents the temperature profile against time at a particular location in the bioreactor. Moving upwards, the curves are for greater and greater heights in the bed. **(b)** Predicted growth profiles at different regions in the bed. Compared with Fig. 24.3, the average growth rate in the bed is much closer to the maximum possible rate

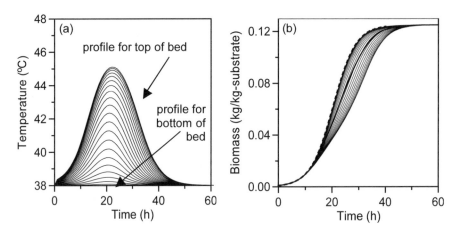

Fig. 24.5. (a) Temperature as a function of time and position for a superficial velocity (V_Z) of 0.1 m s^{-1}, twice as high as the value used to produce Fig. 24.2(a). All other parameters are the same as in Table 24.1. For ease of comparison, the same Y-axis range has been used as was used in Fig. 24.2(a). **(b)** Growth as a function of position for a superficial velocity (V_Z) of 0.1 m s^{-1}

Effect of bioreactor height and fungal specific growth rate. Obviously, with all operating conditions held constant, the height of the bioreactor will affect the maximum temperature reached, due to the unavoidable presence of an axial temperature gradient. In turn, this will affect the performance of the bioreactor. Figure 24.6 shows how the bioreactor height affects the time for the average biomass concentration (i.e., averaged over the whole bed) to reach 90% of the maximum biomass concentration. This time is denoted as t_{90}: the larger the value of t_{90}, the poorer the performance of the fermentation. This criterion is used to compare bioreactors since, for logistic growth kinetics, over a wide range of microbial and system parameters, the productivity of the fermentation, in terms of g-biomass m^{-3}-bioreactor h^{-1}, reaches a maximum when the biomass reaches around 90% of its final value (Mitchell et al. 2002b). In fact, t_{90} is inversely proportional to the productivity. The simulations were done for different specific growth rates.

The value of t_{90} increases approximately linearly with bioreactor height. This occurs because the greater the height, the greater the average deviation of the temperature from the optimum for growth. The value for zero height is the time that it would take for the biomass to reach $0.9X_m$ if the whole of the bioreactor remained at the optimum temperature for growth throughout the entire period.

Summary of strategies for optimizing the operation of traditional packed beds. The previous sections involved only one-by-one changes of variables. Obviously it is possible for more than one variable to be changed simultaneously. Simulations will not be shown for simultaneous changes (readers can use the program supplied with this book to undertake their own explorations), but, in general terms, to improve the performance of a traditional packed-bed, it is necessary to

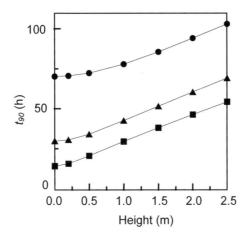

Fig. 24.6. The time taken for the average biomass concentration to reach 90% of X_m (t_{90}), as a function of the bed height within the packed-bed bioreactor, shown for various values for the parameter μ_{opt}. Key: (●) $\mu_{opt} = 0.1$ h^{-1} (▲) $\mu_{opt} = 0.236$ h^{-1} (■) $\mu_{opt} = 0.5$ h^{-1}

decrease the height of the bed within the bioreactor, increase the superficial velocity and use a control system to reduce the temperature of the inlet air in response to temperature increases in the outlet air. Note that, in order to minimize operating costs, it would be preferable not to have to refrigerate the inlet air.

The implications of changes in bed height might need to be considered either at the bioreactor design stage or in an attempt to optimize the performance of a bioreactor that has already been built. At the design stage, decreasing the height while maintaining the superficial velocity constant means that the bioreactor will need to be wider to hold the same amount of substrate, occupying more floor space. Once a bioreactor is built, decreasing the height will mean that the unutilized volume within the bioreactor will increase and therefore the volumetric productivity of the bioreactor will fall, if the calculation is based on the total bioreactor volume and not the bed volume. Therefore, provided the aeration system has the capacity, it would be preferable to increase the superficial velocity than to decrease the bed height, although problems may occur with high pressure drops.

A model similar to the one used in the simulations above was used by Ashley et al. (1999) to investigate whether reversing the airflow direction would help to overcome the problem of overheating at the top of the bed. Figure 24.7 shows that the model predicts that indeed such a strategy will prevent the temperature at the ends of the beds from reaching deleteriously high values. Unfortunately, it is not a useful strategy since the cooling of the middle sections of the bed is very inefficient, allowing them to reach very high temperatures.

Insights into scale-up of traditional packed-beds gained from modeling work. If you are considering using a traditional packed-bed bioreactor due to the inability of your microorganism to tolerate agitation, then, on the basis of the results in

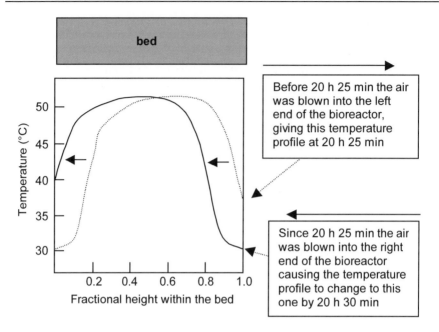

Fig. 24.7. Predicted axial temperature profiles at the time of peak heat production, for a fermentation in a packed-bed, in which there is no heat removal through the side walls and for which the direction of the air flow is reversed every five minutes. The inlet air temperature is 30°C. The bed height is 0.345 m and the superficial air velocity is 2.36 cm s⁻¹. This figure predicts that the temperature at the ends of the beds never exceeds 40°C, however, there is essentially no cooling effect in the central regions of the bed, which reach temperatures of more than 50°C. Adapted from Ashley et al. (1999) with kind permission of Elsevier

the previous section, it is possible to give advice about the research and development program for scale-up.

After the preliminary kinetic investigation in Raimbault columns (Sect. 15.1), experiments should be done in a pilot-scale packed-bed. A reasonable scale would be of the order of 15 cm diameter and as much as 1 m height. The walls of the column should be insulated well, in order to mimic the situation in the large-scale bioreactor, in which radial heat removal will be relatively minor. The 1 m height will allow studies to be done at bed heights that might actually be used in large-scale bioreactors. As such, this pilot bioreactor will represent a vertical section of the full-scale bioreactor (Fig. 24.8). This enables a study of those phenomena that depend on bed height, such as axial temperature profiles and pressure drops, and biomass and product formation as functions of height, and how these are affected by the temperature and velocity of the inlet air.

The advantage of this approach is that you might identify limitations on performance that are not predicted by the mathematical model. For example, the pressure drop may be excessive in your particular system. It is better to identify such problems, and to modify the bioreactor to overcome them, in a pilot-scale bioreactor than it is to build a full-scale bioreactor only to find that it does not work properly.

In the scale-up process, there will be a limit on bed height, in the sense that very tall beds will lead to unacceptably poor performance, due to axial temperature gradients or other considerations. Once this limit is reached, the capacity of the bioreactor can only be increased by making the bed wider. This "critical height" is not a constant, since it depends on the growth rate of the organism and the operating conditions, especially the superficial velocity of the air. An estimate of the "critical height" of a traditional packed-bed can be calculated for logistic growth kinetics as (Mitchell et al. 1999):

$$H = \frac{\rho_a(C_{pa} + f)V_z(T_{out} - T_{in})}{0.25\rho_s(1-\varepsilon)Y\mu_{opt}X_m}, \tag{24.2}$$

where the symbols have the meanings given in Table 24.1. T_{out} is the maximum temperature allowable in the bed while f is an estimate of dH_{sat}/dT.

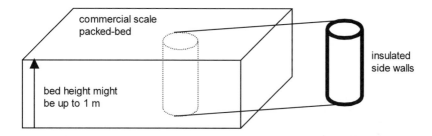

Fig. 24.8. A relatively thin packed-bed can be used for pilot-scale investigations into packed bed design. If its sides are insulated, this will mimic the presence of "hot" substrate around an identical section within the bed of the commercial-scale bioreactor

24.3 A Model of the Zymotis Packed-Bed Bioreactor

The model described here is based on that developed for the Zymotis bioreactor of Roussos et al. (1993) by Mitchell and von Meien (2000). The version used here has been modified by the inclusion of a water balance.

24.3.1 The Model

The model of the Zymotis packed-bed must account for heat transfer in two directions in the substrate bed: (1) the direction that is co-linear with the air flow, which causes convective and evaporative heat removal; and (2) the horizontal conduction to the cooling plates, which is normal to the direction of the air flow. Typically front-to-back gradients will be negligible (Fig. 24.9).

In this model the same growth kinetic equations are used as described by Eqs. (22.1), (22.2), and (22.3) (see Sect. 22.2). The solids and gas phases are assumed to remain in thermal and moisture equilibrium with one another. In other words, the gas phase remains saturated at the temperature of the bed. The differential term dH_{sat}/dT in the energy balance is given by Eq. (19.20) (see Sect. 19.4.1).

The parameter values used in the base-case simulation are the same as those given in Table 24.1 for the traditional packed-bed. There are several extra parameters that do not appear in that model, all of which are associated with the heat transfer plates. The spacing between plates (L) was varied, the overall heat transfer coefficient for heat transfer from the edge of the substrate bed across the plate wall to the cooling water (h) was taken as 95 W m^{-2} °C^{-1} a value that is typical of heat exchangers and, finally, the cooling water temperature (T_w) was set at 38°C in various simulations and varied according to a control scheme in others.

24.3.2 Insights into Optimal Design and Operation of Zymotis Packed-Beds

The model can be used to explore the effect of operating conditions on bioreactor performance. Simulations will not be shown for the effects of superficial air velocity, inlet air temperature, or bioreactor height. The general principles are the same as for the traditional packed-bed discussed in Sect. 24.2, although the effects are not exactly the same, because of the extra heat removal by the heat transfer plates. These parameters are therefore discussed generally, without new simulations being done. Readers with greater interest are encouraged to consult Mitchell and von Meien (2000) and also to use the simulation program provided to explore the performance of Zymotis packed-bed bioreactors in more detail.

After this, the effects of the new design and operating variables introduced by the internal heat transfer plates, namely the spacing between the plates and the temperature of the cooling water, are explored.

General principles (overall trends). The model predictions in Fig. 24.10(a)). show clearly the temperature gradients vertically through the bed (i.e., parallel to the direction of air flow) and horizontally (i.e., normal to the direction of air flow). Figure 24.10(b) compares the central axis temperatures for the Zymotis bioreactor and a traditional packed-bed that is wide enough for heat removal through the side walls to be negligible. The comparison is done for the same microorganism, that is, in both cases μ_{opt} is set at 0.236 h^{-1}, and for the same operating conditions. Along the central axis of the bioreactor, due to the greater heat removal rate in the Zymotis bioreactor, the temperature does not reach such high values in the top half of the bioreactor, although the performance is reasonably similar in the bottom half. Note, however, that for the Zymotis bioreactor the curve represents only the central axis temperature, while for the traditional packed-bed it represents the temperature at all radial positions for that height. In the Zymotis packed-bed the remainder of the bed is cooler than the central axis at the corresponding height, and therefore growth is correspondingly better.

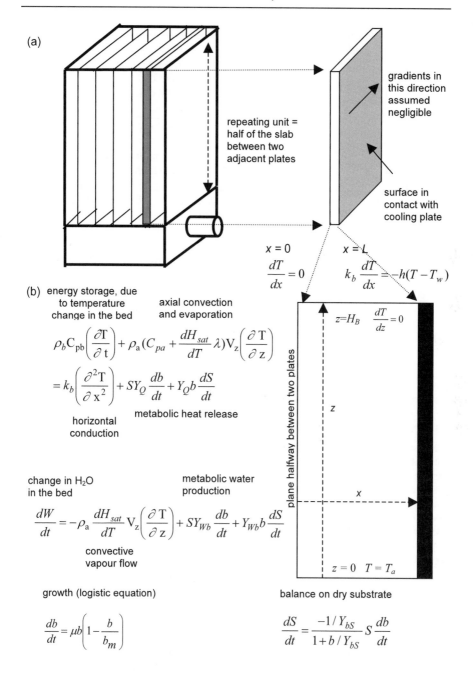

Fig. 24.9. Summary of the model of the Zymotis packed-bed bioreactor used in the second case study. **(a)** The Zymotis bioreactor can be treated as consisting of repeating units. **(b)** Summary of the mathematical model used to model one of the repeating units

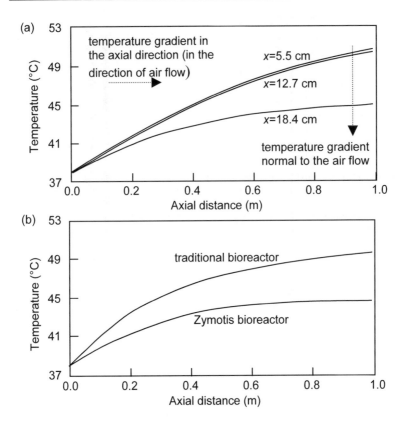

Fig. 24.10. (a) Predictions of the mathematical model about the axial temperature profile at several horizontal positions within the Zymotis packed-bed at the time of peak heat production at 23 h. Note that in the upper regions of the bed the temperature is high from the central plane to 12.7 cm from the central plane (7.3 cm from the plate), only reducing significantly at positions close to the plate (which is at $x=20$ cm). **(b)** Comparison of the central axial temperature profile in a wide traditional packed-bed and the central plane temperature profile of the Zymotis bioreactor, at 20 h, in the case in which $L = 0.03$ cm. In both cases the bioreactor is 1.0 m high, the superficial air velocity is 5 cm s^{-1}, the inlet air temperature is 38°C and the specific growth rate constant (μ_{opt}) is 0.236 h^{-1}

Regarding bioreactor dimensions, the maximum practical value for the front-to-back depth will depend on the size of heat transfer plates that can be constructed. Given that more plates can be added to extend the width of the bioreactor, theoretically there is no limitation on the width. One advantage of the Zymotis bioreactor is that, with the flattening out of the temperature profile, larger heights are theoretically possible than for traditional bioreactors, at least based on temperature considerations, although pressure drop may become problematic at large heights.

In the upper regions of the bed much of the waste metabolic heat is removed by conduction to the heat transfer plates, as evidenced by the flattening out of the temperature profile (Fig. 24.10(b)); in these regions only a relatively minor proportion is removed by axial convection and evaporation. Note that the predicted profiles are similar to the experimental temperature profiles that Saucedo-Castaneda et al. (1990) measured in a water-jacketed packed-bed bioreactor of 6 cm diameter (Fig. 7.5). The position at which the axial temperature profile flattens out in the Zymotis bioreactor depends on the heat production rate and on the various operating parameters, including the superficial air velocity. In fact, due to the acceleration and deceleration in the growth rate during the various phases of a fermentation, and the corresponding changes in the rate of production of waste metabolic heat, the extent of the "flat zone" will fluctuate during the fermentation.

Effect of a cooling water control scheme and of the gap between the plates. In these simulations only data about overall predicted performance is presented; no information is given about the gradients within the bioreactor. However, clearly the best performance must correspond to those operating conditions that minimize the temperature deviations from the optimum temperature, in both space and time.

The spacing between the cooling plates makes a large difference to the predicted performance, with all other parameters and operating variables being held constant (Fig. 24.11). Performance worsens rapidly as the size of the gap increases from 1 to 10 cm. That is, t_{90} increases rapidly over this range. Further increases in gap size worsen performance even further.

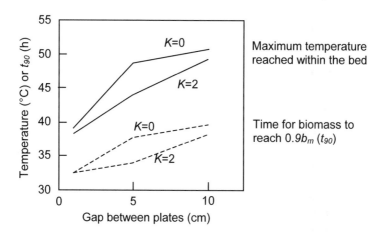

Fig. 24.11. Predictions of the mathematical model about the combined effect on the performance of the Zymotis bioreactor of the gap between the heat transfer plates (the gap is equal to $2L$) and the use of a scheme for the control of the cooling-water temperature according to Eq. (24.3). In this graph the performance of the bioreactor is evaluated on the basis of two criteria: (————) the maximum temperature reached within the bed during the fermentation and (- - - -) the time for the biomass to reach 90% of its maximum value (t_{90})

Two lines are shown on the graph. One corresponds to the case in which the cooling water temperature is maintained at 38°C (K=0). In the other case (K=2) the temperature of the cooling water (T_w) is manipulated in response to the temperature measured at the top of the bed, halfway between the heat transfer plates (this temperature being denoted T^*), according to the following equation:

$$T_w = T_{opt} - K\,(T^* - T_{opt}),\qquad\qquad(24.3)$$

where T_{opt} is the optimum temperature for growth. This equation calculates the number of degrees by which the measured temperature exceeds the optimum for growth and then decreases the temperature of the cooling water by this temperature difference multiplied by a factor K. If possible, the value of K should be chosen so as not to require refrigeration of the cooling water to values below the temperature at which it is normally available. This will avoid the costs of building and operating a water refrigeration system. However, the ability to do this will depend on the optimum growth temperature of the organism in relation to the temperature of the available cooling water.

Whether or not it is advantageous to use this strategy to control the cooling water temperature depends on the spacing between the plates. If the spacing between the plates is large, of the order of 20 cm (L=10 cm), the cooling water has relatively little effect on much of the bed, and therefore the temperature control scheme brings little advantage (Fig. 24.11). If the spacing between the plates is small, of the order of 2 cm (L=1 cm), temperature control is reasonably efficient even without the temperature control scheme, so there is little advantage in having it. The temperature control scheme is most advantageous at intermediate plate spacings.

In fact, intermediate plate spacings are probably preferable. Although a 2 cm gap between plates gives near optimum performance (the minimum possible value of t_{90} for X_o=0.001 kg kg^{-1}, X_m = 0.125 kg kg^{-1}, and μ_{opt} = 0.236 h^{-1} is 29.7 h), it is not a reasonable value, because a significant volume of the bioreactor will be occupied by the plates, leading to a low overall productivity and, additionally, the capital costs of the bioreactor will be much higher. On the other hand, the wider the space between the plates, the less effective they are in cooling the bed, and therefore the higher is the superficial air velocity that is needed to achieve the same cooling effect and, consequently, the higher are the operating costs. Essentially these will need to be balanced against each other. Mitchell et al. (2002b) use a model similar to the one presented here to explore these issues in more depth, identifying a gap of 6 cm (L=3 cm) as optimal in terms of productivity of the bioreactor, calculated per m^3 of overall bioreactor volume, for a microorganism with a specific growth rate of 0.324 h^{-1} and a superficial air velocity of 1 cm s^{-1}. Obviously the optimal plate spacing will differ for different organisms and under different operating conditions. The mathematical model provides a tool that allows the optimum to be determined for any particular combination of growth kinetics, bioreactor design and operating conditions.

24.4 Conclusions on Packed-Bed Bioreactors

Packed-bed bioreactors are necessary in those cases in which the bed must not be mixed during the fermentation. Such bioreactors will always suffer from axial temperature gradients. These can be minimized, although not eliminated, by selection of appropriate design and operating variables. The question as to what combination of design and operating conditions will lead to best performance is not simple, and is best answered with the use of a mathematical model.

The models that were presented in this chapter for the traditional packed-bed bioreactor and Zymotis packed-bed bioreactors could be much more powerful tools for use in the design process if they were improved. Some possible improvements include:

- introduction of a water balance. This would require a description of the effect of the temperature and the water content of the substrate on its water activity and a description of the effect of water activity on the growth kinetics of the microorganism. Note that the model of the intermittently-stirred bioreactor presented in Chap. 25 can be used to explore the water balance during packed-bed operation, simply by suppressing the mixing events;
- incorporation of the effect of growth of the microorganism into the interparticle spaces on the pressure drop through the bed and the resulting effects on air flow and heat transfer phenomena;
- description of changes in the bed due to the shrinkage of substrate particles as dry matter is converted into CO_2 by the microorganism.

Further Reading

Mathematical models of packed-bed bioreactors
Ashley VM, Mitchell DA, Howes T (1999) Evaluating strategies for overcoming overheating problems during solid state fermentation in packed bed bioreactors. Biochem Eng J 3:141–150
Saucedo-Castañeda G, Gutierrez-Rojas M, Bacquet G, Raimbault M, Viniegra-Gonzalez G (1990) Heat transfer simulation in solid substrate fermentation. Biotechnol Bioeng 35:802–808
Gutierrez-Rojas M, Auria R, Benet JC, Revah S (1995) A mathematical model for solid state fermentation of mycelial fungi on inert support. Chem Eng J 60:189–198
Hasan SDM, Costa JAV, Sanzo AVL (1998) Heat transfer simulation of solid state fermentation in a packed-bed bioreactor. Biotechnol Techniques 12:787–791
Oostra J, Tramper J, Rinzema A (2000) Model-based bioreactor selection for large-scale solid-state cultivation of *Coniothyrium minitans* spores on oats. Enz Microb Technol 27:652–663
Mitchell DA, Pandey A, Sangsurasak P, Krieger N (1999) Scale-up strategies for packed-bed bioreactors for solid-state fermentation. Process Biochem 35:167–178

Sangsurasak P, Mitchell DA (1995) Incorporation of death kinetics into a 2-D dynamic heat transfer model for solid state fermentation. J Chem Technol Biotechnol 64:253–260

Sangsurasak P, Mitchell DA (1998) Validation of a model describing two-dimensional heat transfer during solid-state fermentation in packed bed bioreactors. Biotechnol Bioeng 60:739–749

Weber FJ, Oostra J, Tramper J, Rinzema A (2002) Validation of a model for process development and scale-up of packed-bed solid-state bioreactors. Biotechnol Bioeng 77:381–393

Weber FJ, Tramper J, Rinzema A (1999) A simplified material and energy balance approach for process development and scale-up of *Coniothyrium minitans* conidia production by solid-state cultivation in a packed-bed reactor. Biotechnol Bioeng 65:447–458

25 A Model of an Intermittently-Mixed Forcefully-Aerated Bioreactor

David A. Mitchell, Oscar F. von Meien, Luiz F.L. Luz Jr, and Nadia Krieger

25.1 Introduction

True packed-bed operation can only be used in those cases where the bed does not dry out to levels that cause water limitations on growth, because water can only be uniformly distributed within a bed of solids while the solids are being agitated. If the organism tolerates some mixing, then the intermittently-mixed mode of operation can be used, in which the bioreactor operates as a packed-bed during the majority of the fermentation period and undergoes infrequent mixing events, during which water can be added to the bed (Chap. 10). In fact, once the intermittently-mixed mode of operation is selected, the use of dry air to promote evaporative cooling is potentially available as an operating strategy. The current chapter presents a model that can be used to investigate the operation of such bioreactors.

25.2 Synopsis of the Model

The two types of operation, that is, static operation and the mixing event, are modeled separately. During static operation, the system is treated as a packed-bed bioreactor (Fig. 25.1). It is assumed that the bioreactor is sufficiently wide such that heat transfer to the side walls makes a negligible contribution, and therefore only heat and mass transfer in the axial direction are described. Both temperature and water balances are done; in both cases separate balances are written for the air and solid phases. Note that the terms of the balance equations are per cubic meter of bioreactor, and biomass and water contents are expressed on the basis of the total dry solids per cubic meter (S), which is equal to the sum of the dry biomass and dry substrate. The definitions of symbols used are given in Table 25.1.

The model equations are summarized in Fig. 25.2. The balance on the overall mass of dry solids (i.e., the sum of dry biomass and dry residual substrate, Fig 23.2, lower right) is necessary since not all the consumed substrate is converted into biomass; a proportion is lost in the form of CO_2. This is Eq. (16.11), although without the maintenance term.

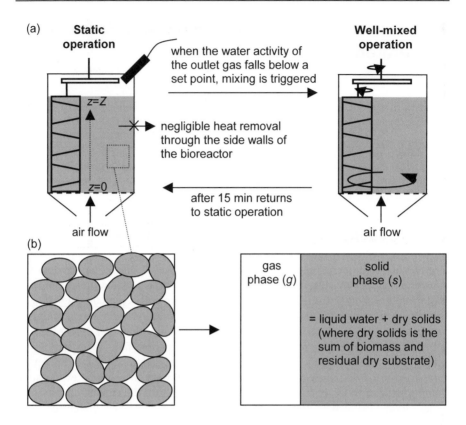

Fig. 25.1. (a) Strategy of operation of an intermittently-mixed forcefully-aerated bioreactor, as modelled in the case study. **(b)** In Fig. 25.2 the solid and gas phases are represented in the manner as shown in the diagram on the right.

Instead of assuming that the solids and air are in thermal and moisture equilibrium, as was done in Chap. 24, the model uses the balance equations, which have driving force equations for heat and mass transfer, to calculate the changes in the dry solids and liquid water, thereby allowing calculation of the water content of the solid phase. Note that the water content of the solid phase in turn affects the rate of evaporation to the gas phase. Given the water content and the temperature of the solid, the model calculates the water activity of the solid. The water activity of the solid, along with its temperature, affects growth. The model calculates and combines fractional growth rates in the same manner as was done in Eqs. (22.1), (22.2), and (22.3) in Sect. 22.2.

Gas phase (g)	Solid phase (s)

Gas phase energy balance (T_g)

$$\varepsilon \rho_g \left(C_{Pg} + H C_{Pv} \right) \frac{\partial T_g}{\partial t}$$

$$= -\left(C_{Pg} + H C_{Pv} \right) G \frac{\partial T_g}{\partial z}$$

$$- ha\left(T_g - T_s \right)$$

G = flux of dry air through the gas phase (mass of dry air per cross-sectional area per second)

Gas phase water balance (H)

$$\varepsilon \rho_g \frac{\partial H}{\partial t}$$

$$= -G \frac{\partial H}{\partial z} + Ka\left(W - W_{sat} \right)$$

H = humidity of the headspace (mass of water vapor per mass of dry air)

flow of dry air and water vapor

Solid phase energy balance (T_s)

$$S\left(C_{Ps} + W C_{Pw} \right) \frac{\partial T_s}{\partial t}$$

$$= ha\left(T_g - T_s \right)$$

$$- Ka\left(W - W_{sat} \right) \lambda$$

$$+ Y_Q \left(S \frac{\partial b}{\partial t} + b \frac{\partial S}{\partial t} \right)$$

W = water content of the bed on a dry basis (mass of liquid water per total mass of dry solids)

Solid phase water balance (W)

$$\frac{\partial (SW)}{\partial t} = Y_{WB} \left(S \frac{\partial b}{\partial t} + b \frac{\partial S}{\partial t} \right)$$

$$- Ka\left(W - W_{sat} \right)$$

b = biomass content of the solids on a dry basis (mass of biomass per total mass of dry solids)

Growth

$$\frac{\partial b}{\partial t} = \mu b \left(1 - \frac{b}{b_m} \right)$$

Consumption of dry solids

$$\frac{\partial S}{\partial t} = \left(1 - \frac{1}{Y_{BS}} \right) \frac{\partial (bS)}{\partial t}$$

S = total mass of dry solids in the bed (dry biomass plus dry residual substrate)

Fig. 25.2. Summary of the model of an intermittently-mixed, forcefully-aerated bioreactor (von Meien and Mitchell 2002). The variables shown in parentheses after the heading in each text box is the variable that is isolated in the differential term on the left hand side of the differential equation before the equation set is solved. Subscripts: s = solids phase, g = interparticle gas phase, sat = saturation. Also indicated are the meanings of several of the symbols representing key system variables

In the gas phase water balance (Fig. 25.2, lower left):

- the left hand side represents the temporal variation in the amount of water vapor in the air phase at a given position;
- the first term on the right hand side represents convective flow of water vapor with the gas phase;
- the second term on the right hand side represents the water exchange between the solid and gas phases.

In the gas phase energy balance (Fig. 25.2, upper left):

- the left hand side represents the temporal variation in the sensible energy of the dry air and water vapor in the air at a given position;
- the first term on the right hand side represents the convective flow of energy in the flowing moist air (i.e., mixture of water vapor and dry air);
- the second term on the right hand side represents the sensible heat exchange between the solid phase and the gas phase.

In the solid phase water balance (Fig. 25.2, middle right):

- the left hand side represents the temporal variation in the water content of the solids phase at a given position;
- the first term on the right hand side represents the metabolic production of water;
- the second term on the right hand side represents the exchange of water between the solid and gas phases.

In the solid phase energy balance (Fig. 25.2, upper right):

- The left hand side represents the temporal variation of the sensible energy within the dry solids and liquid water at a given position;
- the first term on the right hand side represents sensible energy exchange with the gas phase;
- the second term on the right hand side represents the removal of energy from the solid as the latent heat of evaporation;
- the third term on the right hand side represents the liberation of waste metabolic heat in the growth process.

The mixing period is modeled in a simple manner. A 15-minute long mixing event begins whenever the water activity of the outlet air (i.e., percentage relative humidity divided by 100%) falls below a predetermined value (a_{wg}*). This mixing completely inhibits growth during the mixing event, but growth is re-established as soon as static operation is resumed. During the mixing event, the bed temperature is brought back to the optimum temperature for growth and the water activity is brought back to its zero time value. The program calculates how much water needs to be added to reach this water activity. A volume-weighted average biomass content is used as the starting point for a new round of operation in packed-bed mode. It is calculated on the basis of the amount of biomass and dry solids in the various regions of the bioreactor at the time the mixing event is triggered.

Some of the assumptions that are made by the model are:

- there is no change in bed height as dry matter is consumed. Rather, the effect of the loss of solids is to decrease the density of the bed.
- maintenance metabolism is not significant. In other words, water and heat production and dry solids consumption occur only as a result of the production of new biomass.

The values of the various variables and parameters used in the base case simulation are shown in Tables 25.1 and 25.2. The simulations presented here are done for *Aspergillus niger,* using the same growth kinetic parameters as those used in Chap. 22. As in that case, it is possible to specify a combination of parameters either for an *Aspergillus*-type water relation or a *Rhizopus*-type water relation.

As in the well-mixed bioreactor model presented in Chap. 22, the isotherm determined for corn by Calçada (1998), described by Eq. (22.5), is used to calculate W_{sat}, the water content that the solids would have if they were in equilibrium with the gas in the void spaces. The relationship between the water content and the water activity of the solids is as described by Eqs. (19.8) and (19.9).

The coefficients for convective heat transfer and water transfer between the solid and gas phases are given by Eqs. (20.8) and (20.9), determined for the drying of corn (Calado 1993; Mancini 1996).

25.3 Insights the Model Gives into Operation of Intermittently-Mixed Bioreactors

25.3.1 Predictions about Operation at Laboratory Scale

The base case simulation with the model uses the values given in Tables 25.1 and 25.2. This first simulation, although not intended specifically to model the bioreactor of Ghildyal et al. (1994), is for a similar situation. They used a bioreactor of 34.5 cm height and superficial velocities from 0.0047 m s^{-1} to 0.0236 m s^{-1}. In the base case simulation, the bioreactor is 30 cm high, and the air flux used, 0.02 kg-dry-air m^{-2} s^{-1}, corresponds to a superficial velocity of 0.0175 m s^{-1} (i.e., 0.02 kg-dry-air m^{-2} s^{-1}/(1.14 kg-dry-air m^{-3})).

Figure 25.3(a) illustrates the advantages of intermittent mixing over true packed-bed operation. In the absence of any mixing event, the average biomass curve suddenly decelerates at 20 h. This deceleration is due to the bed drying out to water activities that greatly restrict the growth rate. In the intermittent-mixing mode, water is added during the mixing events at 20.35 and 30.8 h, these mixing events being indicated by the sudden drops in bed temperature in Fig. 25.3(c). During a mixing event, the water activity of the solids in the bed is brought back to values that favor fast growth, and the average biomass profile does not deviate too far from the dashed curve in Fig. 25.3(a), which was obtained by plotting the logistic equation with μ always equal to the maximum possible value, 0.236 h^{-1}.

Table 25.1. Values used for the base case simulation of those parameters and variables that can be changed in the accompanying model of an intermittently-mixed forcefully-aerated bioreactor

Symbol	Significance	Base case value and units[a]
Design and operating variables and initial values of key system variables		
Z	Height of the bioreactor	0.3 m
a_{wg}*	Outlet gas water activity set point[b]	0.87
G	Inlet air flux	0.02 kg-dry-air s^{-1} m^{-2}
T_{so}	Initial temperature of the solid phase	38°C
T_{in}	Inlet air temperature[c]	38°C
a_{wso}	Initial water activity of the solids phase[d]	0.99
a_{wgin}	Water activity of the inlet air[e]	0.99
ρ_s	Density of the dry substrate particles[f]	450 kg-dry-substrate m^{-3}-substrate
ε	Void fraction of the bed[f]	0.35 m^3-voids m^{-3}-bed
Microbial parameters		
b_o	Initial biomass concentration	0.002 kg-biomass kg-dry-solids^{-1}
b_m	Maximum possible biomass concentration	0.250 kg-biomass kg-dry-solids^{-1}
μ_{opt}	Optimal specific growth rate constant	0.236 h^{-1}.
T_{opt}	Optimum temperature for growth	38°C
Y_{BS}	Yield of biomass from substrate	0.5 kg-biomass kg-dry-substrate^{-1}
Type	Type of relation of growth with a_w	*Aspergillus*-type (see Fig. 22.3(b))
Parameters related to the mixing event		
t_{mix}	Time taken by the mixing event	0.25 h
μ_{mix}	Fractional value of μ during mixing	0

[a] The program converts all input variables and parameters to a consistent set of units. Note that where "biomass" is mentioned within the units, this represents dry biomass.

[b] When the outlet gas water activity falls below this value, a mixing event is triggered during which water is added to the bed.

[c] The inlet air temperature is also used as the initial temperature of the gas phase within the bed (T_{go} °C).

[d] Used to calculate the initial water content of the solids (W_o, kg-water kg-dry-solids^{-1}), using Eq. (22.5), in which case W_{sat} is replaced with W_o, a_{wg} is replaced with a_{wso}, and T_g is replaced with T_{so}.

[e] The water activity of the inlet air is also used as the initial water activity of the gas phase within the bed (a_{wgo}). This value is used to calculate the humidity of the inlet air (H_{in}) and the initial humidity of the gas phase (H_o), both of which have the units of kg-water kg-dry-air^{-1}.

[f] The initial mass of dry substrate per cubic meter (S_o, kg-dry-substrate m^{-3}) is calculated as $(1-\varepsilon)\rho_s$.

Table 25.2. Values used for the base case simulation of those parameters and variables that cannot be changed in the accompanying model of an intermittently-mixed forcefully-aerated bioreactor

Symbol	Significance	Base case value and units[a]
Parameters related to the growth of the microorganism		
A_1	Constant in the equation describing $\mu=f(T)$	8.31×10^{11}
A_2	Constant in the equation describing $\mu=f(T)$	70225 J mol^{-1}
A_3	Constant in the equation describing $\mu=f(T)$	1.3×10^{47}
A_4	Constant in the equation describing $\mu=f(T)$	283356 J mol^{-1}
Y_Q	Yield of metabolic heat from growth	8.366×10^6 J kg-biomass^{-1}
Y_{WB}	Yield of metabolic water from growth	0.3 kg-water kg-biomass^{-1}
Other parameters and constants		
C_{pg}	Heat capacity of dry air	1006 J kg^{-1} °C^{-1}
C_{pv}	Heat capacity of water vapor	1880 J kg^{-1} °C^{-1}
C_{ps}	Heat capacity of the dry solids	2500 J kg^{-1} °C^{-1}
C_{pw}	Heat capacity of liquid water	4184 J kg^{-1} °C^{-1}
R	Universal gas constant	8.314 J mol^{-1} °C^{-1}
P	Air pressure within the bioreactor	760 mm Hg[b]
λ	Enthalpy of vaporization of water	2.414×10^6 J kg-water^{-1}
ρ_g	Density of the gas phase in the bed	1.14 kg-dry-air m^{-3} [c]

[a] The program converts all variables and parameters to a consistent set of units. Note that where "biomass" is mentioned within the units, this represents dry biomass.
[b] Needed for the calculation of the air water activity and used in the calculation of the solids/gas heat transfer coefficient.

The model predicts that growth will be different at different heights within the bioreactor (Fig. 25.3(b)). This is due to the fact that the conditions for growth at different heights are different. Figure 25.3(c) shows the temporal substrate temperature profiles at various different heights while Fig. 25.3(d) shows the temporal substrate water activity profiles at these heights. The values of μ_{FT} and μ_{FW} are shown in Fig. 25.3(e) and Fig. 25.3(f), respectively. These fractional specific growth rates indicate the relative effects of temperature and water activity in limiting growth. During the fermentation, both the temperature and the water activity significantly affect growth, as indicated by deviations of the fractional specific growth rates below 1.0. The tendency is for the temperature-limitation to occur first, followed by water-limitation as the bed reaches low water activities. For example, temperature limitation dominates before 20 h (Fig. 25.3(e)) while water limitation dominates at 20 h, immediately before the first mixing and water replenishment event (Fig. 25.3(f)).

Figures 25.3(c) and 25.3(d) also indicate that temperature limitations will be greater near the air outlet, that is, at the top of the bed, while water limitations will be greater in the mid sections of the bed.

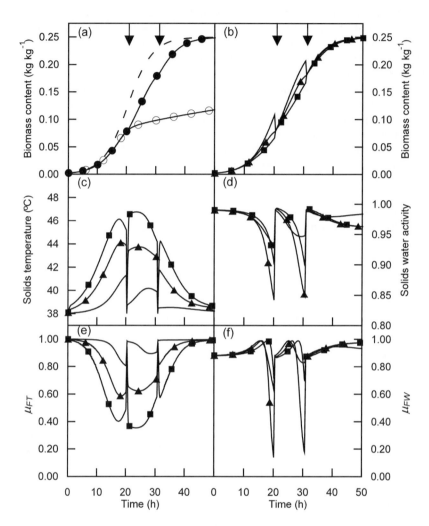

Fig. 25.3. Results of the base case simulation, obtained using the values in Tables 25.1 and 25.2. **(a)** Volume-weighted biomass content. The curve with the solid circles represents the results of the simulation. The curve with the hollow circles represents the results of a simulation undertaken with identical values for all parameters and variables except that no mixing and water addition events were allowed. The dashed line represents growth with the specific growth rate constant equal to μ_{opt} throughout the fermentation; **(b)** Biomass contents predicted at various fractional heights above the air inlet; **(c)** Temperatures of the solids at the various heights; **(d)** Water activities of the solids at the various heights; **(e)** Fractional specific growth rates based on temperature at the various heights; **(f)** Fractional specific growth rates based on water activity at the various heights. Key to graphs (b) to (f): Fractional heights of (———) 0.025; (—▲—) 0.5; (—■—) 1.0. The arrows at the top of the graph denote the mixing events

Reasonable growth and temperature control are predicted for a lab scale bioreactor with a value of G of 0.1 kg-dry-air m^{-2} s^{-1} (which corresponds to a superficial velocity of 0.088 m s^{-1}, calculated as 0.1 kg-dry-air m^{-2} s^{-1}/(1.14 kg-dry-air m^{-3})). Figure 25.4 is plotted using the same axes as Fig. 25.3, to highlight the fact that the temperature remains closer to the optimum of 38°C (Fig. 25.4(c)) and the water activity remains higher (Fig. 25.4(d)). As a result, the temperature-limitation and water-limitation of growth are much less severe (Figs. 25.4(e) and 25.4(f)) and the growth profile is closer to the optimum profile (Fig. 25.4(a)). Due to the fact that the conditions are close to the optimum at all bed positions, there is no significant difference in the growth at different heights. There are two mixing events, one at 17.35 h and the other at 24.20 h.

25.3.2 Investigation of the Design and Operation of Intermittently-Mixed Forcefully-Aerated Bioreactors at Large Scale

To investigate operation at large scale, graphs could be plotted similar to those in the previous section for the operation of a laboratory-scale bioreactor. However, in the current section only graphs of the solids temperature and water activity will be plotted. Better performance can be judged on the basis of how closely the solids temperature is maintained to the optimum for growth of 38°C and how closely the solids water activity is maintained to the optimum for growth of 0.95. Table 25.3 shows the design and operating variables that were changed in these simulations.

Scaling up to a 1 m high bed while maintaining the air flux (G) constant at 0.1 kg-dry-air m^{-2} leads to poor performance with respect to the control of the temperature (Fig. 25.5(a)) and water activity of the solids (Fig. 25.5(b)), as might be expected. On the other hand, good control of the temperature (Fig. 25.5(c)) and water activity of the solids (Fig. 25.5(d)) is predicted when the bioreactor is scaled-up while maintaining the ratio of G/Z at 0.1 kg-dry-air m^{-2} s^{-1}/0.3 m (i.e., 0.33 kg-dry-air m^{-2} s^{-1} for a 1 m high bed). In fact, almost identical performance is expected when the same scale-up strategy is used for a 2 m high bed (for which this strategy gives an air flux of 0.66 kg-dry-air m^{-2} s^{-1}), as shown by Figs. 25.5(e) and 25.5(f).

Table 25.3. Design and operating variables changed in the various explorations of performance of a 1 m bioreactor[a]

Figures	Z (m)	G (kg m^{-2} s^{-2})	a_{wg}*	a_{wgin}	T_{so} & T_{in} (°C)
25.5(a) and (b)	1.0	0.1	0.87	0.99	38
25.5(c) and (d)	1.0	0.33	0.87	0.99	38
25.5(e) and (f)	2.0	0.66	0.87	0.99	38
25.6(a) and (b)	1.0	0.1	0.95	0.99	38
25.6(c) and (d)	1.0	0.1	0.95	0.75	38
25.6(e) and (f)	1.0	0.1	0.95	0.99	35
25.6(g) and (h)	1.0	0.1	0.95	0.99	33

[a] Other conditions are as given in Tables 25.1 and 25.2.

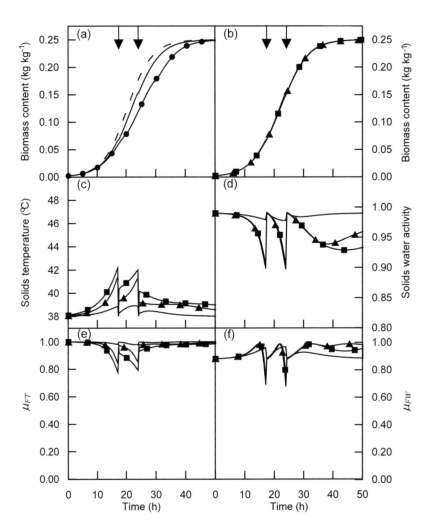

Fig. 25.4. A simulation with a higher air flux of 0.33 in order to give good performance of a laboratory-scale bioreactor. **(a)** Volume-weighted biomass content. The solid line represents the results of the simulation. The curve with the solid circles represents the results of the simulation shown in Fig 25.3(a). The dashed line represents growth with the specific growth rate constant equal to μ_{opt} throughout the fermentation. **(b)** Biomass contents predicted at various fractional heights above the air inlet; **(c)** Temperatures of the solids at the various heights; **(d)** Water activities of the solids at the various heights; **(e)** Fractional specific growth rates based on temperature at the various heights; **(f)** Fractional specific growth rates based on water activity at the various heights. Key to graphs (b) to (f): Fractional heights of (———) 0.025; (—▲—) 0.5; (—■—) 1.0. The arrows at the top of the graph denote the mixing events

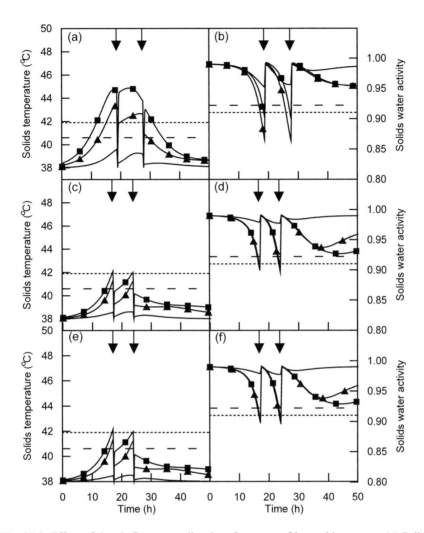

Fig. 25.5. Effect of the air flux on predicted performance of larger bioreactors. **(a)** Solids temperatures and **(b)** solids water activities at various heights within in a 1 m high bioreactor with an air flux of 0.1 kg-dry-air m^{-2} s^{-1}; **(c)** Solids temperatures and **(d)** solids water activities at various heights within in a 1 m high bioreactor with an air flux of 0.33 kg-dry-air m^{-2} s^{-1}; **(e)** Solids temperatures and **(f)** solids water activities at various heights within in a 2 m high bioreactor with an air flux of 0.66 kg-dry-air m^{-2} s^{-1}. In all cases, the other parameters are as given in Tables 25.1, 25.2, and 25.3. The horizontal lines represent temperatures and water activities that give fractional specific growth rates of (- - -) 0.9 and (····) 0.8. Key to graphs (a) to (f): Fractional heights of (——) 0.025; (—▲—) 0.5; (—■—) 1.0. The arrows at the top of each graph denote the mixing events

However, such aeration rates might not be feasible at large scale. The superficial velocity corresponding to an air flux of 0.33 kg-dry-air m^{-2} s^{-1} is 0.29 m s^{-1} while that corresponding to an air flux of 0.33 kg-dry-air m^{-2} s^{-1} is 0.58 m s^{-1}. It may be overly costly to provide such high aeration rates. The simulations presented in Fig. 25.5 represent a search for an operating strategy that will allow good performance in a 1 m high bioreactor but at the lower air flux of 0.1 kg-dry-air m^{-2} s^{-1}. They should be compared with the results presented in Figs. 25.5(a) and 25.5(b), which are for a 1 m high bioreactor with an air flux (G) of 0.1 kg-dry-air m^{-2} but otherwise operated identically to the laboratory-scale bioreactor:

- Increasing the outlet gas relative humidity set point (a_{wg}*) from 0.87 to 0.95 improves performance minimally (Figs. 25.6(a) and 25.6(b)). With the higher set point there are three mixing and water replenishment events instead of two.
- Maintaining the outlet gas relative humidity set point at 0.95 but decreasing the inlet gas relative humidity to 0.75 enables the bed temperature to be maintained at values that give a μ_{FT} of 0.8 or greater for most of the bed and for most of the time (Fig. 25.6(c)). However, the bottom of the bed dries out rapidly (Fig. 25.6(d)). Due to the faster drying, there are four mixing and water replenishment events.
- Maintaining the outlet gas relative humidity set point at 0.95, returning the inlet gas relative humidity to 0.99, but decreasing the initial bed temperature and the inlet gas temperature to 35°C leads to better performance. Note that μ_{FT} is equal to 0.9 at 35°C, so use of this gas temperature does not unduly slow growth. The bed conditions are predicted to remain at near optimal values for growth for the majority of the fermentation (Figs. 25.6(e) and 25.6(f)).
- Maintaining the conditions of the previous simulation but reducing the inlet gas temperature even further, to 33°C, leads to even better performance. Note that μ_{FT} has a value of 0.8 at 33°C, so growth is still acceptable. As shown by Figs. 25.6(g) and 25.6(h), the bed conditions are predicted to give values of μ_{FT} and μ_{FW} of at least 0.8 for almost the entire fermentation.

25.4. Conclusions on Intermittently-Mixed Forcefully-Aerated Bioreactors

It is possible to propose the following general principles for beds that are mixed infrequently so as not to cause overly much damage to the fungus:

- Water limitations will not be an overly large problem in intermittently agitated beds, even with relatively few mixing events. Water limitations can still occur in specific regions of the bed, so care will need to be taken in the selection of the criterion for initiating mixing events, that is, a criterion should not be chosen that allows large regions of the bed to suffer from water limitation for significant periods. In the case study, water limitations could be greatly reduced by increasing the outlet gas relative humidity set point.

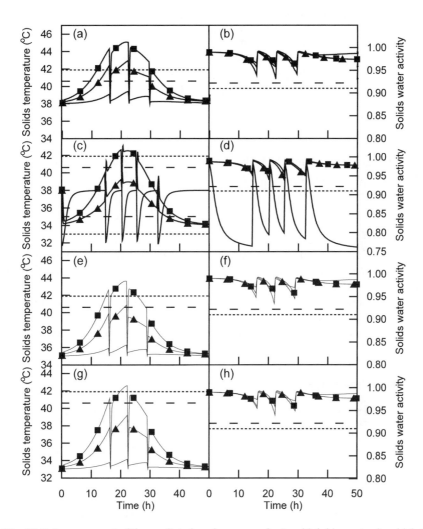

Fig. 25.6. Improvement of the predicted performance of a 1 m high bioreactor in which the air flux is maintained at 0.1 kg-dry-air m^{-2} s^{-1}. The parameters for each simulation are as given in Tables 25.1, 25.2, and 25.3. **(a)** Solids temperatures and **(b)** water activities when the outlet gas relative humidity set point is changed to 0.95; **(c)** Solids temperatures and **(d)** water activities when the outlet gas relative humidity set point is 0.95 and the inlet air humidity is 0.75; **(e)** Solids temperatures and **(f)** water activities when the outlet gas relative humidity set point is 0.95, the inlet air humidity is 0.99, and the inlet air temperature is 35°C; **(g)** Solids temperatures and **(h)** water activities when the outlet gas relative humidity set point is 0.95, the inlet air humidity is 0.99, and the inlet air temperature is 33°C. Key to graphs (a) to (h): Fractional heights of (——) 0.025; (—▲—) 0.5; (—■—) 1.0. The horizontal lines represent temperatures and water activities that give fractional specific growth rates of (- - -) 0.9 and (····) 0.8

- Temperature limitations will more difficult to prevent, since infrequent mixing events do little to control the temperature in the bed, with the temperature gradient that was present before the mixing event being re-established relatively quickly after the mixing event (the speed at which it is re-established depending on the bed height and the flux of air). To minimize temperature limitations, it will be necessary to optimize the combination of air flux, bed height, and inlet air temperature. Of course, the optimal combination will also depend on the specific growth rate of the organism. There appears to be relatively little advantage in blowing unsaturated air into the bioreactor, as this will cause large moisture gradients within the bed during the periods of static operation.

The operation of intermittently-mixed forcefully-aerated bioreactors certainly deserves more attention, as they represent a middle ground between the advantages and disadvantages of continuous mixing. Attention needs to be given to the following issues:

- determination of heat and mass transfer coefficients between the solids and gas phases. For example, in the case study shown in this chapter, the coefficients were borrowed from a study of the drying of corn;
- determination of the deleterious effects of mixing on the organism, and how it recovers after a mixing event;
- optimization of the agitation regime (frequency, duration, and intensity).

Further Reading

Models of intermittently-mixed forcefully-aerated bioreactors
Ashley VM, Mitchell DA, Howes T (1999) Evaluating strategies for overcoming overheating problems during solid state fermentation in packed bed bioreactors. Biochem Eng J 3:141–150
von Meien OF, Mitchell DA (2002) A two-phase model for water and heat transfer within an intermittently-mixed solid-state fermentation bioreactor with forced aeration. Biotechnol Bioeng 79:416–428

26 Instrumentation for Monitoring SSF Bioreactors

Mario Fernández and J. Ricardo Pérez-Correa

26.1 Why Is It Important to Monitor SSF Bioreactors?

Solid-state fermentation plants, like many other processing facilities, are subjected to several kinds of disturbances, such as variations in raw materials, environmental changes, equipment failure, and human errors. If these disturbances are not detected and compensated properly, they will affect the operation and performance of the SSF plant, resulting in accidents, products out of specification, or loss of production. Hence, in order to operate the process reproducibly and safely, some relevant process variables should be measured (measured variables) and some others should be modified (manipulated variables), continuously or at least periodically. In this chapter we give some guidance on how to select relevant measured variables and about which instruments can be used to monitor SSF bioreactors. We will also discuss what to do when no instrument is available to measure a given variable and how to deal with noisy measurements. The selection of manipulation variables for control schemes is left for Chap. 28.

We need to define our specific objectives in putting instrumentation on the bioreactor before selecting the variables to be measured and the instruments to make these measurements. In SSF bioreactors we should ensure that the microorganism grows well and produces the expected quantities of the desired metabolite. To achieve these in the specific case of packed-bed SSF bioreactors operated aseptically in batch mode, for example, we have to stop the process at the right time, detect undesirable heterogeneity in the solid bed, and avoid contamination by other microorganisms. We will emphasize the particular challenges that must be faced when making measurements inside a heterogeneous bed of solid particles.

26.2 Which Variables Would We Like to Measure?

Here we will discuss the variables that must be measured and calculated on-line to achieve what was stated above and how these variables relate with the process objectives.

Bed temperature. This is the most important and easy to measure variable inside the solid bed. The main difference with a submerged fermentation is the heterogeneity of the bed. Therefore, several temperature sensors should be placed inside the solid bed to get a reliable estimation of its average temperature. Growth and production rates can be seriously affected if hot spots inside the bed stay undetected, reducing the performance of the bioreactor.

Bed water content. This variable is related with the water activity of the solid medium, which has a strong influence on the growth and production rates of the microorganism. Water content should be measured periodically to avoid bed over-drying, especially during the exponential growth phase.

Bed porosity. During the fermentation, the growth of filamentous fungi reduces bed porosity, lowering CO_2 and O_2 transfer between the gas and the solid bed and also limiting metabolic heat removal. Hence, detecting inadequate bed porosity is a major concern in operating static or periodically-agitated packed-bed SSF bioreactors.

Inlet air conditions. Practically the only effective way to regulate bed temperature in large SSF bioreactors is through manipulation of the inlet air conditions (see Chap. 29). As part of the control of these conditions the temperature, flow rate, and humidity of the inlet air stream need to be measured.

CO_2 and O_2 concentration (respirometric variables). The evolution of exhaust gases is very rich in information about how well the bioreactor is performing. Respiration gases can tell us about the physiological state of the culture and allow us to detect if the microorganism is under stress. CO_2 and O_2 concentrations and the overall gas flow rate are used to compute important on-line derived variables such as CO_2 Evolution Rate (CER), O_2 Uptake Rate (OUR), and the respiratory quotient (RQ) (Sato et al. 1983; Sato and Yoshizawa 1988; Ooijkaas et al. 1998). With these variables we can estimate the specific biomass growth rate on-line and assess the bioreactor operation.

Volatile metabolites. Monitoring volatile metabolites in the gas outlet stream can provide an indication about the production rate (González-Sepúlveda and Agosin 2000) and metabolic activity or help in defining operating policies to improve process reproducibility (Christen et al. 1997; Bellon-Maurel et al. 2003).

pH. As in submerged fermentation, optimum growth requires a defined pH for each microorganism. However, measuring pH inside solid beds is difficult and unreliable. The heterogeneity of the solid medium poses an extra difficulty. As pH cannot be measured on-line reliably nor controlled, it is common practice to prepare the substrate with a buffered solution, so that pH does not change too much during the fermentation.

Biomass concentration. Despite being the most important variable for assessing the state of the culture, no reliable method is available for continuous measurement of biomass in industrial SSF bioreactors, contrary to the situation in submerged fermentation. Monitoring of the respiratory gases is useful to infer how well the biomass is growing. Note that the heterogeneity of the conditions inside the solid bed that is typical of static SSF processes will cause the microorganisms to grow unevenly across the bed.

Substrate concentration. Establishing the initial concentration of the limiting substrate is usually enough to ensure normal operation in batch bioreactors. However, in the production of secondary metabolites it may be necessary to measure the limiting substrate concentration periodically so that optimal fed-batch policies can be established. In SSF this might be difficult, especially if the substrate contains several carbon or nitrogen sources that the microorganism is capable of using.

Product concentration. As in every industrial process facility, determination of product concentration is critical in SSF plants for optimal productivity and quality assurance. Furthermore, in certain cases the resulting product might be degraded (as is GA_3 within aqueous phases), consequently off-line monitoring of product concentration is necessary to establish the optimum fermentation time.

26.3 Available Instrumentation for On-line Measurements

The most widely used instruments that allow us to measure some of the above variables will be discussed in this section. We will focus on how to measure in order to get the best representation of the bed conditions, on describing the advantages and limitations of the specific instruments, their reliability, expected precision, and relative cost. It is important to remember that any probe located inside the solid bed of a stirred reactor will require special care to avoid damage from the solids movement during the agitation periods.

Temperature. Any SSF control system should include several temperature measurements to monitor the bed temperature distribution and to measure inlet and outlet air temperature so that energy and water balances are kept under control. Several inexpensive temperature sensors are commercially available, thermocouples (TC's) being the most widely used within industrial environments. Their low cost, wide measuring range, and fast and linear response explains their popularity among process engineers; however, thermocouples show poor precision and accuracy. Even though the performance of thermocouples in SSF bioreactors can be significantly improved if they are calibrated frequently, other temperature sensors are superior. Resistance Temperature Detectors (RTD's) are better suited for SSF processing needs. These devices are stable, precise, respond fast, and do not need periodic calibration, even though they are more expensive and fragile than thermocouples. For a complete description of these and other temperature sensors, the reader is encouraged to consult specialized literature (Lipták 1995a; Creus 1998a).

Bed water content or water activity. None of the traditional methods commonly used to measure the water content of solid samples have gained wide acceptance in the SSF field since they are time consuming (needing between 2 and 15 hours)

(Creus 1998b). In addition, capacitance or conductivity-based devices are too sensitive to the electrodes/sample contact area or the apparent density of the sample (Gimson 1989), which normally varies during the fermentation. If properly calibrated and temperature-compensated, such devices may be useful to measure the water content of the solids phase in static packed-bed reactors, but they are limited to a maximum of 50% water content (Creus 1998b) which is too low for the majority of SSF processes. New methods like IR analyzers or IR scales are preferred, since they give a quick and precise indication of the humidity of a sample taken from the process (Fernández et al. 1996; Durand et al. 1997). However, these methods are not appropriate for automatic control of the humidity of the bed of solids, since they require the intervention of an operator to handle the samples. Optic absorbance sensors are a good option for automation in this case, since they provide water activity measurements in 2-3 min with a precision of 0.3% (Bramorski et al. 1998; Bellon-Maurel et al. 2003). Other commercially available humidity measuring devices such as infrared or neutron radiation sensors are not only expensive, but they are also impractical to use in SSF reactors and provide information regarding the surface of the solids only (Brodgesell and Lipták 1995; Creus 1998b). More promising are those based on the emission of radio frequency fields or on Time Domain Reflectometry (TDR) (Bellon-Maurel et al. 2003), since they compute a representative value of the water content in the 3-D zone covered by the electrodes (SCI 1996; SE 1998; Hillen 1999; Atkinson 2000).

Gas flow rate. Several methods exist for measuring both volumetric and mass gas flow rate. Those based in pressure drop (Pitot tube and annubar, Venturi, and orifice flow meters), variable area (rotameters), speed (anemometers, turbines), force (badge meter) and vortex (Von Karman effect) are used to measure volumetric flow rate. These techniques can be adapted to measure mass flow rates also, in which case pressure and temperature compensation is required. In addition, thermal methods (based on the temperature difference between two resistance probes) and Coriolis force (vibrating tube) are specific for mass flow measurements (Creus 1998c). Although there are many options, it is rather difficult to select an adequate flow meter, since not all of them are applicable in a given system due to space, cost, pressure drop, and precision constraints (Lomas and Lipták 1995). Moreover, in some cases a flow rectification device would have to be installed before the sensing probe (Siev et al. 1995). Hence, it is advisable to define the kind of flow meters that cover the measurement range of interest first and then to identify those that better suit the intended application, considering the maximum allowable error, the pressure and temperature that the sensor will be exposed to, and the type of flow (laminar, turbulent, or transition). The lowest cost sensor should then be chosen from this short list. In assessing costs, in addition to the purchase price, the costs of maintenance, spare parts, and sensor operation must be considered, since at times an appropriate instrument can be cheap to buy, but in the long term the other costs can make it impractical. For further details see (Lipták 2002). Table 26.1 (adapted from Cole-Palmer 2003a,b) can be useful in choosing a suitable flow meter.

Table 26.1. Flowmeter characteristics

Attribute→ Flowmeter↓	Accuracy $(\pm\%)^{a,b}$	Repeatability $(\pm\%)^{a,b}$	$P_{max.}$ (psi)	T_{max} (°C)	Reactor size[c]	Average cost (US$)
Pressure drop	2-3[a]	1[a]	100	50	P or I	500-800
Variable area	2-4[a]	0.05-0.15[a]	≥ 200	≥120	L or P	200-600
Turbine	0.25-1[b]	0.1[b]	≥ 5000	≥150	P or I	600-1000
Vortex	0.75-1.5[b]	0.2[b]	300-400	260	L or P	800-2000
Gas mass flow	1.5[a]	0.5[a]	≥500	≥65	L	600-1000
Coriolis	0.05-0.15[b]	0.05-0.10[b]	≥900	≥120	L or P	2500-5000

[a] % of full scale.
[b] % of reading.
[c] L = laboratory, P = Pilot, I = Industrial.

pH. The most accurate and versatile among the many pH measuring methods are the glass electrode and the Ion Sensitive Field Effect Transistor (ISFET). Glass electrodes consist of a glass tube divided by a membrane (also made of glass) that is especially sensitive to hydrogen ions. Their fragility is their main limitation, as the membrane can be broken easily. On the other hand, the transistor electrode ISFET is practically unbreakable, very reliable, and responds quickly (Creus 1998d). The direct measurement of pH in porous solid substrates is not reliable due to poor contact between the solid and the sensitive part of the electrode. Although there are flat electrodes that adapt better to solid samples, their applicability is limited to static beds because agitation may damage the electrode. In this case the measurement cannot be extrapolated to the rest of the bed due to its heterogeneity (Mitchell et al. 1992).

Bed porosity. This effect can be assessed on-line by monitoring the pressure drop through the solid bed (Auria et al. 1993; Villegas et al. 1993; Bellon-Maurel et al. 2003). For example, in static bioreactors pressure drop measurements can be used to keep the inlet air flow rate under control, or in periodically agitated bioreactors, to establish the mixing intervals. Although Bourdon tubes are widely used pressure sensors since they are simple, inexpensive, and reliable, they are neither fast nor precise. On the other hand, piezoelectric sensors can be used to measure within a wider range, their response time is extremely short, and they are insensitive to temperature (Lipták 1995b).

Off-gas analysis. Off-gas analysis can be performed by Gas Chromatography (GC) (Saucedo-Castañeda et al. 1992; Saucedo-Castañeda et al. 1994) or by specific gas analyzers (Smits et al. 1996; Fernández et al. 1997). In both cases, in order to get meaningful results, care must be taken to keep the flow rate regulated and to dry the air sample before it enters the instrument. GC is sometimes preferred over specific analyzers since many compounds, in addition to CO_2 and O_2, can be monitored with the same instrument and over a wider range of values. However, each analysis takes several minutes. On the other hand, gas analyzers are more precise and have fast response times (of a few seconds). These devices make use of chemical or physical properties, like paramagnetism or infrared absorption, which characterize the measured gases. Paramagnetic analyzers, available for O_2, are probably the most effective (Kaminski et al. 1995; Creus 1998e). Although expensive, they are very precise, do not require periodic calibration, present low interference with other gases (if water vapor is removed) and last long. These instruments exploit the property that some gases have of being magnetized when they are exposed to a magnetic field. Electrochemical analyzers are also commonly used to measure O_2 concentrations, since they are low cost and provide good precision; however, the measuring cell must be changed periodically (one or two times per year). CO_2 can be measured reliably with infrared instruments, which are precise and have a long lifetime, though they are expensive and need occasional calibration (Creus 1998e). These analyzers use the capacity that CO_2 has to absorb infrared radiation within a characteristic spectrum.

Volatile metabolites. Traditional analytical methods use special resins to trap volatiles from a gas stream (Sunesson et al. 1995) that are then analyzed by gas chromatography coupled with mass spectrometry (GC/MS). For example, González-Sepúlveda and Agosin (2000) used this technique to record the evolution of ent-kaurene in the outlet gas stream and relate this with the production of gibberellic acid GA_3 in a pilot scale SSF reactor. There are also special devices known as "artificial noses" that are able to detect on-line particular chemical compounds in a mixture of gases. Although some applications of artificial noses in SSF processes have been reported (Wang 1993), their use has not spread yet, mainly due to a poor selectivity, slow response, and high sensitivity to environmental conditions (Bellon-Maurel et al. 2003).

Figure 26.1 shows a periodically agitated SSF bioreactor (Fernández 2001) that incorporates various of the monitoring devices discussed above.

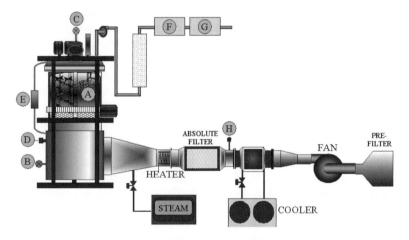

Fig. 26.1. Instrumentation of the SSF bioreactor at Pontificia Universidad Católica de Chile. A: six thermocouples (bed temperature); B: thermocouple (inlet air temperature); C: thermocouple (outlet air temperature); D: Relative humidity transmitter (inlet air relative humidity); E: pressure drop transmitter; F: CO_2 IR detector (CO_2 concentration); G: O_2 paramagnetic detector (O_2 concentration); H: anemometer

26.4 Data Filtering

All measurements are contaminated by some kind of errors (systematic biases, gross errors, and noise) that interfere with process operation and with data analysis for model development and process supervision. When the instrument is not properly calibrated, the average value of a series of measurements does not correspond with the true value of the measured variable. In this case we talk about a systematic bias. Gross errors are caused, for example, by malfunctions of the measuring system, providing unrealistic values of the measured variable. Noise can be classified as (see Fig. 26.2): *high frequency noise,* associated with intrinsic limitations of any instrument that cannot produce exactly the same value after a series of independent measurements, even if the measured variable is kept constant; *medium frequency noise,* due to process heterogeneity (turbulence and poor mixing); and *low frequency noise*, caused by process disturbances (environmental conditions, metabolic heat, bed drying etc.). The latter can be reduced by automatic process control, but high-to-medium noise should be reduced by signal processing, that is, by "data filtering".

In large-scale SSF bioreactors the solid bed is highly heterogeneous and its characteristics are time-varying (water content, biomass content, solid-gas interphase area, porosity etc.), hence it is difficult to infer its average conditions directly from the measurements. These effects mean that typical on-line readings, such as temperatures, gas flow rate or relative humidity, show significant noise and, during the agitation period, many outliers (gross errors), which appear due to the liberation of occluded gas and the electric interference of motor drives. Therefore, data processing is of utmost importance to operate this kind of bioreactor well (Peña y Lillo 2000). This is especially important if advanced control techniques have been implemented, since these control algorithms do not work without reliable process models, which in turn are obtained from good quality process data.

Here we will discuss simple filtering algorithms that can enhance the reliability of noisy measurements significantly.

Although the signals produced by the instruments are continuous (analog), usually control calculations are performed by digital microprocessors that can only operate with digital (discrete) signals. Hence, the analog signal provided by the instrument should be converted to a discrete signal, that is, a signal that has its values reported at regular time intervals. The time interval between two values is known as sampling time and should be provided by the Process Engineer. If the sampling time is too short, the computer control system will be overloaded. Therefore, the sampling rate will be bounded by the processing speed of the control device and by the number of control loops that the control device is handling. On the other hand, if the sampling time is too long, the converted discrete signal will not reproduce the real process dynamics given by the original analog signal. Hence, a compromise must be met and usually the Shannon theorem (Aström and Wittenmark 1984) is used to find a lower limit to the sampling time. Table 26.2 shows common values of sampling times used in practice.

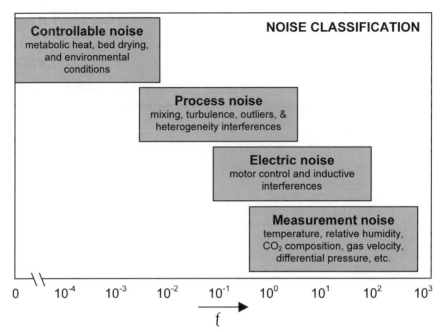

Fig. 26.2. Noise classification

Table 26.2. Typical values of sampling times in the process industries

Variable "y"	Sample Time (s)
Flow	1
Pressure	5
Level	10
Temperature	20

Sampling does not eliminate the noise from the original analog signal; hence digital signals should be filtered. Digital filters are mathematical procedures that process digital signals on-line to reduce their noise and represent better the true dynamics of the measured variable.

In practice, for filtering purposes a high frequency sampling rate is used, known as "time scan". The simplest way to reduce high frequency noise is averaging. If we consider N measurements between two time instants, the average value, $\widetilde{y}_N$, will be defined by,

$$\widetilde{y}_N = \frac{1}{N}\sum_{i=1}^{N} y_i , \tag{26.1}$$

where y_i represents the noisy value measured at time i.

It is easy to implement this formula in a computer using a recursive version,

$$\tilde{y}_i = \tilde{y}_{i-1} + \frac{1}{i}(y_i - \tilde{y}_{i-1}).$$ (26.2)

Here, the correction gain ($1/i$) is much lower for the most recent measurements; hence these will not influence much the average value. This formula is effective if the true signal is constant, however, in practice it is better to use a constant correction gain if the process variables are continuously changing, a situation that is especially true in the batch processes that are typically used in SSF.

$$\tilde{y}_k = \tilde{y}_{k-1} + \frac{1}{\alpha}(y_k - \tilde{y}_{k-1})$$ (26.3)

The gain inverse, α, corresponds approximately to the number of measurements used in the averaging. This filter is also known as a "first-order low-pass filter" or "moving average filter". Obviously, the larger the value of α, the smoother the filtered signal will be. Care must be taken though, since too much filtering can hide the true process dynamics.

Averaging is also a good policy if we are taking samples for off-line analysis. Here, it is convenient to take samples from different places inside the bed, and then mix them all and do the analysis, or analyze each sample independently and then take the average. The advantage of the latter is that we can have an estimation of the heterogeneity of the bed also.

26.5 How to Measure the Other Variables?

When no instrument is available to measure the desired variable on-line at a reasonable cost and within a reasonable time, we have to devise an alternative to keep track of relevant unmeasured variables. For example, it is not possible to measure product concentrations on-line fast enough to be included in a control loop. This is so since these components are typically absorbed in the solid matrix and they must be extracted before analytical determination. Hence the whole measuring procedure can last several hours. *Soft-sensors* or "state observers" are a useful alternative in these cases (Montague 1997). Here, a process model and measurements that are somehow related with the unmeasured variable are used to provide an on-line estimation. These are then processed by a Kalman Filter.

The Kalman Filter is widely used in submerged bioreactors for indirect measurements of specific consumption or production rates, yields, and heat loads. In SSF it has been mainly applied in lab scale bioreactors for biomass, bed water content and heat load estimations. Although this filter was originally developed for linear models, it has been extended to deal with nonlinear systems also.

The Kalman Filter is an on-line data processing algorithm that provides optimal estimation of output process variables, as shown in Fig. 26.3. The main reasons why we need a data processing algorithm are:

- measurements are contaminated by errors and they do not describe completely the state of the process;
- process models provide only an approximation to the real process behavior;
- the process is subjected to disturbances, which cannot be modeled or controlled.

Therefore, given (1) an approximate dynamic linear process model, (2) incomplete and noisy measurements, and (3) statistical information regarding measurement and model errors, the Kalman Filter provides the best linear estimation of the true values of the process outputs, even if some outputs are not measured or the process is subjected to disturbances. The obtained estimation is a compromise between the estimation provided by the model and the measured value. If we have confidence in the model, the optimal estimation will be obtained by putting more weight on the model outputs; on the contrary, if we trust the measurements more, the optimal estimation will be obtained by putting more weight on the measured values. The derivation of the algorithm is rather involved (Welch and Bishop 2003), therefore we will describe only its most basic principles.

Let us consider the following linear dynamic model relating several inputs with several outputs:

$$x_k = A \cdot x_{k-1} + B \cdot u_{k-1} + w_{k-1}$$
$$y_k = C \cdot x_k + v_k$$

(26.4)

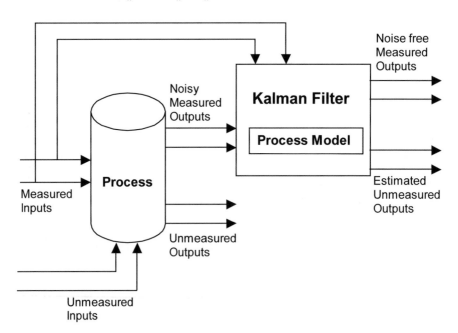

Fig. 26.3. Kalman Filter application

In Eq. (26.4) y_k is the vector of measured variables (outputs), u_k is the vector of manipulated variables (inputs), and w_k and v_k are pure random vector signals, with an average of zero, that represent the process and measurement noise respectively. The above discrete model is written in state space form, where the state vector, x_k, represents the minimum information that we require to predict the system evolution for a given sequence of future inputs. The matrices A, B, and C are of proper dimensions.

The Kalman filter gives the optimum estimation of the true state of the system, given the known covariances, Q and R, of the process and measurement noise vectors. Figure 26.4 summarizes the main algorithm. Here, the corrected state estimate is a compromise, defined by the Kalman gain, between the model estimates and the measured values. The larger the covariance of the process noise, Q, the larger the Kalman gain, therefore the model predictions are weighted less. On the contrary, the larger the measurement noise covariance matrix, R, the smaller the Kalman gain, so model predictions are corrected less.

Even though the Kalman filter is widely used in robotics and related fields to filter measurement noise, in bioprocesses it is more useful for filtering process noise or estimating unmeasured variables like biomass content or metabolite concentrations. Compared with electro-mechanical systems, bioprocesses are slow; hence a high time scan can be used to filter measurement noise with simpler low pass filters. Moreover, to estimate an unmeasured variable once the state estimates

Predictive Equations

(1) Estimate the new state using the model only

$$\hat{x}_k^- = A \cdot \hat{x}_{k-1} + B \cdot u_{k-1}$$

(2) Estimate the new state covariance matrix

$$P_k^- = A \cdot P_{k-1} \cdot A^T + Q$$

Corrective Equations

(1) Compute the correction gain (Kalman gain)

$$K_k = P_k^- \cdot C^T \cdot \left(C \cdot P_k^- \cdot C^T + R \right)^{-1}$$

(2) Correct the state estimate using the new measurement

$$\hat{x}_k = \hat{x}_k^- + K_k \cdot \left(y_k - C \cdot \hat{x}_k^- \right)$$

(3) Correct the state covariance matrix

$$P_k = \left(I - K_k \cdot C \right) \cdot P_k^-$$

Fig. 26.4. Kalman Filter algorithm

are corrected is simple. This is so because the process model is linear and any un-measured variable should be a linear combination of the states. If the process model is nonlinear, a linearized model should be derived and the matrices A, B, and C updated at each sample time. This is known as an Extended Kalman Filter and a detailed discussion is beyond the scope of this chapter. The interested reader is encouraged to visit the web page of Welch and Bishop (Welch and Bishop 2003), where it is possible to find a lot of reading material, Matlab programs to experiment with and also many interesting links.

Further Reading

A deeper treatment of the instruments discussed in this chapter and many more
Lipták BG (1995). Instrument engineers' handbook: process measurement and
    analysis. Radnor, Pennsylvania, Chilton Book Co.

An introduction to the basics of digital signal processing
O'Haver T (2001) An introduction to signal processing in chemical analysis. De-
    partment of Chemistry and Biochemistry, University of Maryland, College
    Park, MD 20742. http://www.wam.umd.edu/~toh/spectrum/TOC.html

A rigorous and updated treatment of digital signal processing
Mitra SJ (2001) Digital signal processing. McGraw Hill, New York

Introduction to Kalman Filter theory and the key ideas that underlie it
Welch G, Bishop G (2003) An introduction to the Kalman Filter. TR 95-04, De-
    partment of Computer Science, University of North Carolina at Chapel Hill.
    http://www.cs.unc.edu/~welch/kalman/
Maybeck PS (1979) Stochastic models, estimation, and control. Vol 1, Chap 1,
    New York, Academic Press.

27 Fundamentals of Process Control

J. Ricardo Pérez-Correa and Mario Fernández

27.1 Main Ideas Underlying Process Control

In this chapter we present the basics that the non-engineer needs to know about process control in order to understand the discussion of the application of process control to solid-state fermentation (SSF) bioreactors that will be presented in Chap. 28. Here, the idea of feedback will be introduced as well as the main components of a control loop. In addition, the most used control algorithms in the process industries will be described and some hints on how to simplify their application will be given. Engineers who understand control can go to the next chapter.

27.1.1 Feedback

F*eedback* is a key concept in automatic control and can be broadly defined as a procedure that uses the past response of the system to compute future corrective actions with the aim of improving the system's performance in the presence of uncertainty. Unmeasured external disturbances, unknown process behavior, and noisy measurements are usual sources of uncertainty. In engineering terms, feedback can be viewed as a procedure by which the values of past process measurements (system past response) are used to deduce or compute the values of future operating variables (corrective actions), in order to keep process measurements as close as possible to predefined values named set points. In many SSF processing plants, corrective actions (cooling, heating, aeration, agitation, water addition etc.) are periodically decided upon and undertaken by the process operator, which represents *manual control*. In a*utomatic control*, an engineering device computes such corrective actions continuously or very frequently, without direct human intervention, providing much better performance than manual control. This improved performance is particularly impressive in complex systems that have many sensors and process inputs, where the severe limitations of even skillful human operators are well known.

27.1.2 Control Loop

A simple automatic feedback control system or control loop, as shown in Fig. 27.1, comprises a measuring device called the sensor, a decision device or algorithm called the controller, and a final control device called the actuator. In the control loop, the sensor measures the past process response (measured variable) and sends a signal to the controller. Here the measured value is compared with the set point and the next corrective action is computed using this difference. The computed corrective action is sent to the actuator, which finally changes the process input (operating variable). Of course, this new value of the operating variable is intended to bring the measured variable back towards the set point.

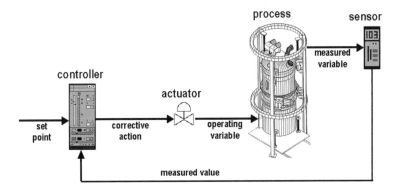

Fig. 27.1. A simple control loop

7.1.3 Computer Control Loop

Usually, the decision device is a computer or digital controller, within which the algorithm that computes the corrective action (control law) is coded. In commercial digital controllers, this algorithm is coded by the manufacturer of the controller and users can only change its characteristic (tuning) parameters. On the other hand, when a computer calculates the control action, the user has the option to code the control algorithm him- or herself, providing much more flexibility. Additional devices are required when a computer is used for automatic control. As seen in Fig. 27.2, the sensor sends a signal (measured value) to an analog-digital (A/D) converter. Here, the continuous electrical signal is transformed into a sequence of equally spaced pulses (discrete signal). The computer periodically compares the pulse value with the set point, performs a calculation, then generates the corrective action as another pulse and sends it to the digital-analog (D/A) converter. Here, the sequence of pulses generated by the computer is transformed into a continuous electric signal. Finally, the analog controller output is sent to the actuator to correct the process behavior.

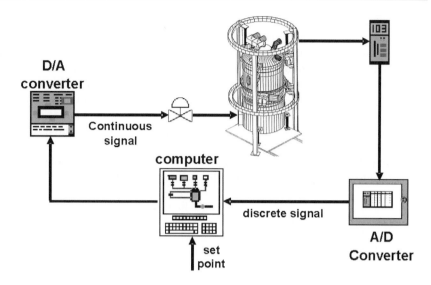

Fig. 27.2. A computer control loop

27.2 Conventional Control Algorithms

On/Off and PID ("proportional integral derivative") are the most frequently used control algorithms in the process industries, and SSF plants are not an exception. Therefore, we will describe below both algorithms in detail and will provide some hints about how to achieve a good performance.

27.2.1 On/Off Control

The simplest algorithm that can be implemented is On/Off control, which explains its widespread use, from home appliances to industrial facilities. This algorithm switches the controller output between two values; the switching is triggered when the measured value crosses the set point, that is, when the error (difference between the measured value and the set point) changes sign. Hence, this algorithm can be written as:

$$If \quad e(t) \geq 0 \quad then \quad u = u_1 \quad else \quad u = u_2$$
$$with \quad e(t) = \text{measured value} - \text{set point value} \qquad (27.1)$$

Here, $e(t)$ is the error computed at time t, u is the controller output, and u_1 and u_2 are the two possible process input values that must be defined by the process engineer. If the difference $u_2 - u_1$ is large, the process will reach the set point fast, but the maximum error will be large and the controller will switch between u_1 and

u_2 very frequently. Hence, proper values of u_1 and u_2 should be tuned on-line to achieve a compromise between speed of response and maximum error.

To illustrate the effect of tuning, here we present two simulations with On/Off control applied to a double pipe heat exchanger as shown in Fig. 27.3. These simulations were generated with a MATLAB™ model provided by Brosilow and Joseph (2002).

In this model, the cold fluid outlet temperature, $T1$, (measured variable) is controlled by manipulating the cold fluid inlet flowrate, $F1$, (operating variable); the controller adjusts the value of $F1$ by moving the valve V-1 (actuator).

We will evaluate the controller performance using a set point step response. Here, the measured variable starts at a given steady state defined by the initial set point value. Next, we suddenly change the set point value to see how the measured value and the operating variable evolve until they reach a new steady state. This is the standard form to evaluate controlled processes in control engineering.

Figure 27.4 shows the process response when the two possible process inputs measured in gallons per minute (GPM) are $u_1 = 0.07$ GPM and $u_2 = 0.12$ GPM, resulting in a difference $u_2 - u_1 = 0.05$ GPM. Although the figure shows that the measured temperature presents small deviations from the reference value, the process takes more than 8 seconds to reach the new steady state, after the set point change at time 15 s. On the other hand, when a difference $u_2 - u_1$ of 0.2 GPM is used, the process takes less than one second to reach the new steady state, however, temperature deviations are larger and the switching between u_1 and u_2 is very frequent (see Fig. 27.5). It is possible to conceive a controller that achieves a fast response and a small maximum deviation, using for example a value of $u_2 - u_1$ of 0.2 when the system is far from the set point and a value of $u_2 - u_1$ of 0.05 when the system is near the set point. However, the controller would not be On/Off anymore, requiring now 4 input values, and being more difficult to implement.

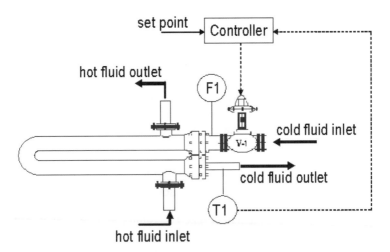

Fig. 27.3. Temperature control in a double pipe heat exchanger

With a little extra cost, as we illustrate below, a proportional (P) controller can achieve a much better performance.

Although effective, low cost, and simple, the On/Off algorithm is not appropriate to control critical variables that should be kept close to the required value (set point) when oscillations of input or output variables are deleterious to the process or when the process responds slowly. In these cases, PID controllers are a good option, therefore they are used in almost all industrial installations; they are the first choice when smooth control action and small deviations from the set point are required.

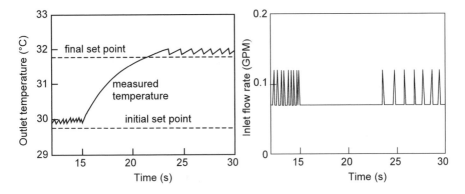

Fig. 27.4. On/Off control with u_1=0.07 and u_2=0.12 GPM. Left: the measured variable slowly reaches the new set point. Right: Operating variable movements are infrequent

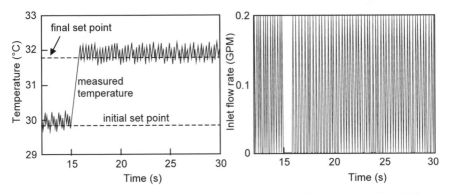

Fig. 27.5. On/Off control with u_1=0.0 and u_2=0.2 GPM. Left: the measured variable responds fast to the set point change and presents large deviations. Right: the operating variable moves very frequently

27.2.2 PID Control

A PID controller produces an output signal that is a linear combination of (1) the error (Proportional action), (2) the integral of the error over time (Integral action), and (3) the time derivative of the error (Derivative action). This algorithm can be represented by the following equation:

$$u(t) = K_c \cdot \left[e(t) + \frac{1}{\tau_i} \cdot \int e(t)dt + \tau_d \cdot \frac{de(t)}{dt} \right].$$

(27.2)

Here, t represents the time, $u(t)$ is the controller output, that is, the signal sent by the controller to the final control element, and $e(t)$ is the controller input, corresponding to the error signal, that is, the set point minus the value of the measured variable. Figure 27.6 represents the terms involved in Eq. (27.2).

The contribution of each term (P, I, and D) to the total control action is shown in Fig. 27.7. Here we can observe that the contribution of the proportional action gets smaller when the error is reduced, therefore, pure proportional control cannot ensure that in the limit (steady state) the measured variable will reach the set point. On the other hand, integral action increases with time; therefore, a non-zero corrective action is applied even if the error disappears. When the system starts moving the derivative action is very high, helping the process to reach the set point sooner.

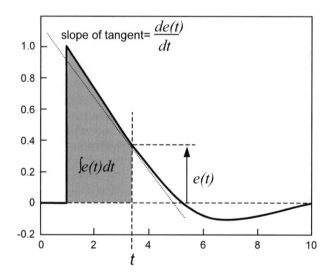

Fig. 27.6. The terms involved in the PID control algorithms. The error, $e(t)$, is the error curve evaluated at time t; the integral term, $\int e(t)dt$, is the (shadowed) area under the error curve; and the derivative term, $de(t)/dt$, is the slope of the error curve at time t

The user can specify the relative contribution of each term in Eq. (27.2) by assigning appropriate values to the parameters K_c (proportional gain), τ_i (integral time), and τ_d (derivative time). This is called *tuning* and many times is not an easy or intuitive task. Control textbooks describe many tuning procedures of varying degrees of difficulty. In this chapter we will present the *relay technique*, which is simple to follow, and we will apply it to the heat exchanger model. This technique, which was proposed by Astrom and Hagglund (1984), uses the process response under On/Off control and the well-known Ziegler and Nichols (1942) tuning rules.

First, the period and amplitude of the process measurement oscillation must be obtained, as seen in Fig. 27.8, which is a magnification of Fig. 27.4.

Then, a parameter called ultimate gain, K_{cu}, must be computed using the following formula:

$$K_{cu} = \frac{4 \cdot (u_2 - u_1)}{\pi \cdot A},\qquad(27.3)$$

where u_2 and u_1 are the On/Off controller values and A is the amplitude of the process output oscillations. From Fig. 27.8 we can see that $A = 0.15°C$, and in Fig 27.4 the value of $(u_2 - u_1)$ is 0.05 GPM, therefore, the ultimate gain is $K_{cu} = 0.42$ GPM °C^{-1}. We also need to get from the figure the ultimate period, Pu.

Ziegler and Nichols (Z-N) tuning rules provide a reasonably good performance for proportional (P), proportional/integral (PI), and proportional/integral/derivative (PID) controllers. We will illustrate their performance below using the heat exchanger model.

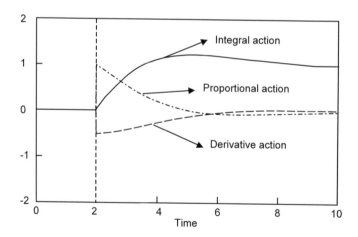

Fig. 27.7. Contribution of each term to the total control action. Proportional action gets small as the error disappears. Integral action takes on large values even when the error is zero. Derivative action only appears when the system is moving

For proportional control (P control), the Z-N rules suggest that $K_c = K_{cu}/2 = 0.21$ and the simulation results of this tuning are shown in Fig. 27.9; here τ_i is infinite and τ_d is zero.

The controller performance shown in the graph can be improved. On the one hand, the measured variable does not reach the new set point; this difference in the steady state is called *offset*. On the other hand, the manipulated variable moves too strongly, which can be deleterious to the actuator or to the process. As seen in Fig 27.10, other K_c values can be tried to improve this performance.

Here we see that, with proportional control alone, we cannot reduce the offset and smooth the control action at the same time. If we reduce the proportional gain, a smooth control action is achieved, but the offset increases. On the contrary, a larger gain reduces the offset but at the expense of an oscillating control action.

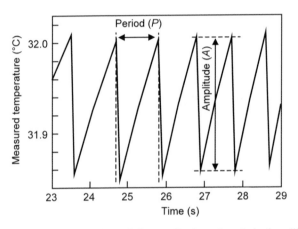

Fig. 27.8. Determination of the amplitude and period of oscillation of the measured temperature under On/Off control

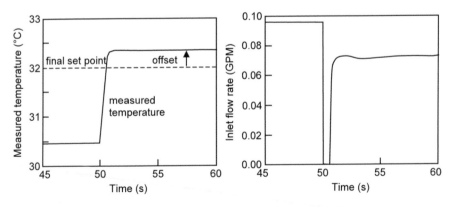

Fig. 27.9. Proportional controller performance using the Z-N tuning rule

Control engineers do not like oscillations, since they reduce the lifespan of the final control elements. In addition, if an even larger proportional gain is applied, the control system will get unstable with oscillations of growing amplitude. The offset can be eliminated with a smooth control, but we need integral control action for this.

The Z-N tuning rules for proportional-integral control (PI control) suggest that $\tau_i = 0.83 \cdot Pu$ and $K_c = 0.45 \cdot K_u$. Therefore, in this case the tuning is given by $K_c = 0.189$ and $\tau_i = 0.955$. Fig. 27.11 shows the response of the controlled system using these values.

Here, the offset is eliminated. This tuning can be improved even further by trial and error (this is called *manual fine tuning*), which is always advisable to do in a real plant to adapt the performance of the controller to the process specifications. For example, a lower proportional gain can be tried in this case to get a smooth control action and reduce the *overshoot* (which is the difference between the maximum value of the process response and the set point), although this will cause a slower response of the controlled process.

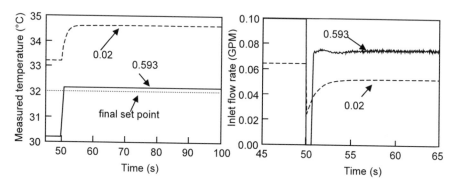

Fig. 27.10. Proportional controller with other K_c values (shown in the figure)

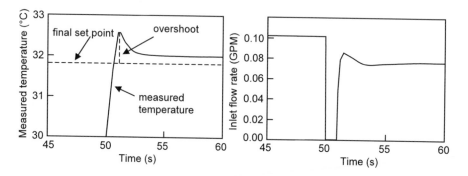

Fig. 27.11. Proportional Integral controller performance using Z-N tuning rules

Derivative control action can improve the controller performance in slow processes. In this example, derivative action is not justified since the process responds fast to changes in the manipulated variable. Derivative action becomes beneficial if we include in the model the dynamics of a temperature sensor with a time constant of 0.3 s. Roughly speaking, the time constant tell us how fast the system responds, which is approximately 5 times the time constant. Therefore, the temperature sensor approximately takes 1.5 s to reach a new steady state after a disturbance. The ultimate gain of the modified model takes on the value 0.212 [GPM °C^{-1}] and the ultimate period the value 1.2 s. Hence, the PI Z-N tuning results in $K_c = 0.0954$ and $\tau_i = 0.996$. For proportional-integral-derivative tuning (*PID tuning*), the Z-N rules state that $K_c = 0.6K_u$, $\tau_i = 0.5$Pu, and $\tau_d = $ Pu/8; for our example this results in $K_c = 0.126$, $\tau_i = 0.6$, and $\tau_d = 0.15$. The response of both settings is shown in Fig. 27.12.

The figure shows that the derivative action reduces the oscillations, so that the measured variable settles at the set point value earlier than with PI control. The difference does not appear very marked in this case, but will be more noticeable in slower processes. However, care must be taken with noisy measurements, since the derivative action can cause undesirable oscillations and instability. In these cases, a low pass filter as described in Chap. 26, is essential.

PID is a good option for industrial control in standard, simple applications. However, the tuning of PID controllers can prove difficult when a process shows complex dynamics, interaction between different control loops, and large and variable time delays, such as occur in large scale SSF bioreactors. In these cases, even periodic attention and retuning do not assure good performance. Model Predictive Control (MPC), also known as Model Based Control, is a popular option nowadays in the process industries since it overcomes most of the limitations of PID control. We will present next the basics of this technique.

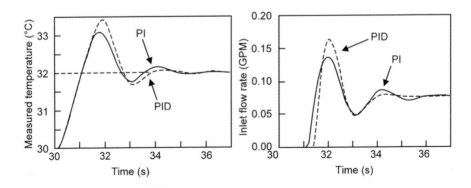

Fig. 27.12. Comparison of PI and PID controllers with Z-N tuning in the case where the temperature sensor has a time constant of 0.3 s

27.2.3 Model Predictive Control

These control algorithms use a process model to predict future outputs, based on past input and output values. At each sampling time, future control movements are calculated that minimize a weighted function of predicted deviations from the set point and control movements. The general algorithm explicitly includes constraints on process inputs and outputs. In addition, MPC can be designed to include a different number of manipulated and controlled variables. While the implementation of these controllers in a real process is difficult and time consuming, they can operate without supervision over long periods. MPC would therefore be particularly useful to control large-scale SSF bioreactors. These algorithms have been widely presented and discussed in standard process control texts (Ogunnaike and Ray 1994) and specialized books (Maciejowski 2002).

The minimization problem can typically be formulated in the following form:

$$\min_{\Delta u_k, \dots, \Delta u_{k+C-1}} \sum_{i=1}^{P} (\hat{y}_{k+i} - r_{k+i}) \cdot W_D^y \cdot (\hat{y}_{k+i} - r_{k+i})^T +$$

$$\sum_{i=1}^{C} \Delta u_{k+i-1} \cdot W_D^{\Delta u} \cdot (\Delta u_{k+i-1})^T$$

s.t.

$$u_L \le u_{k+i-1} \le u_U ; \quad \forall i = 1, \dots, C$$
$$y_L \le \hat{y}_{k+i} \le y_U ; \quad \forall i = 1, \dots, P$$
$$\Delta u_{k+i} = 0 ; \quad \forall i = C, \dots, P$$

(27.4)

Here, $\hat{y}_k$ is the vector of predicted plant outputs at time interval k, r_k is the vector of respective set points, and Δu_k the vector of control moves. The two matrices W_D (one with superscript "y" and the other with superscript "Δu") are diagonal matrices with weights that penalize the output deviations from the set points and the control movements respectively.

The process engineer can make some manipulated variables move more than others and get smaller deviations in some specified plant outputs, by adequately tuning the weights. In addition, the prediction horizon, P, defines the period over which the cost function will be minimized (see Fig. 27.13(a)); a large P assures a smooth and stable performance of the controller and should cover over 80% of the *settling time* (the time taken by the measured variable to settle into a new steady state value after an operating variable is moved). In turn, the control horizon, C, establishes the length of the sequence of future control moves (see Fig. 27.13(b)); heuristics suggests that $C \ll P$. The minimization problem also includes constraints on inputs and outputs, therefore, in the expression above, u_L and y_L represent the lower bounds, and u_U and y_U represent the upper bounds. This optimization problem does not have an analytical solution, except when a linear predictive model is used and no constraints are included, thus in the general case, the problem should be solved numerically.

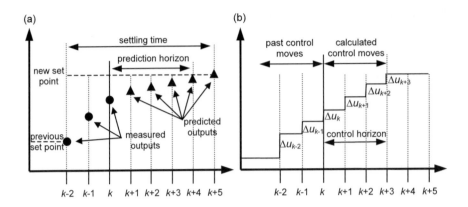

Fig. 27.13. Model predictive control. (a) The deviations between the set point and the future outputs over the prediction horizon are minimized; this minimization is performed at each sample time k. (b) The algorithm calculates all control moves in the control horizon at each time k, although it applies only Δu_k. Next, at time $k+1$, all the calculations are repeated

In next chapter several simulation results are presented with applications of MPC to the control of SSF bioreactors. Some guidance on how to tune the algorithm is given and it is also illustrated how MPC can help to overcome the difficult dynamic behavior of SSF processes.

Further Reading

Intuitive and very simple explanations about PID control can be found in the following Internet addresses
http://abrobotics.tripod.com/ControlLaws/PID_ControlLaws.htm
http://www.control.com/943999705/index_html

The following book provides a simple introduction to process control and at the same time, a reasonably complete treatment of the subject.
Smith CA, Corripio AB (1985) Principles and practice of automatic process control. John Wiley & Sons, New York

28 Application of Automatic Control Strategies to SSF Bioreactors

J. Ricardo Pérez-Correa, Mario Fernández, Oscar F. von Meien, Luiz F.L. Luz Jr, and David A. Mitchell

28.1 Why Do We Need Automatic Control in SSF Bioreactors?

Here, we will discuss what is important to control in SSF bioreactors and how it can be done. The simulations at the end of this chapter show that performance of SSF bioreactors will be better with control schemes; for example, when the temperature in the inlet air is decreased in response to a rise in the average temperature in the bed, or when water is added in response to a drop in the humidity of the off-gases.

As with classical submerged liquid fermentation, in SSF processes there is an optimal set of conditions that will lead to maximal cell growth and metabolite production. As discussed in Sect. 5.1, two of the key process variables are the temperature and water content of the solid bed. In order to maximize the performance of industrial-size SSF bioreactors, these variables must be maintained close to the optimum values for growth and product formation.

In small-size cultures the ratio of the heat transfer capacity of the bioreactor to the metabolic heat generation rate is high; consequently the metabolic heat can be dissipated effectively. Under these conditions, it is straightforward to keep the culture reasonably homogenous. However, this is much more difficult to achieve in industrial or pilot size SSF bioreactors (See Chap. 5). If operated without control, temperature differences between different regions of the solid bed can be as high as 20°C, resulting in disappointingly low productivity, the growth of contaminants, or complete failure of the fermentation run.

Manual control may be useful to regulate the conditions within the solid bed; however, large-scale SSF bioreactors are difficult to operate manually since many variables must be measured and manipulated simultaneously. Under manual control it is not possible to run the bioreactor effectively with just a single operator. In addition, the effects of the many manipulated variables interfere unpredictably with each other throughout the fermentation run. The operators then have a hard job trying to coordinate their respective control actions. Consequently, the bioreactor operation is not reproducible and optimum performance is unattainable.

These difficulties have contributed to the fact that only a minor fraction of the many SSF processes that have been successfully developed at laboratory scale have been scaled up for industrial production.

28.2 How to Control SSF Bioreactors?

As discussed above, no SSF bioreactor can operate efficiently without a good bed temperature control system. The first requirement of a temperature control loop, namely measurement of the temperature, is easy, since Resistance Temperature Detectors (RTDs) or thermocouples can be placed in the bed at desired locations (Sect. 26.3). However, the decision about which operating variable or variables to manipulate in order to try to bring the bed temperature back to the set point value is not so simple. Several control strategies have been designed and tested from laboratory to industrial production scale. These strategies use one or more of the following cooling mechanisms during periods of metabolic high heat production:

- *Conductive cooling*, through cold surfaces like reactor walls or internal plates.
- *Convective cooling*, by forcing cool air through the solid bed.
- *Evaporative cooling*, by forcing partially dry air through the solid bed.

Usually, these mechanisms are enhanced by continuous or periodic bed agitation.

In small size cultures (with beds volumes of the order of 100 cm^3), conduction provides effective cooling due to the large surface-to-volume ratio. Placing the bioreactor (typically a thin column, see Fig. 15.3) inside a thermo-regulated bath ensures good temperature control of the culture. However, in static bioreactors as small as 1 L volume, large temperature differences have been observed inside the solid bed (Saucedo-Castañeda et al. 1990). Therefore, in static pilot-scale or industrial size reactors, metabolic heat must be removed with internal cooling plates, although the design and operation of the reactor becomes complex. Finally, continuous or periodic bed agitation, if not too damaging for the microorganism, may be a convenient alternative to dissipate metabolic heat through the reactor walls. However, Nagel et al. (2001a) showed that this technique becomes insufficient to cool solid beds up to 2 m^3 during the maximum heat production phase, for standard length-to-diameter ratios. Additional advantages of bed agitation are that it reduces temperature gradients, homogenizes the solid bed, and decreases its compactness.

On/Off control (Sect. 27.2.1) performs well with conduction cooling in small bioreactors; however, PID control (Sect. 27.2.2) or Model Based Control (Sect. 27.2.3) are better options for pilot or industrial scale bioreactors.

Effective conductive cooling in large-scale SSF bioreactors may call for excessive agitation, causing a noticeable degradation in process performance due to a low biomass growth rate. Convective cooling can be of great help here, reducing the frequency, intensity, and duration of the mixing events. Of course, bed temperature can be controlled with convection cooling alone, but this is restricted to relatively low bed heights (less than 40 cm height).

Conductive plus convective cooling with agitation cannot remove more than about 50% of the metabolic heat in many types of large-scale SSF bioreactors (see Figs. 22.8 and 23.7). Therefore, the other 50% must be removed by other means. The most effective mechanism for removing metabolic heat in industrial SSF bioreactors is evaporative cooling. Here, the evaporation rate and the consequent cooling rate are determined by the humidity of the partially saturated air that is forced into the solid bed. Therefore, a complete scheme for controlling bed temperature in large-scale SSF bioreactors includes manipulation of inlet air flow rate, temperature, and humidity. The dynamic response of the process and the control configuration can get very complicated when evaporative cooling is used in large-scale SSF bioreactors. Here for example, manipulation of the inlet air temperature affects both controlled variables: bed temperature and bed moisture. Similarly, manipulation of inlet air humidity also affects both controlled variables. This is called *loop interaction*, and it is not usually possible to control this kind of process with PID algorithms alone. In addition, the process can take a long time to respond to changes in the manipulated variables, this is called *time delay* and causes serious difficulties in PID tuning. Moreover, the dynamic response of the system is non-linear, or in other words, the bioreactor does not respond the same at all fermentation times. Hence, PID settings should be changed often since a specific tuning is only effective for a short period. In the face of these various complications, Model Based Control has a better chance of achieving optimum bioreactor performance.

Control of the water content of the bed is also critical to attain good bioreactor performance. For example, an excess of water can cause a reduction of the growth rate due to a low O_2 transfer through the biofilm growing on the surfaces of the solid particles. At the other extreme, a lack of available water can limit growth of the microorganism (in other words, microbial growth is limited by low water activities). If evaporative cooling is used for temperature control, excessive bed drying may occur; however, even if saturated air is fed into the bioreactor some degree of bed drying will occur. Therefore, water content control is usually necessary in large-scale bioreactors. Most large-scale SSF bioreactor types will require periodic addition of fresh water plus agitation to avoid bed over-drying. The amount of water that needs to be added can be established based on humidity measurements of samples removed periodically from the bed. Hence, this control is usually performed manually. The lack of low-cost and reliable on-line sensors for the solids water content makes it difficult to implement an automatic control loop for this variable. If the evaporation rate is high, it is advisable to use a dynamic water balance to get an on-line estimation of the water content, so that the water will be added at the right time (Peña y Lillo et al. 2001).

It may also be beneficial to control the CO_2 or O_2 concentrations in the outlet gas. For example, it has been observed that homogeneous growth in static beds can be attained with a high degree of aeration. Nevertheless, high aeration rates are costly and cause excessive bed drying. Consequently, an optimum aeration rate can be established by regulating either the CO_2 or the O_2 concentration of the outlet gas. This regulation is rather simple and good results can be obtained by switching between different aeration rates or by using PID control coupled with a modulated valve or variable speed blower.

Finally, if a controlled nutrient or pH level in the solid bed is required, appropriate nutrients or acid/base solutions can be sprayed directly over the solid bed or into the inlet air. Typically this is achieved by on/off control. In the case of pH, there is the question of whether reliable on-line measurements can be achieved using pH probes designed for use in SLF bioreactors (Sect. 26.3). However, since pH changes will typically be slow, it may be possible to remove a sample and homogenize it in distilled water and determine the pH, in order to make a decision about whether to implement a control action or not.

The next section will focus on the most difficult aspect of the control of large-scale SSF bioreactors, namely the simultaneous control of bed temperature and water content. The difficulty arises from the fact that evaporation is one of the most effective heat removal mechanisms, meaning that temperature and water control are intrinsically interlinked. Two case studies are presented. The first summarizes practical experience obtained with two pilot-scale SSF bioreactors. The second presents a model-based investigation of control schemes.

28.3 Case Studies of Control in SSF Bioreactors

28.3.1 Control of the Bioreactors at PUC Chile

Two aseptic packed-bed bioreactors (with nominal capacities for 50 kg and 200 kg of fresh solids) with periodic agitation and forced air were built at Pontificia Universidad Católica (PUC) of Chile to scale up the production of gibberellic acid by the filamentous fungus *Gibberella fujikuroi*. This process is particularly difficult to control since *G. fujikuroi* is sensitive to temperatures above 36°C and, due to its slow growth, to contamination by other microorganisms. Due to its filamentous nature, its growth increases bed compactness, which creates a reduction in heat and mass transfer rates, but this filamentous nature also means that it is sensitive to mechanical stresses when the solid bed is agitated. In addition, bed overheating can easily occur during the growth period. Air channelling is also a major problem, causing heterogeneous growth and large temperature differences within the solid bed. Finally, the processing time is long, demanding a robust control system. Key points are outlined below. More information is given in Fernández et al. (1996) and Fernández (2001).

The measured process variables used in these bioreactors are shown in Table 28.1 and the manipulated variables in Table 28.2. Figure 28.1 shows schematically how the instrumentation and control devices presented above were put together to define the control loops used in both bioreactors. The control strategy shown in Fig. 28.1 is described in Table 28.3.

Table 28.1. Measured variables in the PUC SSF bioreactors

Variable	Instrument	Measurement
Inlet and outlet air temperature	Type K thermocouples	On-line
Relative humidity of inlet air	Vaisala HMP 122B (absorption of water on a thin polymer film)	On-line
Solid bed temperature	Type K thermocouples (3 in 50-kg reactor and 6 in 200-kg reactor)	On-line
pH of the solid bed	Schott pH-meter	On-line
Solid bed water content	Precisa IR scale	Off-line (1 sample/hour)
Outlet air CO_2 concentration	IR analyzer (Horiba PIR 2000)	On-line
Outlet air O_2 concentration	SMC Transmitter (electrochemical device)	On-line
Inlet air velocity	Dwyer Inst. 640-0 (Hot wire sensor)	On-line

Table 28.2. Manipulated variables in the PUC SSF bioreactors

Variable	Actuator	Type of control
Inlet air temperature	Electric heater (6 kW) Cooler with fins (6.6 kW)	On/off
Air velocity	Inverter drive Hitachi (50 to 500 $m^3\ h^{-1}$)	Continuous
Steam addition	Solenoid valve + boiler	On/off
Bed rotation	Electric motor with variable frequency (3 to 15 rpm)	On/off
Bed mixing	Local screw stirrers plus inverter drive ABB CDS 150 (for 200-kg reactor only)	On/off
Fresh water addition	Millipore peristaltic pump (60 rpm)	On/off
Pressure drop across the bed	Modus Inst. T30 (Deflection diaphragm)	Continuous

Table 28.3. Control strategies applied in both PUC bioreactors

Controlled variables	Manipulated variables	Type of control	Control algorithm
Bed water content	Fresh water addition plus agitation	Manual	Water balance plus operator experience
Bed porosity and homogeneity	Agitation	Semi-automatic	Operator experience
Inlet air temperature	Heater/Cooler	Automatic	on-off
Inlet air relative humidity	Steam addition	Automatic	on-off
Bed temperature	Inlet air temperature and relative humidity	Semi-automatic	PID plus operator experience
Inlet air flowrate	Air blower velocity	Automatic	PID

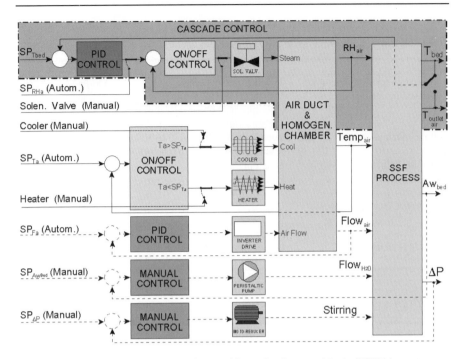

Fig. 28.1. Scheme of the conventional control loops implemented in the PUC bioreactors

A Programmable Logic Controller (PLC) was used for data acquisition and to control the primary actuators. Operator and control calculation interaction were carried out via an IBM-compatible personal computer, linked to the PLC. Project-specific software was developed for the personal computer with a graphic interface to handle the control systems in either automatic or manual mode.

The primary objective of the control system was to regulate the average temperature of the solid bed at a fixed value and to control the bed water content (according to a varying set point). The secondary objectives were to minimize temperature gradients within the bed and also to prevent the bed from becoming overly compact.

Control of the bed temperature was based on evaporative cooling, by manipulating the relative humidity of the inlet air and maintaining its temperature at a low value. During the period of high heat generation, in order to avoid bed overheating, it was necessary to intervene manually to manipulate the inlet air flow rate and temperature and to initiate agitation events. The average bed temperature was fed into a digital PID control algorithm to drive the set point of the inlet air relative humidity. This in turn was controlled with an on/off algorithm with a dead band and hysteresis that commanded a solenoid valve adding steam. The inlet air temperature was further controlled with another on/off algorithm with a dead band and hysteresis that manipulated the heater or cooler, according to process needs.

The water content of the bed was controlled through the periodic addition of fresh water. The reference trajectory of the water content was computed in the laboratory based on water activity studies. A solid sample was taken each hour from the bioreactor for an off-line measurement of the water content. The amount of water to be added was determined by the operator using an approximate water balance and his experience. The bed was agitated upon each addition of water.

In order to keep the bed as homogeneous as possible and to avoid excessive inter-particle aerial growth, which would reduce porosity, a periodic agitation policy was established. The degree of homogeneity was defined by the temperature gradient inside the bed, while inter-particle aerial growth was estimated from the pressure drop through the bed. This involved a semi-automatic loop that employed heuristic logic. The operator could establish the agitation speed, its duration and, in the case of the 200-kg bioreactor, the path of the agitation (left or right).

The control strategy enabled efficient operation of the pilot SSF bioreactors. When the 50-kg bioreactor was run manually, it required the permanent attention of at least two operators. The automated control system required only one operator, even at the 200-kg scale, who intervened relatively little in the process. The system even allowed bioreactor operation without direct supervision at certain times (Fernández 2001). Figure 28.2 shows the performance of the average bed temperature control loop between hours 10 and 40 of a fermentation run in the 200-kg bioreactor. The control system performed reasonably well, since the average bed temperature deviated no more than 4°C from the set point, although most of the time the deviations were smaller than 1°C. However, to achieve this performance, the inlet air temperature had to be changed manually. Moreover, the differences between maximum and minimum temperatures within the solid bed were considerably high (5°C average difference and 15°C maximum differences).

On the other hand, water content control did not perform so well, with deviations of more than 25% with respect to the set point. This was due to the several limitations that this control loop presented, such as manual control, the lack of an on-line sensor and the fact that water had to be added while the bioreactor was being agitated such that control actions could only be taken infrequently. In addition, water content measurements were noisy due to bed heterogeneity.

However, it is noteworthy that the control strategy was successfully scaled up from the 50-kg bioreactor to the 200-kg one with only minor adjustments.

The operation of the bed temperature control loop can be simplified if a model predictive control algorithm is used. When this kind of control was applied in the 200-kg bioreactor, within the same time-span shown in Fig. 28.2, better overall performance was achieved. Temperature differences within the bed and high temperature peaks were reduced when compared with standard control (Fig. 28.4). In addition, the loop was fully automatic therefore no manual operation of the inlet air temperature was necessary.

It should be noted that even better performance could be achieved by tuning the algorithm; however this was not done with the PUC bioreactors since it would have required several fermentation runs, each of which is long and expensive. Despite this, it is probably correct to say that, to attain good performance in industrial-scale SSF bioreactors (2 to 3 tons or more), it is necessary to apply model predictive control in the bed temperature control loop.

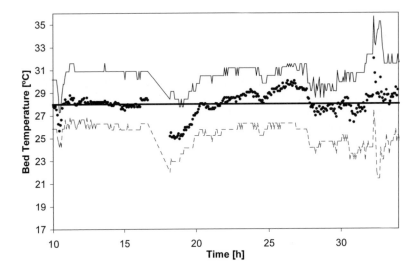

Fig. 28.2. Bed temperature control during a fermentation run with the 200-kg bioreactor. Key: (━━) Bed temperature set point; (——) maximum bed temperature; (•), average bed temperature; (- - -) minimum bed temperature

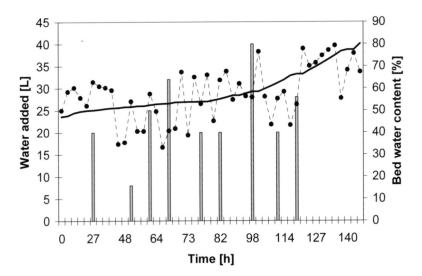

Fig. 28.3. Water content control during a fermentation run with the 200-kg bioreactor. Key: (━━) water content set point; (- • -) measured water content of the bed. The vertical bars represent the volume of water added

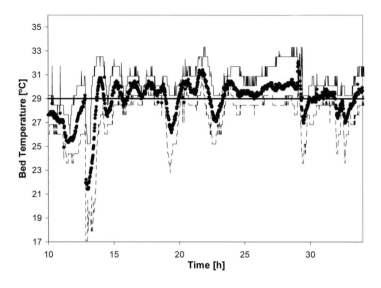

Fig. 28.4. DMC strategy applied to the 200-kg bioreactor. Better temperature control is achieved when two manipulated variables are moved simultaneously, which can be seen by comparing this figure with Fig. 28.2. Key: (—) Bed temperature set point; (—) maximum bed temperature; (•) average bed temperature; (- - -) minimum bed temperature

28.3.2 Model-Based Evaluation of Control Strategies

An intermittently-mixed forcefully-aerated bioreactor, presented in Fig. 25.1, was modeled using the program presented in Chap. 25, the equations of which are shown in Fig. 25.2. Mixing was triggered when the outlet gas water activity fell below a set point. However, unlike the case study in Chap. 25, in which the inlet air conditions were maintained constant during the fermentation, in the present case study a control scheme was implemented to control either or both of the temperature and humidity of the inlet air, based on the average of the temperatures measured at different heights within the bed (Fig. 28.5). The success of the control scheme was evaluated on the basis of the fermentation profile for the average biomass concentration within the bed.

The present case study highlights the main points of interest that were identified in the work of von Meien et al. (2004). Readers interested in a deeper analysis should consult the original paper.

Figure 28.6 shows simulations done with a PID (Proportional-Integral-Derivative) controller, using two different strategies:

- **Humidity control.** In this case the relative humidity of the inlet air is varied by the controller, while the temperature is maintained at 38°C. Figure 28.6(a)) shows that the average temperature in the bed varies significantly from the optimum of 38°C throughout the fermentation.

- **Temperature control.** In this case the temperature of the inlet air is varied by the controller, while the relative humidity is maintained constant. In different simulations the constant relative humidity is maintained at 80% (Fig. 28.6(b)), 90% (Fig. 28.6(c)), and 99% (Fig. 28.6(d)). In this case it is possible to control the average temperature of the bed much better, in other words, the deviations from the optimum temperature are smaller.

The effect of the better temperature control in the case in which the inlet air temperature is manipulated is clear: The growth profiles obtained with "temperature control" (Fig. 28.6(f)) are closer to the optimum than the growth profile obtained with "humidity control" (Fig. 28.6(e)). Note that the relative humidity of the inlet air has no effect on the predicted growth performance in the case of "temperature control". Therefore, it is best to maintain the air saturated since this is easier to achieve in practice than attaining a particular relative humidity set point (see Chap. 29).

Figure 28.7 shows simulations done with a DMC (dynamic matrix control) controller. DMC is a form of Model Predictive Control, which was discussed in Sect. 27.2.3. Again "temperature control" and "humidity control" strategies are compared, with "temperature control" being superior, as it was in the case of PID control (von Meien et al. 2004). Figure 28.7 also shows that DMC control presents an interesting challenge. As discussed in Sect. 27.2.3, model predictive control

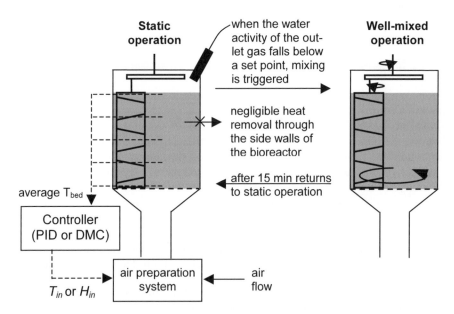

Fig. 28.5. Bioreactor and control scheme for the case study of control of an intermittently-agitated forcefully-aerated bioreactor. The bioreactor simulated is 2.0 m high, with an air flux at the inlet of 0.06 kg dry air s^{-1} m^{-2}. In practice it might be impractical to measure the temperature at many different heights within the bed. In this case a single measurement at a half of the overall bed height will probably be sufficient

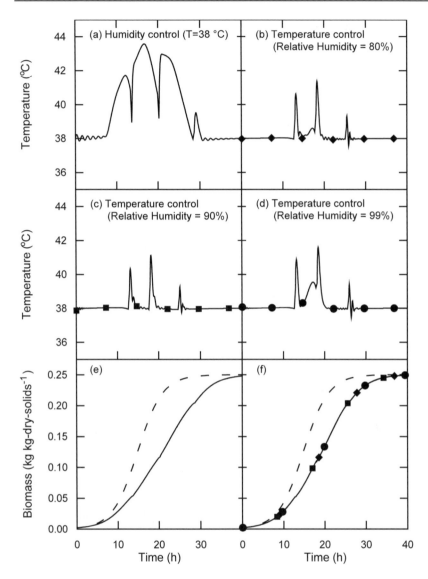

Fig. 28.6. Predictions of performance of an intermittently-agitated, forcefully-aerated bio-reactor when PID control is used. **(a)** to **(d)** Predicted average bed temperature (the average of the temperatures measured at various different heights, as shown in Fig. 28.5) for the various control schemes. **(e)** Predicted average biomass profile for the case of humidity control with the inlet air temperature maintained at 38°C (——). **(f)** Predicted average bio-mass profiles for the case of temperature control with the inlet air humidity maintained at 80% (—◆—), 90% (—■—), and 99% (—●—). In graphs (e) and (f) (- - -) represents the growth curve that would occur if optimum conditions were maintained throughout the en-tire fermentation. Adapted from von Meien et al. (2004) with kind permission of Elsevier

schemes such as DMC use a linear model to calculate the required future changes in the manipulated variable that will result in optimum set point tracking for a specified performance index. This linear model, normally obtained from the initial dynamic response of the bioreactor, is used by the controller to guide the control actions during the whole fermentation run.

However, for this fermentation, when this is allowed to happen, the controller makes large control actions in the latter stages of the process, which cause large and frequent oscillations in the manipulated variable, that is, in the temperature of the inlet air (Fig. 28.7(a)) and of course these oscillations cause similar oscillations in the temperatures within the substrate bed (Fig. 28.7(c)). Such large oscillating control actions are undesirable, especially if they are not necessary. The problem is that the controller had worked out its control strategy based on the initial part of the process during which growth was accelerating and temperature control was becoming ever more difficult. The controller worked out that it is necessary to apply large "preventative" control actions and when growth decelerates in the latter stages of the process, it still applies such large control actions, even though they are not necessary. This problem can be overcome by instructing the controller to establish a new linear model after each mixing event. Since there are three mixing events, four different linear models are used (i.e., the original one plus a new one after each mixing event). In other words, the controller changes its control strategy during the fermentation and this minimizes the oscillations in the inlet temperature (Fig. 28.7(b)) and therefore also the oscillations in the temperatures measured in the bed (Fig. 28.7(d)).

The necessity for the use of multiple linear models can be explained in a different way. The behaviour of the fermentation process is history dependent, that is, the evolution of the system from a particular point relies on what happens before this point. This can be easily understood when it is recognized that the rate of growth at any instant depends to a significant degree on the amount of biomass that was produced in the fermentation from the time of inoculation up until that instant. Since the underlying behaviour of the system (the rate of growth) varies significantly during the process, then the dynamics of the control system need to be changed.

As shown by comparing Figs. 28.7(e) and 28.7(f), it actually makes no difference to the biomass growth curve whether a single linear model or multiple linear models are used, however, it is obvious that multiple linear models should be used since the same performance is achieved, but without large and frequent control actions.

Note that the predicted growth with DMC control is superior to that predicted for PID control (Figs. 28.7(e) and 28.7(f)).

As a final point, this case study has shown that mathematical models of SSF bioreactors are useful tools in the initial stages of the development of control strategies and in the initial tuning of controllers.

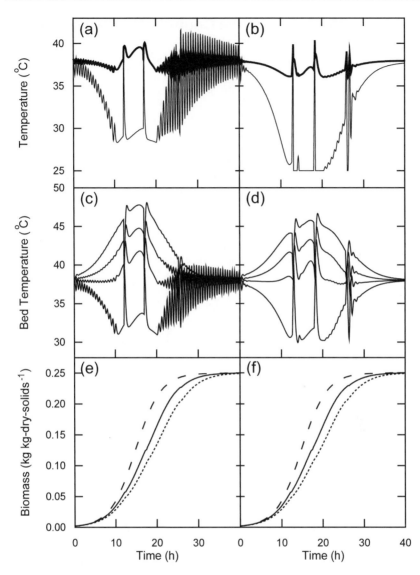

Fig. 28.7. Performance of the bioreactor with DMC control of the inlet air temperature (the relative humidity of the air is fixed at 99%). The results for "a single linear model" and "multiple linear models" are on the left and the right, respectively. **(a)** and **(b)** inlet air temperature (—), which is not allowed to go below 25°C and the average bed temperature (━━), which the controller aims to control at 38°C; **(c)** and **(d)** temperatures at various heights within the bed (from bottom to top the lines are for 0.5, 1.0, 1.5, and 2.0 m); **(e)** and **(f)** biomass profiles, including the predicted average biomass content (—), the growth curve that would occur if optimum conditions were maintained throughout the entire fermentation (- - -) and, for purposes of comparison, growth with PID control of the temperature of the inlet air (·····). Adapted from von Meien et al. (2004) with kind permission of Elsevier

28.4 Future Challenges in the Control of SSF Bioreactors

The control problem is especially challenging in SSF bioreactors. In bioreactors that are not continuously mixed, the problem is not as simple as trying to control the whole of the bed at one set of conditions. Rather, it is a case of accepting the fact that this is not possible and using the control scheme to minimize the average deviations in time and space from the optimal conditions. The challenge is to be able to prevent one part of the bed from reaching deleteriously high temperatures without the control action causing another part of the bed to fall to temperatures that are so low that growth is restricted. Control in distributed systems has received some attention (Christofides 2001), but it is highly mathematical and has not yet proved to be effective in real cases. Further, the distributed nature of the system brings up the question of how many sensors and actuators should be used and where in the bioreactor they should be placed. Such considerations are further complicated by the unpredictable nature of some of the changes that the control scheme might be intended to counteract. For example, it would be desirable for the appearance of channels in the bed to be counteracted by a mixing action. However, it would be difficult to design a measuring device to detect this. The presence of channels might be indicated by local rises in temperature. Such rises would occur in those parts of the bed that do not receive effective aeration due to the preferential flow of the air through the channels. However, if such rises were localized, then they would only be detected if the sensors were in the right place and there is no way of knowing *a priori* where to place the sensors.

Further, SSF bioreactors present an interesting example of what is called "cascade control". For example, it is already an interesting control challenge to provide an air stream of the desired flow rate, temperature, and humidity, especially when the set points for these variables change as the organism proceeds through its growth cycle. For instance, as the metabolic heat generation rate increases early in the process due to the acceleration of growth, the set point of the inlet air temperature would typically be decreased. However, it must be remembered that the final objective of the control action is not the control of the flow rate, temperature, and humidity of the inlet air. Of course this control is important, but the final objective is actually to use the conditions of the inlet air to control the conditions within the bed. However, the intricacies of cascade control are beyond the scope of this book.

Nonlinear model predictive control (NLMPC) may be necessary in SSF systems. This involves the embedding within the control scheme of a set of non-linear differential equations that describe the microbial and heat and mass transfer processes. In this case the optimization problem is in general "non-convex", which means that the solution is hard to find and there is no guarantee that the optimization routine will find it. Furthermore, there are theoretical issues regarding the stability of the control loop in non-convex problems, and these issues have not been completely clarified yet within the area of control theory. In addition, special optimization routines are required. The main problem of NLMPC is that is too sensitive to lack of model accuracy ("model mismatch") and biased model parameters.

The design of a reliable NLMPC system for a SSF bioreactor has as an absolute requirement the development of effective on-line parameter estimation and filtering algorithm, in order to get reliable model parameters and eliminate the process and measurement noise. This also requires the formulation and solution of a non-linear optimization problem. As a result, the design of reliable NLMPC system is very complicated and it is still a developing science even in non-SSF applications.

Further Reading

More information about the control of the PUC bioreactor
Fernandez M, Perez-Correa JR, Solar I, Agosin E (1996) Automation of a solid substrate cultivation pilot reactor. Bioprocess Engineering 16:1–4
Fernandez M (2001). Control automático de un bioreactor piloto de lecho sólido. Ph.D. thesis, University of Chile (http://ing.utalca.cl/~mafernandez)

A theoretical case study of the application of control strategies to a large-scale intermittently-mixed, forcefully-aerated SSF bioreactor
von Meien OF, Luz Jr LFL, Mitchell DA, Pérez-Correa JR, Agosin E, Fernández-Fernández M,. Arcas JA (2004) Control strategies for intermittently mixed, forcefully aerated solid-state fermentation bioreactors based on the analysis of a distributed-parameter model. Chem Eng Sci 59:4493–4504

Control of distributed systems
Christofides PD (2001) Nonlinear and robust control of PDE systems. Birkäuser, Boston

Cascade control
Ogunnaike B, Ray RWH (1994) Process dynamics, modeling and control. Oxford University Press, New York.

Nonlinear Control
Henson MA, Seborg DE (1997) Nonlinear process control. Prentice Hall, New Jersey.

Examples in which control has been attempted in SSF bioreactors
Barstow LM, Dale BE, Tengerdy RP (1988) Evaporative temperature and moisture control in solid substrate fermentation. Biotechnol Techniques 2:237–242
Chinn MS, Nokes SE (2003) Temperature control of a solid substrate cultivation deep-bed reactor using an internal heat exchanger Trans ASAE 46:1741–1749
De Reu JC, Zwietering MH, Rombouts FM, Nout MJR (1993) Temperature control in solid substrate fermentation through discontinuous rotation. Appl Microbiol Biotechnol 40:261–265
Nagel FJI, Tramper J, Bakker MSN, Rinzema A (2001a) Temperature control in a continuously mixed bioreactor for solid state fermentation. Biotechnol Bioeng 72:219–230
Nagel FJI, Tramper J, Bakker MSN, Rinzema A (2001b) Model for on-line moisture-content control during solid-state fermentation. Biotechnol Bioeng 72:231–243

Pajan H, Perez-Correa R, Solar I, Agosin E (1997) Multivariable model predictive control of a solid substrate pilot bioreactor: A simulation study. In: Wise DL (ed) Global Environmental Biotechnology. Kluwer Academic Publishers, Dordrecht, pp 221–232

Ryoo D, Murphy VG, Karim MN, Tengerdy RP (1991) Evaporative temperature and moisture control in a rocking reactor for solid substrate fermentation. Biotechnology Techniques 5:19–24

Sargantanis JG, Karim MN (1994) Multivariable iterative extended kalman filter based adaptive control: case study of solid substrate fermentation. Ind Eng Chem Res 33:878–888

29 Design of the Air Preparation System for SSF Bioreactors

Oscar F. von Meien, Luiz F.L. Luz Jr, J. Ricardo Pérez-Correa,
and David A. Mitchell

29.1 Introduction

All solid-state fermentation (SSF) bioreactor types potentially need an air preparation system. The need is most obvious for forcefully aerated bioreactors, since the conditions of the air at the inlet have a strong influence on the heat and mass transfer phenomena within the bed. However, even those bioreactor types that are not forcefully aerated can benefit from an air preparation system. For example, although it is possible to operate tray bioreactors by circulating air taken directly from the surroundings, it is likely that the process will operate better if conditioned air is circulated through the headspace.

The considerations guiding decisions about the air preparation system are different for SSF and submerged liquid fermentation (SLF). In SLF the rate at which air is blown into the bioreactor is calculated based on the O_2 demand of the microorganism; heat removal considerations do not influence decisions about the aeration rate. In other words, in SLF good temperature control can be attained by circulating hot or cold water through water jackets and cooling coils, without any need to supply or remove heat in the air stream. This is possible due to the diluted nature and favorable heat transfer conditions within the liquid fermentation medium: the medium is typically well agitated and has a high thermal diffusivity.

In contrast, in SSF few alternatives are available for heating or cooling of the substrate bed other than manipulating the temperature, flow rate, and humidity of the inlet air. At first glance, it does not seem to be a difficult task, especially given the high heat removal capacity of evaporative cooling; however, we may be restricted in terms of the values that we can use for these two operating variables, even in a bioreactor operated in the intermittently-mixed mode. For example, if we supply air that is not saturated with water vapor to the bioreactor, this will improve evaporative heat removal but will also accelerate the drying of the bed, increasing the frequency with which water must be added. However, frequent water addition can be undesirable. If the bed is not mixed during the addition of water, then it will be almost impossible to ensure uniform distribution of the water, leading to flooded and dry regions within the bioreactor. Addition of water as a spray while the bed is being agitated may allow uniform distribution of water, however, frequent agitation will typically be deleterious to the performance of processes that involve fungi due to the mechanical damage caused to fungal hyphae by shear and impact forces within the bed. In order to minimize this damage, the frequency of mixing and water addition events should be minimized. However, this means that we should use saturated air at the air inlet in order to minimize the evaporation rate. Therefore the operating strategy must seek to find those conditions that give a reasonable rate of heat removal without causing undue damage to the microorganism. Note that even if saturated air is to be used, it is not necessarily easy to keep the air saturated if one intends to vary its temperature.

In short, in SSF the air stream has a role that goes beyond the supply of O_2. It plays a fundamental role in heat removal and the design of an adequate air conditioning system is essential for good operation and control of an SSF bioreactor.

29.2 An Overview of the Options Available

The air stream not only must be supplied at proper conditions of temperature and humidity as addressed above, it also must overcome the pressure drop caused by the bed, piping, and other accessories. All these aims have to be achieved at low cost, since many SSF processes have low profit margins. At this point, we can examine some alternatives for air preparation, starting with the very simple alternative presented in Fig. 29.1.

The system presented in Fig. 29.1 will only be appropriate if aseptic conditions are not required, since it has only a simple dust collector, like those found in home air conditioning units. A porous plate is necessary for efficient air distribution in the humidifier. Porous stainless steel plates with high permeability are available in the market; they are very efficient but are not cheap. Sintered ceramic plates are also available; they are cheaper than metallic plates, however, they require a mechanical support in order to withstand the mass of the water in the tank. In this system the temperature of the air leaving the humidification tank will be close to the water temperature, therefore it is desirable to control the water temperature, which can be done using electrical resistance heaters and an ordinary thermostat.

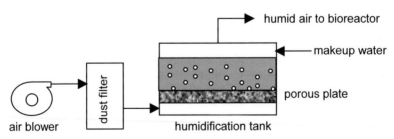

Fig. 29.1. A simple configuration for an air preparation system in which the air is bubbled through a humidification tank

This simple system has three important features. Firstly, it is not practical to obtain air temperatures below the ambient temperature since no cooling unit is included in the system. Secondly, it will not allow a fast change of the temperature of the air fed into the bioreactor due to the high thermal inertia of the water mass. Thirdly, high air flow rates will cause the evaporation of large amounts of water and therefore water must be replenished periodically in order to prevent the tank from drying out.

Since the system presented in Fig. 29.1 does not allow a rapid manipulation of the air temperature, another possible arrangement is presented in Fig. 29.2.

Aseptic cultivation will be easier to achieve with this arrangement. Firstly, the cooling of the air before filtering will eliminate part of the microorganisms with the purged water. Secondly, the micro-porous filter will not allow particles larger than 0.3 μm to pass, which is sufficient to remove fungal spores, bacterial cells, and any larger organisms.

In this system a pair of valves and two sets of electrical resistances can be used to control the air temperature. Typically the process will require saturated air at temperatures higher than the ambient during the lag phase while during the rapid growth phase it will be necessary to supply air at a temperature below the optimum temperature for growth in order to promote heat removal. The highest rates of heat removal will be obtained by supplying cold, dry air in order to promote evaporative cooling, however, this will also promote drying of the bed. It is possible to inject steam into the cold air; but it is not easy to generate steam at temperatures around say 20°C. Finally, it is important to note that it is not a simple matter to produce saturated air by direct mixing of steam and dry air, since it is not easy to design a mixing device that does not produce condensation.

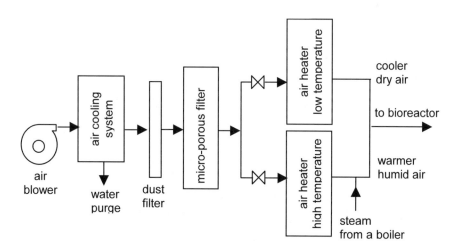

Fig. 29.2. A configuration for an air preparation system that allows control of the temperature and humidity of the air supplied to the bioreactor

A third alternative is presented in Fig. 29.3, where a humidification column re-places the steam-air mixer. A well-designed humidification column will guarantee saturated air. The heater for producing cooler dry air that appeared in Fig. 29.2 is not present since only saturated air will be provided by this system. In this system the temperature of the outlet air will be very close to the water temperature, as was the case for the system presented in Fig. 29.1. It is therefore interesting to work with two reservoirs, one with hot water and other with cold water. A set of syn-chronized solenoid valves can be used to change from circulation of hot water to circulation of cold water and vice-versa, proving saturated air at a higher tempera-ture at the beginning of the fermentation and saturated air at lower temperatures for cooling the reactor during the rapid growth period. It is important to remember that even with the use of saturated air the bed will still dry out, since the air is heated as it passes through the bed and therefore its capacity to carry water in-creases (see Fig. 4.3). In other words, the use of saturated air will reduce the fre-quency with which water must be replenished but water replenishment will still be necessary. Note that aseptic operation is unlikely to be feasible due to the diffi-culty in operating the entire humidification system (reservoirs and column) in an aseptic manner.

Of course, combinations and variations of the alternatives presented in this sec-tion can be worked out. The suggested configurations demonstrate the advantages and disadvantages of selecting a particular combination of devices. The decision for a particular arrangement must be based on criteria of economic performance of the process, which will depend on the capital cost of the selected devices and op-erating costs related to energy consumption for the various unit operations such as blowing the air, heating or cooling the air, producing steam, and heating water. In fact, great care should be taken in computing energy costs, since they can com-promise the feasibility of the project, particularly when working with products that have low profit margins. The following sections give some further advice about various aspects of the design of the air preparation system.

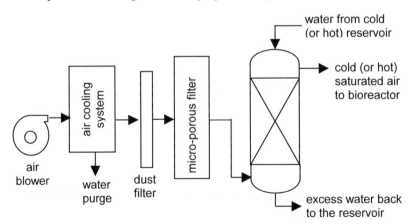

Fig. 29.3. A configuration for an air preparation system that provides saturated air at either a hot or a cold temperature

29.3 Blower/Compressor Selection and Flow Rate Control

The air blower is likely to be responsible for a large proportion of the energy consumed in bioreactor operation and therefore selection of an appropriate blower is very important. In general, SSF bioreactors need high air flow-rates at low pressures. If the pressure drop of the whole system, including piping, accessories, and bioreactor, is equal to or less than 35 cm of water, a fan is the best device. The power consumed by a fan (P, kW) can be calculated as follows:

$$P = 2.72 \times 10^{-5} Fp, \qquad (29.1)$$

where F is the air flow-rate (m^3 h^{-1}) and p is the operating pressure of the fan (cm-H$_2$O). Manufacturers usually supply operational curves and installation details.

For higher pressure-drops, a centrifugal compressor would be the best choice, however, the energy consumption would be prohibitive in many cases. If possible, it is advisable to design and operate the bioreactor and air preparation system in such a manner as to minimize the pressure drop, such that it is possible to use a fan, rather than to work with a compressor.

Usually fans work at fixed velocity whereas the aeration requirements change over time. Therefore a flow control valve (FCV), placed in the air line between the blower and the bioreactor, is required. For air at low pressure a butterfly valve is the best choice. Centrifugal compressors can be operated in a similar manner or, alternatively, their velocity can be controlled with inverter drives.

Both fans and centrifugal compressors work under a "characteristic curve" that gives the flow rate provided by the equipment as a function of the pressure in the air line. Note that the pressure in the air line depends on the pressure drop suffered by the air as it flows from the blower, through the system, to the air outlet (that is, the pressure at the air inlet must be at least equal to the sum of the pressure at the air outlet and the pressure drop within the system). Both these types of blowers will provide larger flow rates at lower pressure, with the flow rate reducing as the pressure in the air line increases. The characteristic curve depends on the type and the design of the blower and should be provided by the manufacturer.

The required air flow rate for the bioreactor will depend on heat removal needs, as was clearly demonstrated in the various modeling case studies presented in Chaps. 22 to 25. The blower must be capable of producing the flow rate required at the time of maximum heat production within the substrate bed. Obviously, models of the type presented in these chapters are useful tools in deciding on the requirements of the blower.

Note that the pressure drop that the system must be capable of overcoming is that which is present in the system at the time of maximum heat production. Potentially, the pressure drop as air flows through the bed can make a significant contribution to the overall pressure drop in the aeration system in those bioreactors in which the bed is forcefully aerated. The pressure drop across the bed depends on the substrate, the microorganism, and their interaction during the process and, due to our relatively poor understanding of these phenomena, it is not possible to use a set of theoretical calculations to predict the magnitude of the pressure drop that can be expected. Therefore, some experimental assays at laboratory scale will be necessary in order to estimate the pressure drops for a particular system. In systems in which the bed is agitated, a maximum allowable pressure drop can be set as a parameter for triggering mixing events, and this should prevent the pressure drop across the bed from reaching high values.

The best strategy is to select the equipment that provides the largest pressure range for the maximum required flow-rate, this flow-rate being deduced from the energy balance model. It is then necessary to check whether the equipment will operate economically in terms of energy consumption at the required combination of pressure and flow rate. If energy consumption is too high then possibly an inferior blower will need to be selected. This may not be capable of meeting the aeration needs during the periods of peak heat generation, so the performance of the process may be deleteriously affected. As stated above, many of the products obtained by SSF have low profit margins; in this case the energy consumption of the aeration system is a crucial factor in determining process profitability.

29.4 Piping and Connections

The specifications for the piping used in the air line will be affected by the sterility requirements of the process. If a high degree of sterility is required then the piping will need to be able to withstand either steam or chemical sterilization before each fermentation: For example, it may be necessary to use stainless steel. If sterility is not a crucial issue, then less resistant materials can be used.

Also, given that saturated air will typically be supplied to the bioreactor, it is possible that condensation will occur within the air line. It is advisable to have strategically placed drains (or "purges") in the air line in order to remove this water intermittently during the process. Rotation of the bioreactor can complicate the aeration system. If the bioreactor rotates while air is introduced into it, then a rotating seal will be necessary between the air line and the bioreactor body. If the inlet and outlet air lines are removed before rotation of the bioreactor, then it is necessary to have a connection that is fully airtight when connected, but simple to remove and re-attach. Also, manually operated butterfly valves may be necessary on the air inlet and air outlet of the bioreactor in order to prevent substrate from flowing out of the bioreactor as it is rotated.

29.5 Air Sterilization

The selection of an appropriate filter type will be based on considerations of minimizing the pressure drop for a desired air flow rate while removing very small particles. It may also be influenced by the types of filters that local suppliers actually have available.

At laboratory scale, contamination is typically somewhat easier to control while at large scale sterilization and prevention of contamination is more complicated. In fact, contamination may be one of the most frequent operational problems for those processes that require aseptic operation of the bioreactor. During the production of gibberellic acid by *Gibberella fujikuroi* in the 200-kg capacity pilot scale bioreactor of PUC, Chile (see Sect. 10.3.1.3 and Fig. 10.4) there were frequent contamination problems when the air system contained only a pre-filter and an absolute filter with a cut-off of 0.3 μm. These contamination problems were reduced significantly by (1) the installation of UV lamps in the air duct between the filter system and the bioreactor and (2) chemical sterilization of the air duct system prior to each fermentation run.

In designing an appropriate system, several factors will need to be considered, such as capital and operating costs, the effectiveness of removal of microorganisms, the potential for failures in the system (such as the rupturing of filters), and the pressure drop contributed by the air sterilization system. It is also important to consider at what stage of the air preparation system the sterilization should occur. If the air is dry at the time of sterilization, then there should be no problem of wetting of filters, however, any subsequent humidification steps will need to be done aseptically if aseptic bioreactor operation is required for the particular process in question.

29.6 Humidification Columns

Usually, it is hard to find a supplier for humidification columns such as that shown in Fig. 29.3, so they are typically custom-made from a design supplied by the purchaser. Within the interior of humidification columns the water flows downwards through a bed of packing that ensures a high superficial area of contact between the air and water, in order to guarantee saturation at the air outlet.

The column diameter is chosen obeying the criteria of minimum pressure drop and avoidance of flooding. Flooding, namely the accumulation of water at the top of the column, happens in packed humidification columns when the air velocity through the column becomes large enough to stop the liquid from flowing down the bed. The height of the column necessary to ensure saturation can be calculated from mass and energy balances. Once the dimensions are determined then the pressure drop across the column can also be calculated.

Sources of advice on how to design humidification columns are given in the Further Reading section at the end of the chapter.

29.7 Case Study: An Air Preparation System for a Pilot-Scale Bioreactor

The air preparation system presented in Fig. 29.3 represents a compromise between technical specifications and costs in the sense that, while it does not allow as flexible a control of the conditions of the air supplied to the bioreactor as the system shown in Fig. 29.2, it will be much cheaper to build and operate than that system and it will allow better control than the system presented in Fig. 29.1. On the basis of these considerations, the system shown in Fig. 29.4 was recently constructed for a pilot-scale SSF bioreactor with a 200-L substrate bed. Although the bioreactor has not yet gone into operation, it is worthwhile to describe briefly the calculations that were used to design the system.

Maximum air flow rate requirement. The maximum air flow rate that would be needed was calculated on the basis of heat removal considerations. Assuming logistic growth kinetics, the maximum heat generation rate (R_Q, kJ h^{-1}) is given by (Mitchell et al. 1999):

$$R_Q = \rho_b X_{max} Y_q \mu_{max} V_b / 4. \tag{29.2}$$

The substrate packing density (ρ_b) was estimated as 350 kg-dry-substrate m^{-3}, the maximum biomass content (X_{max}) as 0.125 kg-dry-biomass kg-dry-substrate^{-1}, the metabolic heat yield coefficient (Y_q) as 10^7 J kg-dry-biomass^{-1}, the maximum value of the specific growth rate constant (μ_{max}) as 0.324 h^{-1}, and the bed volume (V_b) as 0.2 m^3. The calculation gave a value of R_Q of 7.1 MJ h^{-1}.

A conservative estimate of the capacity of the air to remove heat from the bed was made as Q_{rem} = 5 kJ kg-air^{-1} °C^{-1}. This represents the sum of the heat capacity of humid air (~1.0 kJ kg-dry-air^{-1} °C^{-1}) and the contribution of evaporative cooling of "$\lambda.(dH_{sat}/dT)$" where λ is the enthalpy of evaporation of water (2500 kJ kg-water^{-1}) and dH_{sat}/dT (kg-vapor kg-dry-air^{-1} °C^{-1}) is the change in the water-carrying capacity of air with a change in temperature (see Sects. 18.5.2.2 and 19.4.1). Use of Eq. (19.20) shows that dH_{sat}/dT varies from 0.0016 kg-vapor kg-dry-air^{-1} °C^{-1}at 30°C to 0.0048 kg-vapor kg-dry-air^{-1} °C^{-1} at 50°C. Using the value of dH_{sat}/dT at 30°C therefore leads to a more conservative estimate and with this value "$\lambda.(dH_{sat}/dT)$" is calculated as 4 kJ kg-air^{-1} °C^{-1}.

The mass flow rate of air required (W_{air}, kg h^{-1}) was then calculated as:

$$W_{air} = \frac{R_Q}{Q_{rem} \Delta T}, \tag{29.3}$$

where ΔT is the maximum allowable rise in temperature of the air as it flows through the bed. This was taken as 5°C. Substituting the values of R_Q and Q_{rem}, W_{air} was calculated as 283.5 kg h^{-1} (235 m^3 h^{-1} at 15°C and 1 atm). Since a conservative value was used for dH_{sat}/dT, this is probably an overestimate, but will therefore allow a reasonable margin for error.

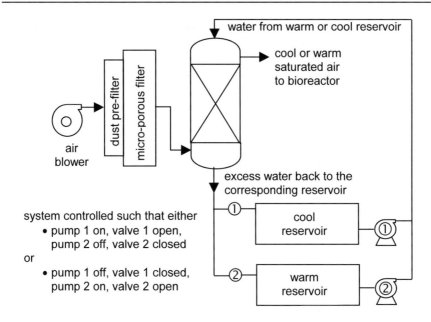

Fig. 29.4. Configuration used for an air preparation system for a 200-L bed capacity pilot-scale SSF bioreactor

Filter. The filter selected is a HEPA (High Efficiency Particulate Air filter), with a minimum efficiency of 99.97% for particles larger than 0.3 μm. The dimensions of the filter cartridge are 305 mm by 305 mm, with an overall thickness of 78 mm. The pressure drop caused by this filter is equal to a 25 mm water column. In order to protect the filter, a dust pre-filter is included in the cartridge.

Reservoirs. The reservoirs of warm and cool water are large (1 m^3 each) in order to make the regulation of their temperature easier, that is, the reservoirs have a large thermal inertia. Note that the intention is to maintain the water temperature of each reservoir constant at the desired set point during the whole fermentation; manipulation of the reservoir temperatures is not part of the control strategy. For the saturation, at 40°C, of 283.5 kg-air h^{-1}, the evaporation rate will be 7.7 kg-H$_2$O h^{-1} (an inlet relative humidity of 80% was assumed, based on local weather information; obviously it can vary significantly with location). If the make-up water is provided at this rate but at 10°C, then heat must be added at the rate of 270 W in order to maintain the temperature of the water in the warm reservoir at 42°C. By placing a resistance heater of 1000 W in the warm water tank and one of 700 W in the cool water tank, the water temperature can be controlled easily (this can be assured even without making specific calculations about heat losses since the reservoirs are made of polypropylene, which has insulating properties, and also the tanks are covered to prevent evaporation to the surrounding air). This extra capacity also allows a faster warm-up of the reservoir at the beginning of a fermentation.

Humidification column. A computer program was used to determine the minimum design necessities for the humidification column. For a column diameter of 40 cm, a water flow rate through the column of 1.5 m^3 h^{-1}, a packing consisting of 25 mm Raschig rings, an inlet water temperature of 42°C and an inlet air temperature of 20°C, Fig. 29.5 shows the resulting predictions for the air and water temperature and the air humidity as functions of height within the column.

A 35 cm high column is sufficient to saturate the air at 40°C. For this height the pressure drop is equivalent to 2 mm of water. The air temperature can be manipulated by changing the water temperature or flow rate, although any such change will be done between different fermentation runs and not during a given fermentation (temperature change during a run is impractical due to the large thermal inertia of the reservoir). With this particular set of parameters, the air velocity through the column is 0.45 m s^{-1}, far below the air velocity that would cause flooding (1.5 m s^{-1}). Our column was designed with a packing height of 70 cm in order to guarantee saturation of the air even if we use different operating conditions.

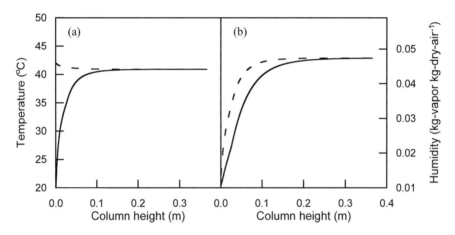

Fig. 29.5. Predicted performance of the humidification column operating under the conditions given in the text. **(a)** Temperatures of the (——) air and (- - -) water as a function of height within the column; **(b)** Humidity of the air as a function of height in the column. (- - -) saturation humidity, which changes due to the change in air temperature as the air passes through the column; (——) actual humidity of the air. This graph is used to decide on the height of the bed in the humidification column

Further Reading

Humidification
Coulson JM, Richardson JF (1990) Chemical engineering, 4th edn, Vol. 1, Pergamon, Oxford, Chap 13, pp 621-672
McCabe WL Smith JC (1976) Unit operations of chemical engineering, 3rd edn, McGraw-Hill, New York, Chap 19, pp 596-621

30 Future Prospects for SSF Bioreactors

David A. Mitchell, Marin Berovič, and Nadia Krieger

30.1 The Increasing Importance of SSF

Solid-state fermentation will become of ever increasing importance. The need to use rationally the organic solid wastes that we generate will increase as the increasing world population puts increasing pressure on environmental resources.

In the same manner as the world is coming to understand that, in dealing with liquid wastes, "dilution is NOT the solution to pollution", in the case of organic solid wastes, we can say (even if it does not rhyme so well) that "land-filling or disposal in the environment are NOT the solution to dealing with organic solid wastes". Unfortunately the economic models used in the majority of countries do not adequately penalize the generation and inappropriate disposal of organic solid wastes. Organic solid wastes are often treated as "somebody else's problem". The late Douglas Adams, in the book "Life, The Universe and Everything", one of the books of the "Hitchhiker's Guide to the Galaxy" series, defined "somebody else's problem" as "something that we can't see, or don't see, or our brain doesn't let us see, because we think that it's somebody else's problem" (Adams 1982). Unfortunately, in this case, when we say "somebody else" we mean "our descendants". We bury large quantities of organic solid wastes in the ground or dispose of them inadequately in the environment and wait for the real problem of dealing with them adequately to be faced by future generations. This seems neither sensible nor fair. However, currently, we seem unable or even unwilling to recognize this as a society. As a society we should be dealing adequately with organic solid wastes now and we should feel guilty that we are not doing it. What legacy are we leaving for future generations, not to mention the rest of the ecosystems on this earth?

It may not currently be "economic" to reuse or treat organic solid wastes properly. However, this is a question of the point of view that we use to look at the problem. We need to change our economic models. What price do we put on environmental quality? If we are not careful, sooner or later we are going to be knee-deep in putrefying rubbish. Maybe our difficulty in appreciating the problem is that this will probably not happen in our own lifetimes, but is this really a reason not to worry? We should start thinking harder about organic solid wastes now. We should put a realistic cost on their generation and disposal. The revenue raised can be used to develop and perfect various technologies, including SSF. Of course, the rational processing and treatment of these solids will require a multi-faceted approach. SSF in itself will be only one technology amongst many that will be necessary to cope with the huge volumes that we generate.

Although Holker et al. (2004) do point out correctly that we should be interested in small-scale SSF for specialty products that are induced by the physiological conditions to which the organism is subjected in SSF systems, SSF will also be one of the technologies applied to dealing with the large volumes of organic solid wastes that we generate, so we will need efficient large-scale SSF bioreactors and optimized operating strategies.

30.2 Present State and Future Prospects

It is clear that SSF is not a simple technology. In order for large-scale bioreactors to operate efficiently, we need to base their design and operation on engineering principles. We hope that we have convinced readers of this book that mathematical models of bioreactor operation will be important tools in the design and optimization of performance of large-scale SSF bioreactors. It will also be necessary to extend process control theory to deal adequately with the many complexities that SSF bioreactors present.

Despite our attempts in the present book to bring together the fundamental principles of SSF bioreactor design and operation, we are quite aware that much more work is required. The range of organisms and substrates and the particular challenges that they bring is so great that it is not currently possible to give a generally applicable key to bioreactor selection and operation. In fact, at the moment it would be quite misleading to present anything more specific than the simple key that was presented in Fig. 3.3. Perhaps the best we can do at the moment is to say:

- Do not expect the SSF bioreactor selection and design process to be easy. We hope that your ability to make good judgments in the design and development process has been improved by your having read this book.
- Take advantage of the mathematical models that accompany the present book in order to help you make design and operation decisions. However, do realize that these models are still imperfect tools that need many refinements.
- Despite the advances over the last three decades, the complexity of the SSF system and the phenomena occurring within it are so great that there are significant gaps in our understanding. For example: (1) Although experiments in simplified systems have given us insights into the role of intra-particle diffusion in controlling system performance, we understand relatively little about how these phenomena operate within real solid substrate particles, which are often quite heterogeneous at the microscale; (2) Many SSF processes involve filamentous fungi and we do not sufficiently understand the shear forces within agitated solid beds, the damage they cause to fungal hyphae and the resulting effects on growth of the fungus on the particle; (3) Although particle technology is a well-established area, we do not understand enough about particle be-

havior in solid beds in which fungal hyphae bridge the gaps between particles. As a result of these gaps in our knowledge, there is no guarantee that the application of our best currently available knowledge will lead us to perfectly optimized bioreactors.

We are aware that our book has left many loose ends. Each new large-scale SSF process that is established through the use of a rational approach (that is, through the application of engineering principles rather than through best-guessing or trial-and-error) will contribute to refining our knowledge base.

We envisage a future in which a single computer program can have fast-solving models of several SSF bioreactors embedded, requiring a minimum of information to be input (key bed parameters, heat and mass transfer coefficients, selection of an appropriate kinetic equation for growth and for product formation and input of kinetic parameters, and also parameters of equations describing environmental effects on growth). Such a program might be able to explore optimal design and operational variables for each of various bioreactors, giving outputs of the best performing fermentations for each bioreactor type. Such a program might also enable estimation of capital and operating costs. However, it might be as much as 20 years before a useful version of such a model is available. The amount of work required, not only to advance the theoretical framework of SSF bioreactor design and operation, but also to expand the database through experimental research and development, should not be underestimated!

So SSF bioreactor design remains a challenge. We hope that this book has not only contributed to your understanding of the issues and your ability to apply them, but has also stimulated you to contribute to further developments in this area.

References

Adams D (1982). Life, the universe and everything. Pan Books, London

Agosin E, Aguilera JM (1998) Industrial production of active propagules of *Trichoderma* for agricultural uses. In: Harman GE (ed) *Trichoderma & Gliocladium*, vol. 2. Taylor & Francis, London, pp 205–227

Ashley VM, Mitchell DA, Howes T (1999) Evaluating strategies for overcoming overheating problems during solid state fermentation in packed bed bioreactors. Biochem Eng J 3:141–150 (Figs. 6,7, and 8 adapted with permission from Elsevier)

Astrom KJ, Hägglund T (1984) Automatic tuning of simple regulators with specifications on phase and amplitude margins. Automatica 20:645-651

Aström KJ, Wittenmark B (1984) Computer controlled systems: Theory and design. Prentice-Hall, Englewood Cliffs

Atkinson JM (2000) Technical surveillance counter measures. Available in http://www.tscm.com/. Consulted: September, 2000

Auria R, Hernandez S, Raimbualt M, Revah S (1990) Ion exchange resin: A model support for solid state growth fermentation of *Aspergillus niger*. Biotech Techniques 4:391–396

Auria R, Morales M, Villegas E, Revah S (1993) Influence of mold growth on the pressure drop in aerated solid state fermentors. Biotech Bioeng 41:1007–1013

Auria R, Ortiz I, Villegas E, Revah S (1995) Influence of growth and high mould concentration on the pressure drop in solid state fermentations. Process Biochem 30:751–756

Auria R, Palacios J, Revah S (1991) A simple diffusion reactor to evaluate kinetic parameters in solid state fermentation. 4th World Cong Chem Eng, Karlsruhe, Germany, Abstract No.7.2-65

Bahr D, Menner M (1995) Solid-state-fermentation of starter cultures in fluidized bed. Part 2: Mathematical modeling. Bioforum 18:366–372

Bandelier S, Renaud R, Durand A (1997) Production of gibberellic acid by fed-batch solid state fermentation in an aseptic pilot-scale reactor. Process Biochem 32:141–145

Barstow LM, Dale BE, Tengerdy RP (1988) Evaporative temperature and moisture control in solid substrate fermentation. Biotechnology Techniques 2:237–242

Bell TA, Couch S W, Krieger TL, Feise HJ (2003) Screw feeders: A guide to selection and Use. Chem Eng Progress 99:44–51

Bellon-Maurel V, Orliac O, Christen P (2003) Sensors and measurements in solid state fermentation: a review. Process Biochem 38:881–896

Berovic M, Ostroversnik H (1997) Production of Aspergillus niger pectolytic enzymes by solid state bioprocessing of apple pomace. J Biotech 53:47–53

Blumberg W, Schlünder E-U (1996) Transversale Schuettgutbewegung und konvektiver Stoffübergang in Drehrohren. Teil 1: Ohne Hubschaufeln, Chem Eng Proc 35:395–404

Bramorski A, Christen P, Ramirez M, Soccol C, Revah S (1998) Production of volatile compounds by the edible fungus *Rhizopus oryzae* during solid state cultivation on tropical agro-industrial substrates. Biotech Lett 20:359–362

Brodgesell A, Lipták BG (1995) Analytical instrumentation (moisture in solids). In: Lipták BG (ed) Instrument engineers' handbook: process measurement and analysis. Chilton Book Company, Radnor, Pennsylvania, pp 1079–1084

Brosilow C, Joseph B (2002) Techniques of model-based control. Prentice Hall, Upper Saddle River, New Jersey

Calado VMA (1993) Secagem de cereais em leito fixo e fluxos cruzados. Ph.D. thesis. Universidade Federal de Rio de Janeiro

Calçada LA (1998) Secagem de materiais granulares porosos. Ph.D. Thesis. Universidade Federal de Rio de Janeiro

Chamielec Y, Renaud R, Maratray J, Almanza S, Diez M, Durand A (1994) Pilot-scale reactor for aseptic solid-state cultivation. Biotech Techniques 8:245–248

Chisti Y (1999) Solid substrate fermentation, enzyme production, food enrichment. In: Flickinger MC, Drew SW (eds) The encyclopedia of bioprocess technology: fermentation, biocatalysis and bioseparation, vol. 5. John Wiley, New York, pp 2446–2462

Christen P, Meza JC, Revah S (1997) Fruity aroma production in solid state fermentation by *Ceratocystis fimbriata*: influence of the substrate type and the presence of precursors. Mycological Research 101:911–919

Christofides PD (2001) Nonlinear and robust control of PDE systems. Birkäuser, Boston

Churchill SW, Chu HHS (1975) Correlating equations for laminar and turbulent free convection from a vertical plate. Int J Heat Mass Transfer 18:1323

Cole-Palmer Co (2003a) Selecting the right flowmeter – Part 1. Available in http://www.coleparmer.com/techinfo/techinfo.asp?openlist=D,E,C,A,D5,A1&htmlfile=SelectingFlowmeter1%2Ehtm&Title=Selecting+the+Right+Flowmeter%3A+Part+1. Consulted: July 2003

Cole-Palmer Co (2003b) Selecting the right flowmeter – Part 2. Available in http://www.coleparmer.com/techinfo/techinfo.asp?openlist=D,E,C,A,D5,A1&htmlfile=SelectingFlowmeter2%2Ehtm&Title=Selecting+the+Right+Flowmeter%3A+Part+2. Consulted: July 2003

Cooney CL, Wang DIC, Mateles RI (1968) Measurement of heat evolution and correlation with oxygen consumption during microbial growth. Biotech Bioeng 11:269–281

Costa JAV, Alegre RM, Hasan SDM (1998) Packing density and thermal conductivity determination for rice bran solid-state fermentation. Biotech Techniques 12:747–750

Creus A (1998a) Medida de Temperatura. In: Marcombo A (ed) Instrumentación industrial, 6th edn. Boixareu Editores, Colombia pp 223–300

Creus A (1998b) Humedad en sólidos. In: Marcombo A (ed) Instrumentación industrial, 6th edn. Boixareu Editores, Colombia pp 327–329

Creus A (1998c) Medida de Caudal. In: Marcombo A (ed) Instrumentación industrial, 6th edn. Boixareu Editores, Colombia pp 90–192

Creus A (1998d) Otras variables. In: Marcombo A (ed) Instrumentación industrial, 6th edn. Boixareu Editores, Colombia pp 301–363

Creus A (1998e) Concentración de gases. In: Marcombo A (ed) Instrumentación industrial, 6th edn. Boixareu Editores, Colombia pp 358–363

Cuero RG, Smith JE, Lacey J (1985) A novel containment system for laboratory scale solid particulate fermentations. Biotech Lett 7:463–466

Dalsenter FDH, Viccini, G, Barga MC, Mitchell DA, Krieger N (2005) A mathematical model describing the effect of temperature variations on the kinetics of microbial growth in solid-state culture. Process Biochem 40:801–807

de Reu JC, Zwietering MH, Rombouts FM, Nout MJR (1993) Temperature control in solid substrate fermentation through discontinuous rotation. Appl Microb Biotech 40:261–265 (Figs. 1 and 3 adapted with permission from Springer Science and Business Media)

Dominguez A, Rivela I, Couto SR, Sanroman MA (2001) Design of a new rotating drum bioreactor for ligninolytic enzyme production by *Phanerochaete chrysosporium* grown on an inert support. Process Biochem 37:549–554 (Fig. 1 adapted with permission from Elsevier)

dos Santos MM, da Rosa AS, Dal'Boit S, Mitchell DA, Krieger N (2004) Thermal denaturation: Is solid-state fermentation really a good technology for the production of enzymes? Bioresource Technol 93:261–268

Durand A (2003) Bioreactor designs for solid state fermentation. Biochem Eng J 13:113–125

Durand A, Vergoignan C, Desgranges C (1997) Biomass estimation in solid state fermentation. In: Roussos S, Lonsane BK, Raimbault M, Viniegra-González G (eds) Advances in solid state fermentation. Kluwer Academic Publishers, Dordrecht, pp 23-37

Durand A, Chereau D (1988) A new pilot reactor for solid-state fermentation: Application to the protein enrichment of sugar beet pulp. Biotech Bioeng 31:476–486 (Fig. 1 adapted with permission of John Wiley & Sons, Inc.)

Durand A, Pichon P, Desgranges C (1988) Approaches to K_La measurements in solid state fermentation. Biotech Techniques 2:11–16

Ellis SP, Gray KR, Biddlestone AJ (1994) Mixing evaluation of a Z-blade mixer developed as a novel solid state bioreactor. Food and Bioproducts Processing: Trans IChE, Part C. 72(C3), 158–162

Fernández M (2001) Control automático de un bioreactor piloto de lecho sólido. Ph.D. thesis, University of Chile

Fernández M, Pérez-Correa JR, Solar I, Agosin E (1996) Automation of a solid substrate cultivation pilot reactor. Bioprocess Eng 16, 1–4

Fernández M, Ananías J, Solar I, Perez R, Chang L, Agosin E (1997) Advances in the development of a control system for a solid substrate pilot bioreactor. In: Roussos S, Lonsane BK, Raimbault M, Viniegra-González G (eds) Advances in solid state fermentation. Kluwer Academic Publishers, Dordrecht, pp 155-168

Fogler HS (1999) Elements of chemical reaction engineering, 3rd edn. Prentice-Hall, New Jersey

Fung CJ, Mitchell DA (1995) Baffles increase performance of solid-state fermentation in rotating drums. Biotech Techniques 9:295–298 (Fig. 1 adapted with permission from Springer Science and Business Media)

Geankoplis CJ (1993) Transport processes and unit operations, 3rd edn. Prentice-Hall International, Englewood Cliffs

Ghildyal NP, Gowthaman MK, Raghava Rao KSMS, Karanth NG (1994) Interaction between transport resistances with biochemical reaction in packed-bed solid-state fermentors: Effect of temperature gradients. Enzyme Microb Technol 16:253–257 (Figs. 2 and 3 adapted with permission from Elsevier)

Gibbons WR, Westby CA, Dobbs TL (1984) A continuous, farm-scale, solid-phase fermentation process for fuel ethanol and protein feed production from fodder beets. Biotech Bioeng 26:1098–1107

Gibbons WR, Westby CA, Dobbs TL (1986) Intermediate-scale, semicontinuous solid-phase fermentation process for production of fuel ethanol from sweet sorghum. Appl Environ Microb 51:115–122

Gimson C (1989) Using the capacitance charge transfer principle for water content measurement. Measurement and Control 22:79

Glenn DR, Rogers PL (1988) A solid substrate process for an animal feed product: Studies on fungal strain improvement. Aust J Biotech 2(1):50–57 (Fig. 1 adapted with permission from the authors)

González-Sepúlveda M, Agosin E (2000) Capture of volatile metabolites for tracking the evolution of gibberellic acid in a solid-state culture of *Gibberella fujikuroi*. Biotech Lett 22:1849–1854

Gowthaman MK, Ghildyal NP, Raghava Rao KSMS, Karanth, N.G. (1993a) Interaction of transport resistances with biochemical reaction in packed-bed solid state fermenters: The effect of gaseous concentration gradients. J Chem Technol Biotech 56:233–239

Gowthaman MK, Raghava Rao, KSMS, Ghildyal NP, Karanth NG (1993b) Gas concentration and temperature gradients in a packed-bed solid-state fermentor. Biotech Adv 11:611–620 (Figs. 3 and 5 and Table 1 adapted with permission from Elsevier)

Gowthaman MK, Raghava Rao KSMS, Ghildyal NP, Karanth NG (1995) Estimation of $K_l a$ in solid-state fermentation using a packed-bed bioreactor. Process Biochem 30:9–15

Gumbira-Sa'id E, Greenfield PF, Mitchell DA, Doelle HW (1993) Operational parameters for packed-beds in solid-state cultivation. Biotech Adv 11:599–610 (Fig. 6 adapted with permission from Elsevier)

Gutierrez-Rojas M, Amar Aboul Hosn S, Auria R, Revah S, Favela-Torres E (1996) Heat transfer in citric acid production by solid state fermentation. Process Biochem 31:363–369

Gutierrez-Rojas M, Auria R, Benet JC, Revah S (1995) A mathematical model for solid state fermentation of mycelial fungi on inert support. Chem Eng J 60:189–198

Habijanic J, Berovic M (2000) The relevance of solid-state substrate moisturing on *Ganoderma lucidum* biomass cultivation. Food Technol Biotech 38:225–228

Hallstrom B, Skjoldebrand C, Tragardh C (1988) Heat transfer and food products. Elsevier Applied Science, New York

Hamidi-Esfahani Z, Shojaosadati SA, Rinzema A (2004) Modelling of simultaneous effect of moisture and temperature on *A. niger* growth in solid-state fermentation. Biochem Eng J 21:265–272

Hardin MT, Howes T, Mitchell DA (2001) Residence time distributions of gas flowing through rotating drum bioreactors. Biotech Bioeng 74:145–153 (Figs. 6 and 7 adapted with permission from John Wiley & Sons, Inc.)

Hardin MT, Howes T, Mitchell DA (2002) Mass transfer correlations for rotating drum bioreactors. J Biotech 97:89–101 (Figs. 3 and 5 and Table 1 adapted with permission from Elsevier)

Hasan SDM, Costa JAV, Sanzo AVL (1998) Heat transfer simulation of solid state fermentation in a packed-bed bioreactor. Biotech Techniques 12:787–791

Hesseltine CW (1977) Solid state fermentation. Part II. Process Biochem 12(9), 24-27

Hillen G (1999) A Complete Moisture Sensing and Control System. Available in http://www.aquapro-sensors.com/. Consulted: February, 2001

Himmelblau DM (1982) Basic Principles and calculations in chemical engineering, 5th ed. Prentice Hall, Englewood Cliffs

Holker U, Hofer M, Lenz J (2004) Biotechnological advantages of laboratory-scale solid-state fermentation with fungi. Appl Microb Biotech 64:175–186

Hrubant GR, Rhodes RA, Orton WL (1989) Continuous solid substrate fermentation of feedlot waste grain. Biological Wastes 28:277–291

Ikasari L, Mitchell DA (1998) Oxygen uptake kinetics during solid state fermentation with *Rhizopus oligosporus*. Biotech Techniques 12:171–175

Ikasari L, Mitchell DA, Stuart DM (1999) Response of *Rhizopus oligosporus* to temporal temperature profiles in a model solid-state fermentation system. Biotech Bioeng 64:722–728

Ishida H, Hata Y, Kawato A, Abe Y, Suginami K, Imayasu S (2000) Identification of functional elements that regulate the glucoamylase-encoding gene (glab) expressed in solid-state culture of *Aspergillus oryzae*. Curr Genet 37:373–379

Ishikawa Y, Kase M, Sasaki M, Satoh K, Sasaki S (1982) Recent progress in the sintering technology – high reproducibility and improvement of fuel consumption. Ironmaking Conference Proceedings 41, 80–89

Kalogeris E, Fountoukides G, Kekos D, Macris BJ (1999) Design of a solid-state bioreactor for thermophilic microorganisms. Bioresource Technol 67:313–315

Kaminski RK, Razaq M, Lipták BG (1995) Analytical instrumentation (oxygen in gases). In: Lipták BG (ed) Instrument engineers' handbook: process measurement and analysis. Chilton Book Company, Radnor, Pennsylvania pp 1116-1123

Kays WM, Bjorklund IS (1958) Heat transfer from a rotating cylinder with and without crossflow. Trans ASME 80:70–78

Kossen NWF, Oosterhuis NMG (1985) Modelling and scaling up of bioreactors. In: Rehm HJ, Reed G (eds) Biotechnology, vol 2. Verlag Chemie, Weinheim, pp 571–605

Lai MN, Wang HH, Chang FW (1989) Thermal diffusivity of solid mash of sorghum brewing - a solid state fermentation. Biotech Bioeng 34:1337–1340

Lekanda JS, Pérez-Correa JR (2004) Energy and water balances using kinetic modeling in a pilot-scale SSF bioreactor. Process Biochem 39:1793–1802

Lipták BG (1995a) Temperature measurement. In: Lipták BG (ed) Instrument engineers' handbook: process measurement and analysis. Chilton Book Company, Radnor, Pennsylvania pp 399–522

Lipták BG (1995b) Pressure measurement. In: Lipták BG (ed) Instrument engineers' handbook: process measurement and analysis. Chilton Book Company, Radnor, Pennsylvania pp 523–601

Lipták B (2002) How to select the right flowmeter. Available in http://www.controlmagazine.com/Web_First/ct.nsf/ArticleID/PSTR-5B7RG3/. Consulted: July, 2003

Lomas DJ, Lipták GB (1995) Flow measurement (application and selection). In: Lipták BG (ed) Instrument engineers' handbook: process measurement and analysis. Chilton Book Company, Radnor, Pennsylvania pp 76–87

Lotong N, Suwanarit P (1983) Production of soy sauce koji mold spore inoculum in plastic bags. Appl Env Microb 46:1224–1226

Lüth P, Eiben U (1999) Solid-state fermenter and method for solid-state fermentation. World Patent No. WO 99/57239

Maciejowski JM (2002) Predictive control with constraints. Pearson Education Limited, Prentice Hall. Harlow, UK

Mancini MC (1996) Transferência de massa em secadores de grãos. PhD thesis. Universidade Federal de Rio de Janeiro

Matsuno R, Adachi S, Uosaki H (1993) Bioreduction of prochiral ketones with yeast cells cultivated in a vibrating air-solid fluidized bed fermentor. Biotech Adv 11:509–517

McCabe WL, Smith JC, Harriott P (1985) Unit operations of chemical engineering. McGraw-Hill, Sydney.

Milner RJ (2000) Current status of *Metarhizium* as a mycoinsecticide in Australia. Biocontrol News and Information 21(2) 47N–50N

Mitchell DA, Lonsane BK, Durand A, Renaud R, Almanza S, Maratray J, Desgranges C, Crooke PS, Hong K, Tanner RD, Malaney GW (1992) General principles of reactor design and operation for SSC. In: Doelle HW, Mitchell DA, Rolz CE (eds), Solid substrate cultivation. Elsevier Applied Science, London. pp 115–139

Mitchell DA, Pandey A, Sangsurasak P, Krieger N (1999) Scale-up strategies for packed-bed bioreactors for solid-state fermentation. Process Biochem 35:167–178

Mitchell DA, Berovic M, Krieger N (2000) Biochemical engineering aspects of solid state bioprocessing. Adv Biochem Eng/Biotech 68:61–138

Mitchell DA, von Meien OF (2000) Mathematical modeling as a tool to investigate the design and operation of the Zymotis packed-bed bioreactor for solid-state fermentation. Biotech Bioeng 68:127–135 (Fig. 5 reproduced with permission from Springer Science and Business Media)

Mitchell DA, Tongta A, Stuart DM, Krieger N (2002a) The potential for establishment of axial temperature profiles during solid-state fermentation in rotating drum bioreactors. Biotech Bioeng 80:114–122 (Fig. 3 reproduced with permission from John Wiley & Sons, Inc.)

Mitchell DA, von Meien OF, Luz Jr LFL, Krieger N (2002b) Evaluation of productivity of Zymotis solid-state bioreactor based on total reactor volume. Food Technol Biotech 40:135–144

Montague G (1997) Monitoring and control of fermenters. Institution of Chemical Engineering, UK

Moo-Young M, Moreira AR, Tengerdy RP (1983) Principles of solid-substrate fermentation. In: Smith JE, Berry DR, Kristiansen B (eds) The filamentous fungi, vol 4. Edward Arnold, London, pp 117–144

Moyers CG, Baldwin GW (1999) Psychrometry, evaporative cooling, and solids drying. In: Perry RH, Green DW (eds) Perry's chemical engineers' handbook, 7th edn. McGraw-Hill, New York, Section 12, pp 12-1–12-90

Nagel FJI, Tramper J, Bakker MSN, Rinzema A (2001a) Temperature control in a continuously mixed bioreactor for solid state fermentation. Biotech Bioeng 72:219–230 (Figs. 5, 6, 7 , 8, and 10 adapted with permission from John Wiley & Sons. Inc.)

Nagel FJI, Tramper J, Bakker MSN, Rinzema A (2001b) Model for on-line moisture-content control during solid-state fermentation. Biotech Bioeng 72:231–243

Nandakumar MP, Thakur MS, Raghavarao KSMS, Ghildyal NP (1994) Mechanism of solid particle degradation by *Aspergillus niger* in solid state fermentation. Process Biochem 29:545–551

Ogunnaike B, Ray RWH (1994) Process dynamics, modeling, and control. Oxford University Press, New York

Ooijkaas LP, Tramper J, Buitelaar RM (1998) Biomass estimation of *Coniothyrium minitans* in solid-state fermentation. Enzyme Microb Technol 22:480–486

Ooijkaas LP, Buitelaar RM, Tramper J, Rinzema A (2000) Growth and sporulation stoichiometry and kinetics of *Coniothyrium minitans* on agar media. Biotech Bioeng 69:292–300

Oostra J, Tramper J, Rinzema A (2000) Model-based bioreactor selection for large-scale solid-state cultivation of *Coniothyrium minitans* spores on oats. Enzyme Microb Technol 27:652–663

Oostra J, le Comte EP, van den Heuvel JC, Tramper J, Rinzema A (2001) Intra-particle oxygen diffusion limitation in solid-state fermentation. Biotech Bioeng 74:13–24

Pajan H, Perez-Correa R, Solar I, Agosin E (1997) Multivariable model predictive control of a solid substrate pilot bioreactor: A simulation study. In: Wise DL (ed) Global environmental biotechnology. Kluwer Academic Publishers, Dordrecht, pp 221–232

Pandey A (1991) Aspects of fermenter design for solid-state fermentations. Process Biochem 26:355–361

Peña y Lillo M, Perez-Correa R, Latrille E, Fernandez M, Acuna G, Agosin E (2000) Data processing for solid-substrate cultivation bioreactors. Bioprocess Eng 22:291–297

Peña y Lillo M, Pérez-Correa R, Agosin E, Latrille E (2001) Indirect measurement of water content in an aseptic solid substrate cultivation pilot-scale bioreactor. Biotech Bioeng 76:44–51

Pérez-Correa JR, Agosin E (1999) Automation of solid-substrate fermentation processes. In: Flickinger MC, Drew SW (eds) The encyclopedia of bioprocess technology: fermentation, biocatalysis and bioseparation, vol. 5. John Wiley, New York, pp. 2429–2446

Perry RH, Green DW, Maloney JO (1984) Perry's Chemical Engineer's Handbook, 5th ed. McGraw-Hill, New York.

Ragheva Rao KSMS, Gowthaman MK, Ghildyal NP, Karanth NG (1993) A mathematical model for solid state fermentation in tray bioreactors. Bioprocess Eng 8:255–262

Rahardjo YSP, Weber FJ, le Comte EP, Tramper J, Rinzema A (2002) Contribution of aerial hyphae of *Aspergillus oryzae* to respiration in a model solid-state fermentation system. Biotech Bioeng 78:539–544

Raimbault M, Alazard D (1980) Culture method to study fungal growth in solid fermentation. Eur J Appl Microb Biotech 9:199–209

Rajagopalan S, Modak JM (1994) Heat and mass transfer simulation studies for solid-state fermentation processes. Chem Eng Sci 49:2187–2193 (Figs. 2 and 3 and Table 2 adapted with permission from Elsevier)

Rajagopalan S, Modak JM (1995) Evaluation of relative growth limitation due to depletion of glucose and oxygen during fungal growth on a spherical solid particle. Chem Eng Sci 50:803–811

Rajagopalan S, Rockstraw DA, Munson-McGee SH (1997) Modeling substrate particle degradation by *Bacillus coagulans* biofilm. Bioresource Technol 61:175–183

Ramos-Sánchez LB (2000) Aplicación de la modelación matemática al desarrollo de la tecnología de fermentación del BAGARIP. Ph.D. Thesis. University of Camagüey, Cuba

Ramos–Sánchez LB, Julian-Ricardo MC, Suárez-Rodríguez Y, Chacón-Reyes M, Rodríguez AG (2003) Process development for food processing by solid-state fermentation. Current Studies of Biotechnology, in press

Rathbun BL, Shuler ML (1983) Heat and mass transfer effects in static solid-substrate fermentations: Design of fermentation chambers. Biotech Bioeng 25:929–938

Reid RC, Prausnitz JM, Sherwood TK (1977) The properties of gases and liquids, 3rd edn. McGraw-Hill, New York

Rottenbacher L, Schossler M, Bauer W (1987) Modelling a solid-state fluidized bed fermenter for ethanol production with *Saccharomyces cerevisiae*. Bioprocess Eng 2:25–31

Roussos S, Raimbault M, Prebois J-P, Lonsane BK (1993) Zymotis, a large scale solid state fermenter. Appl Biochem Biotech 42:37–52

Ryoo D, Murphy VG, Karim MN, Tengerdy RP (1991) Evaporative temperature and moisture control in a rocking reactor for solid substrate fermentation. Biotech Techniques 5:19–24

Sangsurasak P, Mitchell DA (1995) Incorporation of death kinetics into a 2-D dynamic heat transfer model for solid state fermentation. J Chem Technol Biotech 64:253–260

Sangsurasak P, Mitchell DA (1998) Validation of a model describing two-dimensional heat transfer during solid-state fermentation in packed bed bioreactors. Biotech Bioeng 60:739–749

Sargantanis J, Karim MN, Murphy VG, Ryoo D, Tengerdy RP (1993) Effect of operating conditions on solid substrate fermentation. Biotech Bioeng 42:149–158

Sastry KVS, Cooper H, Hogg R, Jespen TLP, Knoll F, Parekh B, Rajamani RK, Sorenson T, Wechsler I (1999) Solid-solid operations and equipment. In: Perry RH, Green DW, Maloney JO (eds), Perry's chemical engineers' handbook, 7th ed. McGraw-Hill, New York, Section 19, pp. 19-1–19-66

Sato K, Sudo S (1999) Small-scale solid-state fermentations. In: Demain AL, Davies JE (eds) Manual of industrial microbiology and biotechnology, 2nd edn. ASM Press, Washington DC, pp 61–79

Sato K, Yoshizawa K (1988) Growth and growth estimation of *Saccharomyces cerevisiae* in solid-state ethanol fermentation. J Ferment Technol 66:667–673

Sato K, Nagatani M, Nakamura K, Sato S (1983) Growth estimation of *Candida lipolytica* from oxygen uptake in a solid state culture with forced aeration. J Ferment Technol 61:623–629

Saucedo-Castañeda G, Gutierrez-Rojas M, Bacquet G, Raimbault M, Viniegra-Gonzalez G (1990) Heat transfer simulation in solid substrate fermentation. Biotech Bioeng 35:802–808. (Figs. 2, 4, and 6 adapted with permission from John Wiley & Sons, Inc.)

Saucedo-Castañeda G, Lonsane BK, Navarro JM, Roussos S, Raimbault M (1992) Control of carbon dioxide in exhaust air as a method for equal biomass yields at different bed heights in a column fermentor. Appl Microb Biotech 37:580–582

Saucedo-Castañeda G, Trejo-Hernández MR, Lonsane BK, Navarro JM, Roussos S, Dufour D, Raimbault M (1994) On-line automated monitoring and control system for CO_2 and O_2 in aerobic and anaerobic solid-state fermentation. Process Biochem 29:13–24

Savage SB (1979) Gravity flow of cohesionless granular materials in chutes and channels. J Fluid Mech 92:53–96

Schutyser MAI, Padding JT, Weber FJ, Briels WJ, Rinzema A, Boom RM (2001) Discrete particle simulations predicting mixing behavior of solid substrate particles in a rotating drum fermenter. Biotech Bioeng 75:666–675 (Table II adapted with permission from John Wiley & Sons, Inc.)

Schutyser MAI, Weber FJ, Briels WJ, Boom RM, Rinzema A (2002) Three-dimensional simulation of grain mixing in three different rotating drum designs for solid-state fermentation. Biotech Bioeng 79:284–294 (Fig. 2 adapted with permission from John Wiley & Sons, Inc.)

Schutyser MAI, Briels WJ, Boom RM, Rinzema A (2004) Combined discrete particle and continuum model predicting solid-state fermentation in a drum fermentor. Biotech Bioeng 86:405-413

Schutyser MAI, de Pagter P, Weber FJ, Briels WJ, Boom RM, Rinzema A (2003a) Sub-strate aggregation due to aerial hyphae during discontinuously mixed solid-state fermentation with *Aspergillus oryzae*: Experiments and modeling. Biotech Bioeng 83:503–513

Schutyser MAI, Briels WJ, Rinzema A, Boom RM (2003b) Numerical simulation and PEPT measurements of a 3D conical helical-blade mixer: A high potential solids mixer for solid-state fermentation Biotech Bioeng 84:29–39 (Figs. 2, 6, 8, and 9 adapted with permission of John Wiley & Sons, Inc.)

Schutyser MAI, Weber FJ, Briels WJ, Rinzema A, Boom RM (2003c) Heat and water transfer in a rotating drum containing solid substrate particles. Biotech Bioeng 82:552–563

SCI (1996) Smells light up artificial nose. Available in http://pharma.mond.org/9617/ 961710.html. Consulted: December, 2000

SE (1998) Soilmoisture Equipment Corp. Website. Available in http://www.soilmoisture.com/. Consulted: December, 2000

Siev R, Arant J.B, Lipták GB (1995) Flow measurement (laminar flow-meters). In: Lipták BG (ed) Instrument engineers' handbook: process measurement and analysis. Chilton Book Company, Radnor, Pennsylvania, pp 104–109

Silva EM,Yang ST (1998) Production of amylases from rice by solid-state fermentation in a gas-solid spouted-bed bioreactor. Biotech Progress 14:580–587

Smits JP, Rinzema A, Tramper J, Schlosser EE, Knol W (1996) Accurate determination of process variables in a solid-state fermentation system. Process Biochem 31:669–678

Smits JP, Rinzema A, Tramper J, van Sonsbeek HM, Hage JC, Kaynak A, Knol W (1998) The influence of temperature on kinetics in solid-state fermentation. Enzyme Microb Technol 22:50–57

Smits JP, van Sonsbeek HM, Tramper J, Knol W, Geelhoed W, Peeters M, Rinzema A (1999) Modelling fungal solid-state fermentation: the role of inactivation kinetics. Bioprocess Eng 20: 391–404

Stuart DM (1996) Solid-state fermentation in rotating drum bioreactors. Ph.D. thesis, The University of Queensland

Stuart DM, Mitchell DA, Johns MR, Litster JD (1999) Solid-state fermentation in rotating drum bioreactors: Operating variables affect performance through their effects on transport phenomena. Biotech Bioeng 63:383–391 (Fig. 4 adapted with permission from John Wiley & Sons, Inc.)

Stuart DM, Mitchell DA (2003) Mathematical model of heat transfer during solid-state fermentation in well-mixed rotating drum bioreactors. J Chem Technol Biotech 78:1180–1192

Sunesson AL, Vaes WHJ, Nilsson CA, Blomquist G, Andersson B, Carlson R (1995) Iden-tification of volatile metabolites from five fungal species cultivated on two media. Appl Environ Microb 61:2911–2918

Suryanarayan S (2003) Current industrial practice in solid state fermentations for secondary metabolite production: the Biocon India experience. Biochem Eng J 13:189–195

Suryanarayan S, Mazumdar K (2000) Solid state fermentation. World Patent No. WO 00/29544

Sweat VE (1986) Thermal properties of foods. In: Rao MA, Rizvi SS (eds) Engineering properties of foods. Marcel Dekker, New York, pp 49–132

Szewczyk KW (1993) The influence of heat and mass transfer on solid state fermentation. Acta Biochimica Polonica 40:90–92

Takamine J (1914) Enzymes of *Aspergillus oryzae* and the application of its amyloclastic enzyme to the fermentation industry. Ind Eng Chem 6:824–828

Thibault J, Pouliot K, Agosin E, Perez-Correa R (2000a) Reassessment of the estimation of dissolved oxygen concentration profile and K_La in solid-state fermentation. Process Biochem 36:9–18

Thibault J, Acuna G, Perez-Correa R, Jorquera H, Molin P, Agosin E (2000b) A hybrid representation approach for modelling complex dynamic bioprocesses. Bioprocess Eng 22:547–556

Toyama N (1976) Feasibility of sugar production from agricultural and urban cellulosic wastes with *Trichoderma viride* cellulase. Biotech Bioeng Symp 6:207–219

Tscheng SH, Watkinson AP (1979) Convective heat transfer in a rotary kiln. Canadian J Chem Eng 57:433–443

Underkofler LA, Fulmer EI, Schoene L (1939) Saccharification of starchy grain mashes for the alcoholic fermentation industry. Ind Eng Chem 31:734–738

van de Lagemaat J, Pyle DL (2001) Solid state fermentation and bioremediation: Development of a continuous process for the production of fungal tannase. Chem Eng J 84:115–123

Viccini G, Mitchell DA, Boit SD, Gern JC, da Rosa AS, Costa RM, Dalsenter FDH, von Meien OF, Krieger N (2001) Analysis of growth kinetic profiles in solid-state fermentation. Food Technol Biotech 39:271–294

Villegas E, Aubague S, Alcantara L, Auria R, Revah S (1993) Solid state fermentation: acid protease production in controlled CO_2 and O_2 environments. Biotech Adv 11:387–397

von Meien OF, Mitchell DA (2002) A two-phase model for water and heat transfer within an intermittently-mixed solid-state fermentation bioreactor with forced aeration. Biotech Bioeng 79:416–428

von Meien OF, Luz Jr LFL, Mitchell DA, Pérez-Correa JR, Agosin E, Fernández-Fernández M, Arcas JA (2004) Control strategies for intermittently mixed, forcefully aerated solid-state fermentation bioreactors based on the analysis of a distributed parameter model. Chem Eng Sci 59:4493–4504 (Figs. 2, 5, and 6 adapted with permission from Elsevier)

Wang HH (1993) Assessment of solid state fermentation by a bioelectronics artificial nose. Biotech Adv 11:701–710

Weber FJ, Oostra J, Tramper J, Rinzema A (2002) Validation of a model for process development and scale-up of packed-bed solid-state bioreactors. Biotech Bioeng 77:381–393 (Figs. 3, 7, and 8 adapted with permission from John Wiley & Sons, Inc.)

Weber FJ, Tramper J, Rinzema A (1999) A simplified material and energy balance approach for process development and scale-up of *Coniothyrium minitans* conidia production by solid-state cultivation in a packed-bed reactor. Biotech Bioeng 65:447–458

Welch G, Bishop G (2003) An introduction to the Kalman filter. (TR 95-04, Department of Computer Science, University of North Carolina at Chapel Hill). Available in: http://www.cs.unc.edu/~welch/kalman/. Consulted: May, 2003

Wightman C, Muzzio FJ (1998) Mixing of granular material in a drum mixer undergoing rotational and rocking motions I. Uniform particles. Powder Technol 98:113–124 (Fig. 1 adapted with permission from Elsevier)

Xue, M., Liu, D., Zhang, H., Hongyan, Q. & Lei, Z. (1992) A pilot process of solid state fermentation from sugar beet pulp for the production of microbial protein. J Ferment Bioeng 73:203–205

Yokotsuka T (1985) Traditional fermented soybean foods. In: Moo-Young M, Drew S, Blanch HW (eds) Comprehensive biotechnology. Pergamon Press, Oxford, New York, pp 395–427

Ziegler JG, Nichols NB (1942) Optimum settings for automatic controllers. Trans ASME 64:759-768

Ziffer J (1988) Wheat bran culture process for fungal amylase and penicillin production. In: M. Raimbault (ed) Solid state fermentation in bioconversion of agro-industrial raw materials. ORSTOM, Montpellier, pp 121–128

Appendix: Guide to the Bioreactor Programs

A.1 Disclaimer

The models that accompany this book represent tools developed in research. They have not yet been fully proven. They are provided with the intention of helping workers in the area of solid-state fermentation (SSF) to understand the complex types of behaviors that can be expected to be shown by SSF bioreactors. Please note that no guarantees or warranties are given as to their suitability for designing bioreactors. We fully recognize that much work is required to improve the models before they will become truly useful tools in SSF bioreactor development.

A.2 General Information and Advice

Five different programs can be obtained by sending an email to David Mitchell <davidmitchell@ufpr.br>. Each program is available in a zipped format (Table A.1):

Table A.1. The program files

Related to	File (.zip)	The program simulates:
Chap. 22	WellMixed	A well-mixed bioreactor
Chap. 23	RotatingDrum	A rotating drum bioreactor
Chap. 24.2	PackedBed	A traditional packed-bed bioreactor
Chap. 24.3	Zymotis	A Zymotis packed-bed bioreactor
Chap. 25	Intermittent	An intermittently-mixed forcefully-aerated bioreactor

Each of these zipped files contains the two files that you need to run the corresponding program. These two files are:

- an input file that specifies the values of some of the parameters and operating variables that the program will use. A default input file (named "input.txt" in all cases) is provided for each program.
- an executable file.

These programs should run on IBM-compatible PCs with Microsoft Windows®. The general procedure for running each program is to:

- save the ".zip" file to a directory on your hard disk;
- extract the zip file, which will produce an .exe file;
- edit the input file (changing values of parameters and variables as appropriate);
- run the program by clicking twice on the executable file. A DOS window will appear. It will ask you for the input and output file names. Answer with file names in DOS format, pressing RETURN after each file name. If you make a mistake, it is not possible to go back and change file names. In this case, press CTRL-C to close the DOS window and then open another one by clicking twice on the executable file). Please make sure that you type the input file name exactly, otherwise you will get an error message. After entering the last file name the program will run. Also, the input file must be in the same directory as the executable file that you are running. As the program runs, it will show within the DOS window the time of fermentation. When the program has finished a message will appear saying "Press any key to continue". After pressing a key the DOS window will disappear. The output files should have appeared in the directory (the same directory as the executable program file);
- inspect the output files that appear in the directory, importing them into a spreadsheeting program if you wish to make graphs.

More detail about running each program is provided in the following subsections for each of the individual bioreactor programs. When prompted for file names, you should use DOS-type filenames (i.e., a filename of eight letters or numbers followed by a point and an extension of three letters), without using any special characters. Note also that you can save various input data files containing different information with different file names. When you run the program you can then give, when prompted, the name of any one of the saved input files.

Each input file contains the following instructions within it "The program reads one number per line before skipping to the next line. (That is, it doesn't even see the text that is to the right of each number). So please leave this input file with exactly the appearance that it has now. It would be a good idea to save a copy of this as "backup.dat", so that you always have a file in the correct format to come back to, if it becomes necessary! Also, please put decimal points in all numbers EXCEPT those marked as integers." Note that scientific notation is used for very large or very small numbers. For example, "23245000" would be written as "2.3245D7", while "0.000023245" would be written as "2.3245D-5".

The various output files are text files, designed to be easily read by a spreadsheeting program. Most spreadsheeting programs can be used for the rapid construction of graphs, which will help you to visualize the trends in the data. In most cases you will need to open your spreadsheeting program and then, once within the spreadsheet, you should import the ".txt" file. Note that there may potentially be a problem if your spreadsheeting program uses commas as the decimal separator, since the output of the bioreactor will have the decimal point. Some of the most recent spreadsheeting programs allow you to specify whether the file that you are importing uses the comma or decimal point, at the time that you import the data. If there are problems, one strategy is to open the ".txt" file in a text editor or word processor and then use the "substitute all" function to substitute all the decimal points with commas. You should make sure that you save the file as a simple "text only" file before exiting the text editor/word processor.

The remaining sections assume that you have followed the above instructions and have already extracted the ".zip" file for each bioreactor program to different directories on your hard disk.

Please note that all the figures and tables in this appendix show the variables, parameters, and explanations in the format that they appear in the text files that either accompany or are generated by the corresponding computer program (i.e., in normal font and without superscripts or subscripts).

A.3 Use of the Well-Mixed Bioreactor Model

This section gives advice about the use of the model for the well-mixed bioreactor that was presented in Chap. 22. The following steps should be followed.

First use a text editor to edit the input data file. An input data file (text file) has been supplied with the name "input.txt". The contents and format of this file are shown in Fig. A.1. Change the values as appropriate, save the file (as a text file), and exit the file editor.

Then click twice on "WellMixed.exe". A "DOS window" will appear on the screen. Several questions will appear, one after the other.

- What is the name of the INPUT file?
- What is the name of the OUTPUT file?
- What is the name of the COOLING CONTRIBUTIONS file?
- What is the name of the ECHO file?

After you answer the last question, the program will run automatically. When the instruction "Press any key to continue" appears, press the space bar.

Inspect the output files. Assuming that you have called the three output files "output.txt", "cooling.txt", and "echo.txt", respectively:

- Output.txt will have 15 columns with the headings that are explained in Table A.2. The number of rows will depend on the "number of outputs to be reported" that was selected in the input file.
- Cooling.txt will have 6 columns with the headings that are explained in Table A.3. The number of rows will depend on the "number of outputs to be reported" that was specified in the input file.
- Echo.txt contains an echo of various input values and other values calculated in the program. The content is self-explanatory.

2.0	height = bed height (m)
2.0	diameter = bioreactor diameter (m)
5.0	wallthickness (mm)
1.0	vvm = volumes of air per total bed volume per minute
0.87	outlet gas water activity setpoint for triggering water addition
1.0	fold increase in transfer due to mixing
35.0	initial system temperature (deg C) Wall solids gas
35.0	Tin = temperature of air at the air inlet (deg C)
35.0	Tsetpoint = setpoint temperature used in water jacket temperature control scheme (C)
0.0	gain = gain factor used in the water jacket temperature control scheme
0.99	awgo = initial water activity of the gas phase
0.50	awgin = water activity of the inlet air
0.99	awso initial water activity of the solid phase
0.002	Y(1) = initial biomass concentration (kg biomass/kg solid)
0.250	Xmax (kg X /kg solids)
0.236	Uopt (h-1)
0.5	Yxs
1	Organism type 1=Aspergillus 2=Rhizopus (MUST BE AN INTEGER)
0.5	Effective bed porosity during mixing (volvoids/voltotal)
450.0	Density of dry solid particles (the particles in themselves, not a bed of particles)
60.0	Total time to be simulated (h)
60	Number of outputs to be reported (MUST BE AN INTEGER)
200.	heat transfer coefficient (wall to cooling water)(W/(m2.K))
200.	heat transfer coefficient (solid phase in bed to wall)(W/(m2.K))
200.	heat transfer coefficient (gas phase in bed to wall)(W/(m2.K))

Fig A.1 Appearance of the input file (input.txt) for the well-mixed bioreactor model

Table A.2. Significance of the column headings in the "OUTPUT" file generated by the well-mixed bioreactor model

No.	Heading	Significance
1	t(h)	Time (h)
2	X(kg)	Total mass of dry biomass in the bioreactor (kg)
3	M(kg)	Total mass of dry solids (residual substrate and biomass) in the bioreactor (kg)
4	W(kg/kg)	Water content of the solids phase (kg-water kg- dry- solids^{-1})
5	H(kg/kg)	Gas phase humidity (kg-vapor kg-dry-air^{-1})
6	Ts(C)	Temperature of the solids (°C)
7	Tg(C)	Temperature of the inter-particle gas phase (°C)
8	Tb(C)	Temperature of the bioreactor wall (°C)
9	Xopt(kg)	Total mass of dry biomass in the bioreactor (kg) that would be obtained if conditions were ideal throughout the process
10	Awgas	Water activity of the inter-particle gas phase
11	Awsolid	Water activity of the solids phase
12	mu(T)	Fractional specific growth rate based on temperature (μ_{FT})
13	mu(W)	Fractional specific growth rate based on water activity (μ_{FW})
14	mu(1/h)	Value of the specific growth rate constant (h^{-1})
15	Twater	Temperature of the cooling water in the water jacket (°C)

Table A.3. Significance of the column headings in the "COOLING CONTRIBUTIONS" file generated by the well-mixed bioreactor model

No.	Heading	Significance
1	t(h)	Time (h)
2	growth	Rate of waste heat generation by the growth reaction that is taking place in the solids phase (W)
3	toWALL	Rate of sensible heat transfer from the solids phase to the bioreactor wall (W). A negative number means transfer from the wall to the solids
4	toGAS	Rate of sensible heat transfer from the solids phase to the gas phase (W). A negative number means transfer from the gas phase to the solids phase)
5	byEVAP	Rate of heat removal from the solids phase by evaporation (W). A negative number means that condensation is occurring, bringing energy to the solids phase
6	growth-others	Difference between the rate of waste metabolic heat and heat removal, calculated as "growth – (toWall+toGAS+byEVAP)"

A.4 Use of the Rotating-Drum Bioreactor Model

This section gives advice about the use of the model for the rotating-drum bioreactor that was presented in Chap. 22. The following steps should be followed.

First use a text editor to edit the input data file. An input data file (text file) has been supplied with the name "input.txt". The contents and format of this file are shown in Fig. A.2. Change the values as appropriate, save the file (as a text file) and exit the file editor.

Then click twice on "RotatingDrum.exe". A "DOS window" will appear on the screen. Several questions will appear, one after the other.

- What is the name of the INPUT file?
- What is the name of the OUTPUT file?
- What is the name of the COOLING CONTRIBUTIONS file?
- What is the name of the ECHO file?

After you answer the last question, the program will run automatically. When the instruction "Press any key to continue" appears, press a key (for example, the space bar).

Inspect the output files. Assuming that you have called the three output files "output.txt", "cooling.txt", and "echo.txt", respectively:

- Output.txt will have 15 columns with the headings shown in Table A.4. The number of rows will depend on the "number of outputs" that was selected in the input file.

5.8	Length of the drum (m)
1.2	Diameter of drum (m)
1.	Air flow in vvm (m3-air/m3-total-bioreactor-volume/minute)
30.	%fill = percentage of total volume of the bioreactor occupied by the bed
.99	Awso - initial water activity of the solids (0 to 1)
0.95	AwSP - solids water activity that triggers water addition
0.5	Initial water activity of the inlet air (0 to 1)
0.9	Water activity of the inlet air when T>38 celsius (0 to 1)
35.	Tin - temperature of air entering the headspace (deg C) (in absence of control scheme)
35.	Tsurr = temperature of the bioreactor surroundings
0.001	binoc kg of dry biomass per kg of dry solids
0.25	bmax = maximum concentration of biomass (kg-biomass/kg INITIAL dry solids)
0.236	Uopt = value of the specific growth rate constant at Topt (h-1)
38.0	Topt = temperature at which the specific growth rate is optimal (deg C)
387.	rhobed = packing density of the bed (must know for the given water content!!!) (kg/m3)
0.005	Thickness of the metal wall of the drum (m)
35.	Initial temperature of the bed (deg C)
35.	Initial temperature of the headspace gases (deg C)
35.	Initial temperature of the drum body (deg C)
0.5	Ybd = kg of dry substrate produced per kg of dry substrate consumed
0.0	emd = kg of dry solids eaten per kg of dry biomass present per SECOND (maintenance)
8.37D6	Yqb = J liberated per kg of dry biomass produced
0.0	emq = J liberated per kg of dry biomass present per SECOND (maintenance)
0.3	Ywb = kg of water produced per kg of dry biomass produced
0.0	emw = kg of water liberated per kg of dry biomass present per SECOND (maintenance)
10.0	enfactor = increase in mass and heat transfer (substrate to headspace) due to rotation
100.0	Time at which the simultion is to finish (h)
100	Number of outputs INTEGER
1	Type of organism (integer) "1"= A. niger type "2" = R.oligosporus

Fig A.2. Appearance of the input file (input.txt) for the rotating-drum bioreactor model

Table A.4. Significance of the column headings in the "OUTPUT" file generated by the rotating-drum bioreactor model

No.	Heading	Significance
1	t(h)	Time (h)
2	X(kgtotal)	Total mass of dry biomass in the bioreactor (kg)
3	Solids(kg)	Total dry solids (substrate + biomass) in the bioreactor (kg)
4	Wbed	Water content of the solids phase (kg-water kg-dry-solids^{-1})
5	Whead	Gas phase humidity (kg-vapor kg-dry-air^{-1})
6	Tbed(C)	Temperature of the bed (°C)
7	Thead(C)	Temperature of the headspace gases (°C)
8	Tdrum(C)	Temperature of the bioreactor wall (°C)
9	Wsat	Water content that the solids would have to have in order to be in equilibrium with the headspace gases (kg-water kg-dry-solids^{-1})
10	Xmit	Fractional specific growth rate based on temperature (μ_{FT})
11	Hsat	Saturation gas phase humidity (kg-vapor kg-dry-air^{-1})
12	awhead	Water activity of the headspace gases
13	Xmir	Fractional specific growth rate based on water activity (μ_{FW})
14	awbed	Water activity of the solids in the bed
15	U(1/h)	Fractional specific growth rate parameter (h^{-1})

- Cooling.txt will have 5 columns with the headings shown in Table A.5. The number of rows will depend on the "number of outputs to be reported" that was selected in the input file.
- Echo.txt contains an echo of various input values and other values calculated in the program. The content is self-explanatory.

Table A.5. Significance of the column headings in the "COOLING CONTRIBUTIONS" file generated by the rotating-drum bioreactor model

No.	Heading	Significance
1	t(h)	Time (h)
2	growth	Rate of waste heat generation by the growth reaction thati is taking place in the bed (W)
3	toHEAD	Rate of sensible heat transfer from the bed to the headspace wall (W) A negative number means headspace-to-bed transfer
4	toWALL	Rate of sensible heat transfer from the bed to the bioreactor wall (W). A negative number means wall-to-bed transfer
5	byEVAP	Rate of heat removal from the bed by evaporation of water to the headspace (W). A negative number means that condensation is occurring, bringing energy to the bed

A.5 Use of the Traditional Packed-Bed Bioreactor Model

This section gives advice about the use of the model for the traditional packed-bed bioreactor that was presented in Chap. 24.2. The following steps should be followed.

First use a text editor to edit the input data file. An input data file (text file) has been supplied with the name "input.txt". The contents of this file are shown in Fig. A.3. Change the values as appropriate, save the file (as a text file), and exit the file editor.

Then click twice on "PackedBed". A "DOS window" will appear on the screen. Several questions will appear, one after the other.

- What is the name of the INPUT file?
- Filename for temperature profile?
- Filename for biomass?
- Filename for echo?

After you answer the last question, the program will run automatically. When the instruction "Press any key to continue" appears, press the space bar.

Inspect the output files. Assuming that you have called the three output files "temp.txt", "biomass.txt", and "echo.txt", respectively:

- temp.txt will have various columns without headings. The number of columns will be 3 greater than the number of "internal collocation points" that was entered into the file "input.txt". The first column represents the time of the output (h). The other columns represent the temperatures (°C) at the various heights within the bioreactor. The locations of these heights (as a fraction of the total height) are given, in the same order, at the end of the echo file. The number of rows within temp.txt will depend on the values for "Number of times that output is to be reported per hour " and "Time at which the simulation is to finish (h)" that were entered in the input file.

- biomass.txt will have the same organization as "temp.txt". The first column represents the time of the output (h). The entries in the other columns will represent biomass concentrations (kg-biomass kg-substrate^{-1}) at the various fractional heights within the bioreactor.
- echo.txt contains an echo of various input values and other values calculated in the program. The content is self-explanatory.

38.0	Initial bed temperature (deg C)
38.0	Initial inlet air temperature (deg C)
0.	K (gain factor that controls the inlet air temperature)
3.0	Superficial velocity of the air (cm/s)
1.0	Bed height (m)
1.	Initial biomass (g dry biomass per kg moist solids)
125.	Final biomass (g dry biomass per kg moist solids)
.236	Maximum value of the specific growth rate constant (1/h)
38.0	Optimum temperature for growth (deg C)
21	Number of internal points along height
1.	Number of times that output is to be reported per hour
101.	Time at which the simulation is to finish (h)

Fig A.3. Appearance of the input file (input.txt) for the traditional packed-bed bioreactor model

A.6 Use of the Zymotis Packed-Bed Bioreactor Model

This section gives advice about the use of the model for the Zymotis packed-bed bioreactor that was presented in Chap. 24.3. The following steps should be followed.

First use a text editor to edit the input data file. An input data file (text file) has been supplied with the name "input.txt". The contents of this file are shown in Fig. A.4. Change the values as appropriate, save the file (as a text file), and exit the file editor.

Then click twice on the executable file. A "DOS window" will appear on the screen. Several questions will appear, one after the other.

- What is the name of the INPUT file?
- Name of file for output at top of bioreactor (Z=H)?

- Name of file for 2-D output of temperatures?
- Name of file for 2-D output of water activities?
- Name of file for 2-D output effect of T on growth?
- Name of file for 2-D output effect of Aw on growth?
- Name of file for echo?

After you answer the last question, the program will run automatically. When the instruction "Press any key to continue" appears, press the space bar.

Inspect the output files. Assuming that you have called the six output files "output.txt", "temp.txt", "water.txt", "tempgro.txt", "watergro.txt", and "echo.txt", respectively.

- output.txt will have nine columns, with the headings shown in Table A.6. These values represent the conditions at the top of the bioreactor; at a position midway between two heat transfer plates.
- temp.txt, water.txt, tempgro.txt, and watergro.txt contain output data tabulated as a function of both horizontal and vertical position. Fig. A.5 shows how these files are organized.
- echo.txt contains an echo of various input values and other values calculated in the program. The content is self-explanatory.

1.0	H = Height of the bioreactor (m)
.10	xL = half of the distance between plates (m)
.03	Vz = superficial velocity of the air (m/s)
38.	Tair = temperature of the air fed into the bioreactor (deg C)
2.0	Gain factor for the temperature control scheme for the cooling water(-)
700.	RhoS0 = density of moist subsrate particles at t=0 (kg/m3)
.35	Epsilon = void fraction of the bed (dimensionless)
.002	Bo = initial biomass concentration (kgX/kg DRY S)
.25	Bm = maximum biomass concentration (kgX/kg DRY S)
95.	hwall (W/m2/K) (heat transfer coefficient at the wall)
0.5	Ybiom/sub = kg biomass produced per kg of substrate consumed
0.3	Yh20/biom = kg H2O produced per kg of biomass produced
0.5	FMo Fraction of moisture at t=0 (kg H2O/kg total moist substrate)
0.236	Specific growth rate constant at Topt (1/h)
38.	Topt = optimum temperature for growth (deg C)
1	Organism (MUST BE AN INTEGER) 1 = Aspergillus 2 = Rhizopus
50.	Final time for the simulation (h)
50	Number of outputs (MUST BE AN INTEGER)

Fig A.4. Appearance of the input file (input.txt) for the Zymotis packed-bed bioreactor model

Table A.6. Significance of the column headings in the "OUTPUT" file generated by the Zymotis packed-bed bioreactor model

No.	Heading	Significance
1	t(h)	Time (h)
2	T(degC)	Temperature (°C)
3	X(kg/kg)	Biomass concentration (kg-dry-biomass kg-dry-substrate^{-1})
4	S(kg/m3)	Dry substrate in the bed (kg-dry-substrate m^{-3})
5	H2O(kg/m3)	Water in the bed (kg-water m^{-3})
6	W(kg/kg)	Water content of the bed solid phase (kg-water kg-dry-solids^{-1})
7	xMiR	Fractional specific growth rate based on water activity (μ_{FW})
8	xMiT	Fractional specific growth rate based on temperature (μ_{FT})
9	Tw(degC)	Temperature of cooling water circulated through the plates (°C)

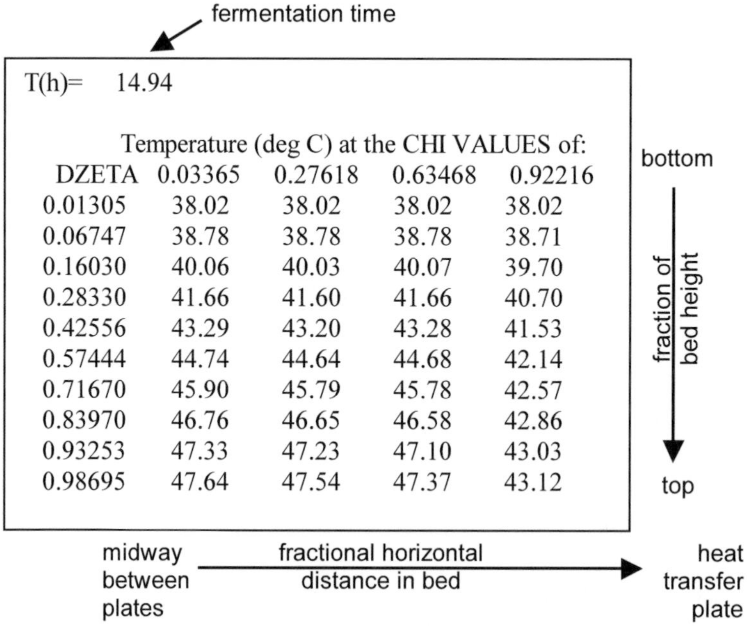

Fig A.5. Appearance of the output file "temp.txt" that will be generated by the Zymotis packed-bed bioreactor model (shown within the box). At each of the output times, the temperatures are tabulated for various values of the "fractional horizontal distance in the bed" (DZETA values) and "fraction of bed height" (CHI values). The files "water.txt", "tempgro.txt", and "watergro.txt" will be similar in format, but will contain bed water activities, the factional specific growth rates based on temperature (μ_{FT}) and the fractional specific growth rates based on water activity (μ_{WT}), respectively

A.7 Use of the Model of an Intermittently-Mixed Forcefully-Aerated Bioreactor

This section gives advice about the use of the model for the intermittently-mixed, forcefully-aerated bioreactor that was presented in Chap. 25. The following steps should be followed.

First use a text editor to edit the input data file. An input data file (text file) has been supplied with the name "input.txt". The contents of this file are shown in Fig. A.6. Change the values as appropriate, save the file (as a text file), and exit the file editor.

Then click twice on the executable file. A "DOS window" will appear on the screen. Several questions will appear, one after the other.

- What is the name of the INPUT file?
- Filename for output of average biomass? (BI)
- Filename for output of biomass as f(Z)? (BD)
- Filename for gas temperature? (TG)
- Filename for solid temperature? (TS)
- Filename for gas water activity? (WG)
- Filename for solid water activity? (WS)
- Filename for actual specific growth rate? (MU)
- Filename for specific growth rate as f(T)? (MT)
- Filename for specific growth rate as f(W)? (MR)
- Filename for humidity of gas? (FG)
- Filename for water ratio in the solid? (FS)
- Filename for eqm water ratio in the solid? (FE)
- Filename for mixing event output? (EV)
- Filename for 3-dimensional temperature output (TD)
- Name of file for echo? (EC)

The letters in parentheses after the question represent suggestions as to the final letters that you might use within the filename in order to help organize your files. After you answer the last question, the program will run automatically. When the instruction "Press any key to continue" appears, press a key (e.g., the space bar).

Inspect the output files. Assuming that you have used the suggested endings for the file names and called the fourteen output files "outbi.txt", "outbd.txt", "outtg.txt", "outts.txt", "outwg.txt", "outws.txt", "outmu.txt", "outmt.txt, "outmr.txt", "outfg.txt", "outfs.txt", "outfe.txt", "outev.txt", and "echo.txt", respectively:

- outbi.txt will have three columns with the time (h), the biomass concentration (kg-dry-biomass kg-dry-solids^{-1}), and the solids per unit volume (kg-dry-solids m^{-3}). The biomass and solids values are weighted averages calculated over the whole bed.

- outbd.txt, outts.txt, outwg.txt, outws.txt, outmu.txt, outmt.txt, outmr.txt, outfg.txt, outfs.txt, and outfe.txt will each have 6 columns. In all cases, the first column represents the time (h). The other five columns are various variables at different heights within the bed. "Z~0+" represents one-fortieth of the overall bed height, "Z=H/4" represents one quarter of the overall bed height, "Z=H/2" represents a half of the overall bed height, "Z=3H/4" represents three quarters of the overall bed height, and "Z=H" represents the top of the bed. The specific information that will appear within each file is shown in Table A.7.
- "outev.txt" gives information about the mixing events. For each mixing event are listed the time (h) at which it began, the time (h) at which it ended, the average biomass (kg-dry-biomass kg-dry-solids^{-1}) at the end of the mixing event, the average solids per unit volume (kg-dry-solids m^{-3}), the amount of energy that needed to be removed from the solids to bring them back to the initial bed temperature (J kg-dry-solids^{-1}), and the amount of water that needed to be added to the solids in order to bring their water activity back to the initial water activity (kg-water kg-dry-solids^{-1});
- outec.txt contains an echo of various input values and other values calculated in the program. The content is self-explanatory.

1.0	Bioreactor height (m)
.1	Mass flowrate of dry air (kgA/s/m2)
.95	Outlet relative humidity at which the mixing event initiated
38.	Tso Initial temperature of the solids in the bed (deg C)
38.	Tin Inlet air temperature (deg C) (also = Toa initial gas phase temperature)
0.99	Water activity of solid at time zero
0.90	Water activity of inlet air (also = awgo)
450.	rhos Density of dry substrate (kg/m3)
.35	epsilon Porosity of the bed (-)
.002	bo Initial biomass content (kgX/kgS)
.250	bm Maximum biomass content (kgX/kgS)
0.236	xMiopt maximum specific growth rate (1/h)
38.0	Topt optimum temperature for growth (deg C)
.5	Ybs Yield of biomass from substrate (kgX/kgS)
1	Organism type 1=Aspergillus 2=Rhizopus
.25	RELATED WITH MIXING EVENT Time taken by mixing event (h)
0.	RELATED WITH MIXING EVENT Specific growth rate (fractional) during mixing
50.	Number of hours of simulation time
500	Number of times that output will be reported

Fig A.6. Appearance of the input file (input.txt) for the model of the intermittently-mixed, forcefully-aerated bioreactor

Table A.7. Contents of some of the files that are generated by the model of the intermittently-mixed, forcefully-aerated bioreactor

Filename	Content of the file (as a function of height, as explained in the text)
outbd.txt	Biomass concentration (kg-dry-biomass kg-dry-solids^{-1})
outtg.txt	Temperature of the gas in the inter-particle phase (°C)
outts.txt	Temperature of the solids phase (°C);
outwg.txt	Water activity of the gas phase (dimensionless)
outws.txt	Water activity of the solids phase (dimensionless)
outmu.txt	Value of the specific growth rate constant (h^{-1})
outmt.txt	Fractional specific growth rate based on temperature (μ_{FT})
outmr.txt	Fractional specific growth rate based on water activity (μ_{WT})
outfg.txt	Gas phase humidity (kg-vapor kg-dry-air)
outfs.txt	Solids phase moisture content (kg-water kg-dry-solids^{-1})
outfe.txt	Moisture content that the solids would have to have in order to be in equilibrium with the gas phase (kg-water kg-dry-solids^{-1})

Index